# Topics in Boundary Element Research

Edited by C. A. Brebbia

## Volume 1: Basic Principles and Applications

With 144 Figures and 11 Tables

Springer-Verlag
Berlin Heidelberg GmbH 1984

Editor:
Dr. Carlos A. Brebbia

Department of Civil Engineering
The University
Southampton SO9 5NH
England

ISBN 978-0-387-13097-2

Library of Congress Cataloging in Publication Data
Main entry under title: Topics in boundary element research.
Contents: v. 1. Basic principles and applications. 1. Boundary value problems. I. Brebbia, C. A.
TA347.B69T67 1984 620'.001'51535 84-10644
ISBN 978-0-387-13097-2 ISBN 978-1-4899-2877-1 (eBook)
DOI 10.1007/978-1-4899-2877-1


Originally published by Springer-Verlag Berlin Heidelberg New York Tokyo in 1984
Softcover reprint of the hardcover 1st edition 1984



Typesetting: Graphischer Betrieb Konrad Triltsch, Würzburg

2061/3020-5 4 3 2 1 0

# Contributors


| | | |
|---|---|---|
| M.C. Au | University of Calgary, Canada | (Chap. 5) |
| G. Beer | University of Queensland, Australia | (Chap. 8) |
| C.A. Brebbia | Southampton University and Computational Mechanics Institute Southampton, England | (Chaps. 0, 5, 7) |
| J.J. Connor | MIT, Cambridge, MA, U.S.A. | (Chap. 0) |
| J.L.M. Fernandes | Instituto Superior Tecnico, Lisbon, Portugal | (Chap. 2) |
| I. Herrera | Universidad Nacional Autónoma de México, Mexico | (Chap. 10) |
| M.A. Jaswon | City University, London, England | (Chap. 1) |
| N. Kamiya | Mei University, Japan | (Chap. 9) |
| M. Kikuta | Nippon Sheet Glass Co., Ltd., Itami, Japan | (Chap. 3) |
| J.A. Liggett | Cornell University, Ithaca, N.Y., U.S.A. | (Chap. 4) |
| P.L.-F. Lui | Cornell University, Ithaca, N.Y., U.S.A. | (Chap. 4) |
| J.L. Meek | University of Queensland, Australia | (Chap. 8) |
| C. Patterson | University of Sheffield, England | (Chap. 6) |
| H.L. Pina | Instituto Superior Tecnico, Lisbon, Portugal | (Chap. 2) |
| Y. Sawaki | Mei University, Japan | (Chap. 9) |
| M.A. Sheikh | University of Sheffield, England | (Chap. 6) |
| M. Tanaka | Shinshu University, Nagano, Japan | (Chap. 3) |
| H. Togoh | Nippon Sheet Glass Co., Ltd., Itami, Japan | (Chap. 3) |
| W. Venturini | University of São Paulo, Brazil | (Chap. 7) |

# Introduction to the Series "Topics in Boundary Element Research"

The continuing interest in the application of Boundary Element Methods in engineering has generated a series of books and numerous scientific papers, not least those regularly presented at the International Conferences on Boundary Elements which have been held under my direction since 1978. Most recently a new journal, "Engineering Analysis", has been launched which concentrates on new developments in this important area. In spite of all this activity, the need exists for a serial publication in which the most recent advances in the method are documented in a more complete form than is usually the case in papers presented at conferences or scientific gatherings. This unfulfilled need prompted me to launch the present series.

Each volume in this series will comprise chapters describing new applications of the method. The emphases will be on contributions which are self-contained and explain a particular topic in sufficient detail for the analytical engineer or scientist to be able to understand the theory and in due course to write the relevant computer software. All chapters are written by scientists who are actively involved in Boundary Element research, the internationally best known names being balanced with those of new researchers who have recently made significant contributions in this area.

Another objective of the series is to report work for direct application by the practising engineer. Furthermore, I feel that it is important to include sections which discuss the modelling strategies and presentation of results as well as theoretical chapters. The relationship between Boundary Element analysis codes and computer aided design packages will be discussed in subsequent volumes to achieve the right perspective on the application of Boundary Elements. It is all too easy when dealing with these types of analytical techniques to forget that they exist within the framework of the final engineering product.

It is my intention that the series should be open to all those researchers who have made significant contributions to the advancement of the new method. In this regard I shall be happy to receive any suggestions that such members of the scientific community may wish to make, in an effort to produce a publication that is indispensable to all concerned with the advancement of Boundary Elements.

**Carlos A. Brebbia**
Editor

# Preface

As the Boundary Element Method develops into a tool of engineering analysis more effort is dedicated to studying new applications and solving different problems. This book contains chapters on the basic principles of the technique, time dependent problems, fluid mechanics, hydraulics, geomechanics and plate bending.

The number of non-linear and time dependent problems which have become amenable to solution using boundary elements have induced many researchers to investigate in depth the basis of the method. Chapter 0 of this book presents an approach based on weighted residual and error approximations, which permits easy construction of the governing boundary integral equations. Chapter 1 reviews the theoretical aspects of integral equation formulations with emphasis in their mathematical aspects.

The analysis of time dependent problems is presented in Chap. 2 which describes the time and space dependent integral formulation of heat conduction problems and then proposes a numerical procedure and time marching algorithm.

Chapter 3 reviews the application of boundary elements for fracture mechanics analysis in the presence of thermal stresses. The chapter presents numerical results and the considerations on numerical accuracy are of interest to analysts as well as practising engineers.

Boundary elements are becoming an accepted tool for the solution of many fluid mechanic problems as demonstrated in Chaps. 4 and 5. Chapter 4 describes the use of the method for solving problems in aerodynamics and hydrodynamics, free body problems and water waves. The chapter starts with some historical remarks linking the boundary elements to be more classical techniques such as panel methods. Chapter 5 is dedicated to the application of boundary elements to find water waves forces on fixed, free-floating or moored offshore structures. The authors point out the convenience of using the simple fundamental solution of Laplace's equation together with radiation type boundary conditions, in preference to more sophisticated fundamental solutions. The main advantage of this treatment is that it gives accurate solutions which can be used for general types of structures without presenting any convergence problems.

One of the advantages of boundary elements which is now beginning to be fully exploited in commercial programs[1] is the possibility of using discontinuous elements (Chap. 6). These elements have the advantage of being able to represent dis-

---

[1] "BEASY, A Boundary Element Analysis System", "Boundary Element Methods in Engineering". (Ed. C. Brebbia) Springer-Verlag, Berlin, Heidelberg, New York, 1982

continuities such as singularities better and their meshes can be refined in a much simpler way.

The next two chapters deal with problems in geomechanics. Chapter 7 presents the basic formulation for no-tension and problems with joints and discontinuities such as those frequently appearing in geomechanics. The chapter also uses the viscoplastic formulation as presented by Telles and Brebbia. Chapter 8 is addressed more specifically to the applications of the method to mining, a field in which boundary elements have been very successfully applied. The authors present the displacement discontinuity formulation and discuss the combination of the method with other techniques. The chapter contains a number of interesting examples which lend weight to the authors' conclusions regarding the suitability of boundary elements for problems in mining engineering.

Very little has been done on the topic of geometrically non-linear problems using boundary elements. The authors of Chap. 9 present work carried out on finite deflections of plates using different integral formulations corresponding to diverse plate theories. Numerical illustrations are presented for the approximate Berger equations and their validity is discussed in this chapter. A section on extension of the method to deal with non-linear problems in sandwhich plates and shallow shells is also included.

The final chapter (Chap. 10) is dedicated to the interpretation of boundary integral methods using Trefftz original ideas. The chapter concentrates on the basic mathematical definitions but is written in a way that the engineer can easily interpret. The last chapter reviews the theory and in particular its applications in potential and elastostatic problems, including the indirect formulations.

The book is a testimony to the strength and vitality of the new method which has made considerable advances in a period of 7 years or so, giving as its 'official' date of birth the 1st Conference in 1978 for which the name boundary elements was used[2] and the first book published on the topic also in 1978[3]. Judging by a rule of thumb that a method takes 15 years or so of gestation and a similar time for development, the editor is certain that the need for this series will continue. New volumes will continue to provide up to date information on the state of the art of the technique and its applications at the same high standard.

**Carlos A. Brebbia**

---

[2] "Recent Advances in Boundary Element Methods" Pentech Press, London, 1978

[3] Brebbia, C.A. "The Boundary Element Method for Engineers" Pentech Press, London and Halstead Press, New York, 1978

# Contents

# Chapter 0

# Boundary Integral Formulations

*by C. Brebbia and J. J. Connor*

## 0.1 Fundamentals of Functional Analysis

An operator is a process which applied to a function or a set of functions produces another function, i.e.,

$$\mathscr{L}(u) = b \tag{0.1}$$

where $\mathscr{L}(u)$ is the operator which applied to $u$ produces $b$; $u$ and $b$ may be scalars or vectors: $\mathscr{L}(\,)$ may be an ordinary differential operator such as

$$\mathscr{L}(\,) = a_0 \frac{d^2(\,)}{dx^2} + a_1 \frac{d(\,)}{dx} + a_2(\,) \tag{0.2}$$

a partial differential operator such as

$$\mathscr{L}(\,) = \frac{\partial}{\partial x}\left(k \frac{\partial(\,)}{\partial x}\right) + \frac{\partial}{\partial y}\left(k \frac{\partial(\,)}{\partial y}\right) \tag{0.3}$$

or an integro operator

$$\mathscr{L}(\,) = \int_0^t v(t-\tau)\,(\,)\, d\tau \tag{0.4}$$

which can also be written in terms of convolution notation as

$$L(\,) = v(t) * (\,) \tag{0.5}$$

Most of the operators encountered in engineering formulations are differential but for some problem areas, such as visco-elasticity and transient response, one needs to consider integro or integro-differential operators.

Another important definition of functional analysis is the generalized inner product. Given two vector functions $u$ and $v$, the product is denoted as

$$\langle u, v\rangle = \langle u \cdot v\rangle \tag{0.6}$$

where the dot represents the scalar product of $u, v$ at a point and the brackets may indicate different types of operations. The most common operation is an integration over a domain $\Omega$, i.e.,

$$\langle u, v\rangle = \int_\Omega u \cdot v \, d\Omega \tag{0.7}$$

Other operations are the convolution,

$$\langle u, v\rangle = \int_0^t u(t-\tau) \cdot v(\tau)\, d\tau \tag{0.8}$$

and integration over a certain region of the domain. In particular, if the region is the boundary, $\Gamma$, enclosing the domain, the operation is denoted by

$$\langle u, v\rangle_\Gamma = \int_\Gamma u \cdot v \, d\Gamma \tag{0.9}$$

Having defined the operator $\mathscr{L}(\,)$ as in (0.1), one can now introduce the concept of its *adjoint*. The adjoint of an operator is similar, conceptually, to the transpose of a matrix. Let $u$ and $v$ be two arbitrary functions and multiply (0.1) by $v$. We wish to determine how the operations can be transferred to the $v$ function, i.e.,

$$[\mathscr{L}(u)]\, v \Rightarrow [\mathscr{L}^*(v)]\, u + \ldots \tag{0.10}$$

where $\mathscr{L}^*(\,)$ is referred to as the adjoint of $L$. Forming the inner product over the domain,

$$\langle \mathscr{L}(u), v\rangle = \int_\Omega \mathscr{L}(u)\, v \, d\Omega \tag{0.11}$$

and integrating by parts until all derivatives (we are talking about differential operators) are passed to $v$ gives

$$\langle \mathscr{L}(u), v\rangle = \langle u, \mathscr{L}^*(v)\rangle + \langle P(u,v), 1\rangle_\Gamma \tag{0.12}$$

where $P(u, v)$, a result of the partial integration, is called the "bilinear concomitant". We will shortly examine the character of this term. The inner product of a single function say $\langle f, 1\rangle$ is written as $\langle f\rangle$ for convenience, when no confusion as to meaning arises.

**Example 1.** Consider the following second order operator;

$$\mathscr{L}(u) = a_0 \frac{d^2u}{dx^2} + a_1 \frac{du}{dx} + a_2 u\,. \tag{a}$$

Forming the inner product,

$$\langle \mathscr{L}(u), v\rangle = \int_{x_1}^{x_2} \left( a_0 \frac{d^2u}{dx^2} + a_1 \frac{du}{dx} + a_2 u \right) v \, dx \tag{b}$$

and integrating, one finds

$$\langle \mathscr{L}(u), v\rangle = \int_{x_1}^{x_2} \left\{ - \frac{du}{dx} \frac{d}{dx} (a_0 v) - u \frac{d}{dx} (a_1 v) + a_2 u v \right\} dx$$

$$+ \left[ v \left\{ a_0 \frac{du}{dx} + a_1 u \right\} \right]_{x_1}^{x_2} \tag{c}$$

Operating on the first term to eliminate $du/dx$ gives;

$$\langle \mathscr{L}(u), v\rangle = \int_{x_1}^{x_2} \left\{ \frac{d^2}{dx^2} (a_0 v) - \frac{d}{dx} (a_1 v) + a_2 v \right\} u \, dx$$

$$+ \left[ v \left\{ a_0 \frac{du}{dx} + a_1 u \right\} - u \left\{ \frac{d}{dx} (a_0 v) \right\} \right]_{x_1}^{x_2} \tag{d}$$

It follows that

$$\mathscr{L}^*(u) = \frac{d^2}{dx^2}(a_0 u) - \frac{d}{dx}(a_1 u) + a_2 u$$
$$P(u,v) = a_0 v \frac{du}{dx} - u \frac{d}{dx}(a_0 v) + a_1 u v \tag{e}$$

**Example 2.** Let us consider

$$\mathscr{L}(u) = \frac{d^2}{dx^2}\left(b_0 \frac{d^2u}{dx^2}\right) + \frac{d}{dx}\left(b_1 \frac{du}{dx}\right) + b_2 u \tag{a}$$

One obtains after *two* integrations of the inner product,

$$\int_{x_1}^{x_2} v \mathscr{L}(u)\, dx = \int_{x_1}^{x_2} \left[ b_0 \frac{d^2u}{dx^2}\frac{d^2v}{dx^2} - b_1 \frac{du}{dx}\frac{dv}{dx} + b_2 u v \right] dx$$
$$+ \left[ v \left\{ \frac{d}{dx}\left(b_0 \frac{d^2u}{dx^2}\right) + b_1 \frac{du}{dx} \right\} + \frac{dv}{dx}\left( - b_0 \frac{d^2u}{dx^2}\right) \right] \tag{b}$$

This result corresponds to the point where we have converted the integral such that all the terms have, individually, the same order derivatives in $x$ for $u$ and $v$. Continuing for two more integrations gives the final result,

$$\int_{x_1}^{x_2} \mathscr{L}(u)\, v\, dx = \int_{x_1}^{x_2} \left[ \frac{d^2}{dx^2}\left(b_0 \frac{d^2v}{dx^2}\right) + \frac{d}{dx}\left(b_1 \frac{dv}{dx}\right) + b_2 v \right] u\, dx$$
$$+ \left[ v \left\{ \frac{d}{dx}\left(b_0 \frac{d^2u}{dx^2}\right) + b_1 \frac{du}{dx} \right\} + \frac{dv}{dx}\left( - b_0 \frac{d^2u}{dx^2}\right) \right]$$
$$- \left[ u \left\{ \frac{d}{dx}\left(b_0 \frac{d^2v}{dx^2}\right) + b_1 \frac{dv}{dx} \right\} + \frac{du}{dx}\left( - b_0 \frac{d^2v}{dx^2}\right) \right] \tag{c}$$

This expression can be written as,

$$\langle \mathscr{L}(u), v \rangle = \langle u, \mathscr{L}^*(v) \rangle + \langle (R(u,v) - R(v,u)), 1 \rangle_\Gamma \tag{d}$$

where $\mathscr{L}^*(\,) = \mathscr{L}(\,)$, i.e. the operator is *self adjoint*, and the $R(u,v)$ term has the form

$$R(u,v) = v \left\{ \frac{d}{dx}\left(b_0 \frac{d^2u}{dx^2}\right) + b_1 \frac{du}{dx} \right\} + \frac{dv}{dx}\left( - b_0 \frac{d^2u}{dx^2}\right) \tag{e}$$

Interchanging $u$ and $v$ in $R(u,v)$ gives $R(v,u)$.

## 0.2 Generalized Green's Formula

Formula (0.12) specialized for the one dimensional case is sometimes called Lagrange's identity. Green extended the derivation to partial differential operators and also expressed the 'bilinear concomitant' $P(u,v)$ as a product of four sets of terms, represented in vector form for generality here as

$$\begin{aligned} &\mathbf{F}(u), \ \mathbf{F}^*(v) \\ &\mathbf{G}(u), \ \mathbf{G}^*(v) \end{aligned} \tag{0.13}$$

We write (12) as,

$$\langle v, \mathscr{L}(u)\rangle - \langle u, \mathscr{L}^*(v)\rangle = \langle \mathbf{F}^*(v), \mathbf{G}(u)\rangle_\Gamma - \langle \mathbf{F}(u), \mathbf{G}^*(v)\rangle_\Gamma \tag{0.14}$$

This is equivalent to requiring

$$\langle P(u,v)\rangle_\Gamma = \langle \mathbf{F}^*(v), \mathbf{G}(u)\rangle_\Gamma - \langle \mathbf{F}(u), \mathbf{G}^*(v)\rangle_\Gamma \tag{0.15}$$

Equation (0.15) is called Green's formula and we will shortly see that the $F$ and $G$ vector differential operators are related to the boundary conditions for the scalar differential operator, $\mathscr{L}$.

For the case of self-adjoint operators, such as the one encountered in example 2, we find that $\mathscr{L}^* = \mathscr{L}$ and $\mathbf{F}^* = \mathbf{F}$, $\mathbf{G}^* = \mathbf{G}$. This occurs because the boundary terms for the second phase of the integration are similar to the terms during the first phase; $u$ and $v$ are interchanged and $a-1$ is introduced. Hence (0.14) becomes

$$\langle v, \mathscr{L}(u)\rangle - \langle u, \mathscr{L}(v)\rangle = \langle \mathbf{F}(v), \mathbf{G}(u)\rangle_\Gamma - \langle \mathbf{F}(u), \mathbf{G}(v)\rangle_\Gamma . \tag{0.16}$$

**Example 3.** We return to Example 2. Comparing (0.16) with (c) leads to

$$\begin{aligned} \mathbf{F}(v) &= \left\{v, \frac{dv}{dx}\right\} \\ \mathbf{G}(u) &= \left\{\frac{d}{dx}\left(b_0 \frac{d^2u}{dx^2}\right) + b_1 \frac{du}{dx}, \ -b_0 \frac{d^2u}{dx^2}\right\} \end{aligned} \tag{a}$$

It is conventional to define the forms of **F** and **G** after the first phase of the integration. Notice that the order of **F** and **G** is equal to *half* the order of $\mathscr{L}(\ )$. Hence is $\mathscr{L}(\ )$ is of order $2n$, **F** and **G** will have terms of order $n$. **F** will involve derivatives up to order $n-1$ according to our convention for defining **F** (i.e. result after the first $n$ integrations). The boundary conditions are defined as

$$\begin{aligned} &\text{Essential} \\ &\qquad \mathbf{F}(u) = \mathbf{f} \text{ (a prescribed vector)} \\ &\text{Natural} \\ &\qquad \mathbf{G}(u) = \mathbf{g} \text{ (a prescribed vector).} \end{aligned} \tag{b}$$

Applied to Example 2, one needs to specify,

$$u = f_1 \quad \text{or} \quad \frac{d}{dx}\left(b_0 \frac{d^2u}{dx^2}\right) + b_1 \frac{du}{dx} = g_1$$

and

$$\frac{du}{dx} = f_2 \quad \text{or} \quad -b_0 \frac{d^2u}{dx^2} = g_2$$

at each point on the boundary.

## 0.3 Variational Formulation

When the operator is *self-adjoint*, one can transform the differential equation and boundary conditions to a variational statement. We just outline the procedure here to stress its relationship to integral equation methods, but our primary interest is in weighted residual techniques which allow for non-self-adjoint operators, including non-linear cases.

Consider the following differential equation,

$$\mathscr{L}(u) + b = 0 \quad \text{in } \Omega \tag{0.17}$$

and boundary conditions,

$$\begin{aligned} \mathbf{F}(u) &= \mathbf{f} \quad \text{on } \Gamma_1 \\ \mathbf{G}(u) &= \mathbf{g} \quad \text{on } \Gamma_2 \end{aligned} \tag{0.18}$$

We multiply equation (0.17) by $v$ and integrate by parts until we have balanced derivatives of $u$ and $v$. This yields

$$\int [\mathscr{L}(u) + b]\, v\, d\Omega = \int \mathbf{F}(v) \cdot \mathbf{G}(u)\, d\Gamma + \int [b\, v + \mathscr{D}(u) \cdot \mathscr{D}(v)]\, d\Omega \tag{0.19}$$

where $\mathscr{D}(\,)$ is a vector containing derivatives of $u$ or $v$. We can also write (0.19) in terms of scalar product notation,

$$\langle \mathscr{L}(u) + b, v \rangle = \langle \mathbf{F}(v), \mathbf{G}(u) \rangle_\Gamma + \langle b, v \rangle + \langle \mathscr{D}(u), \mathscr{D}(v) \rangle \tag{0.20}$$

As an illustration, the equation discussed in examples 2 and 3 has the following $\mathscr{D}(\,)$ operator,

$$\mathscr{D}(\,) = \left\{ b_0^{1/2} \frac{d^2(\,)}{dx^2},\ i\, b_1^{1/2} \frac{d(\,)}{dx},\ b_2^{1/2}(\,) \right\} \tag{0.21}$$

To interpret (0.19) as a stationary requirement for a functional, we define $v$ as the variation of $u$, i.e.

$$v \equiv \delta u \tag{0.22}$$

and (0.20) then takes the form

$$\langle \mathscr{L}(u) + b, \delta u \rangle = \langle \mathbf{F}(\delta u), \mathbf{G}(u) \rangle_\Gamma + \langle b, \delta u \rangle + \langle \mathscr{D}(u), \mathscr{D}(\delta u) \rangle \tag{0.23}$$

Next, we transform the last inner product by defining a functional $H$ such that

$$\langle H(u) \rangle = \tfrac{1}{2} \langle \mathscr{D}(u), \mathscr{D}(u) \rangle \tag{0.24}$$

This step is possible only when $\mathscr{L}(\,)$ is a self-adjoint. Since $H$ is a function of $(u^2)$, $(du/dx)^2$ etc. the 1/2 is needed when carrying out the variation. For the case discussed in example 2,

$$H(u) = \tfrac{1}{2} b_0 \left( \frac{d^2 u}{dx^2} \right)^2 - \tfrac{1}{2} b_1 \left( \frac{du}{dx} \right)^2 + \tfrac{1}{2} b_2 u^2 \tag{0.25}$$

Noting (0.24), equation (0.23) can be written as

$$\langle \mathscr{L}(u) + b, \delta u \rangle - \langle \mathbf{F}(\delta u), \mathbf{G}(u) \rangle = \delta \langle H(u) + b\, u \rangle \tag{0.26}$$

It remains to incorporate the boundary conditions in the variational statement. As a starting point, we form the variation of the inner product of $\mathbf{F}(u)$ and $\mathbf{G}(u)$

on the boundary. Since **F** and **G** are linear in $u$, the variational operator can be shifted inside the bracket, i.e., $\delta[\mathbf{F}(u)] = \mathbf{F}(\delta u)$, and the result is

$$\delta\langle \mathbf{F}(u), \mathbf{G}(u)\rangle_\Gamma = \langle \mathbf{F}(\delta u), \mathbf{G}(u)\rangle_\Gamma + \langle \mathbf{F}(u), \mathbf{G}(\delta u)\rangle_\Gamma \tag{0.27}$$

Now we require $u$ to satisfy the essential boundary conditions on $\Gamma_1$, i.e.

$$\mathbf{F}(u) = \mathbf{f} \quad \text{on } \Gamma_1 \tag{0.28}$$

where **f** is prescribed. In methods such as finite elements – but not in boundary elements – the variation is usually required to satisfy the homogeneous form of the essential boundary conditions

$$\mathbf{F}(\delta u) = \mathbf{0} \quad \text{on } \Gamma_1 \tag{0.29}$$

The remaining boundary conditions that have to be enforced are the natural conditions in $\Gamma_2$.

$$\mathbf{G}(u) = \mathbf{g} \quad \text{on } \Gamma_2 \tag{0.30}$$

One way of preceeding is to require $\delta u$ to be constrained by (0.29) and set $\mathbf{G}(u) = \mathbf{g}$ in (0.27). This yields

$$\delta\langle \mathbf{F}(u), \mathbf{g}\rangle_\Gamma = \langle \mathbf{F}(\delta u), \mathbf{g}\rangle_{\Gamma_2} \tag{0.31}$$

Combining (0.31) with (0.26) gives the desired result,

$$\delta[\langle H(u) + b\,u\rangle + \langle \mathbf{F}(u), \mathbf{g}\rangle_{\Gamma_2}] = \langle \mathscr{L}(u) + b, \delta u\rangle + \langle \mathbf{F}(\delta u), (\mathbf{g} - \mathbf{G}(u))\rangle_{\Gamma_2} \tag{0.32}$$

with $\mathbf{F}(\delta u) = \mathbf{0}$ on $\Gamma_1$.

Equation (0.32) shows that the solution of (0.17), (0.18) corresponds to a stationary value of the functional,

$$I_1 = \langle H(u) + b\,u\rangle + \langle \mathbf{F}(u), \mathbf{g}\rangle_{\Gamma_2} = \int_\Omega (H(u) + b\,u)\,d\Omega + \int_{\Gamma_2} \mathbf{F}(u)\,\mathbf{g}\,d\Gamma \tag{0.33}$$

A more general approach which is followed for the boundary element method and mixed variational formulations is to include the *essential boundary conditions* on $\Gamma_1$ as a constraint, as well as the *natural* boundary conditions on $\Gamma_2$. In this case, the expanded form of (0.26) is

$$\begin{aligned} &\delta[\langle H(u) + b\,u\rangle + \langle \mathbf{F}(u), \mathbf{g}\rangle_{\Gamma_2}] + \langle (\mathbf{F}(u) - \mathbf{f}), \mathbf{G}(u)\rangle_{\Gamma_1} \\ &\quad = \langle \mathscr{L}(u) + b, \delta u\rangle + \langle \mathbf{F}(\delta u), (\mathbf{g} - \mathbf{G}(u))\rangle_{\Gamma_2} + \langle (\mathbf{F}(u) - \mathbf{f}), \mathbf{G}(\delta u)\rangle_{\Gamma_1} \end{aligned} \tag{0.34}$$

where now $\mathbf{F}(\delta u)$ is not constrained (i.e. $\mathbf{F}(\delta u) \neq 0$ on $\Gamma_1$). Defining the functional $I$ as

$$I_2 = \langle H(u) + b\,u\rangle + \langle (\mathbf{F}(u) - \mathbf{f}), \mathbf{G}(u)\rangle_{\Gamma_1} + \langle \mathbf{F}(u), \mathbf{g}\rangle_{\Gamma_2} \tag{0.35}$$

it follows that the statement $\delta I_2 = 0$ for arbitrary $\delta u$ is equivalent to the complete set of conditions,

$$\begin{aligned} \mathscr{L}(u) + b &= 0 \quad && \text{in } \Omega \\ \mathbf{F}(u) &= \mathbf{f} && \text{on } \Gamma_1 \\ \mathbf{G}(u) &= \mathbf{g} && \text{on } \Gamma_2 \end{aligned} \tag{0.36}$$

This approach to incorporating all the boundary conditions is followed in establishing the general weighted residual form. The important difference is that while the above approach is only valid for self-adjoint operators, the weighted residual technique allows us to produce boundary element type statements for any problem.

**Example 4.** Consider the operator of Example 2.

$$\mathscr{L}(u) = \frac{d^2}{dx^2}\left(b_0 \frac{d^2u}{dx^2}\right) + \frac{d}{dx}\left(b_1 \frac{du}{dx}\right) + b_2 u$$

We have shown that,

$$\mathbf{F}(u) = \left\{u, \frac{du}{dx}\right\}$$

$$\mathbf{G}(u) = \left\{\frac{d}{dx}\left(b_0 \frac{d^2u}{dx^2}\right) + b_1 \frac{du}{dx}, -b_0 \frac{d^2u}{dx^2}\right\}$$

$$H(u) = \tfrac{1}{2} b_0 \left(\frac{d^2u}{dx^2}\right)^2 - \tfrac{1}{2} b_1 \left(\frac{du}{dx}\right)^2 + \tfrac{1}{2} b_2 u^2$$

For the case where $\mathbf{F}(u)$ is specified on $\Gamma_1$, the expanded form is

$$I = \int (H(u) + b\,u)\, d\Omega$$
$$- \left[ (u - f_1) \left\{\frac{d}{dx}\left(b_0 \frac{d^2u}{dx^2}\right) + b_1 \frac{du}{dx}\right\} + \left(\frac{du}{dx} - f_2\right)\left(-b_0 \frac{d^2u}{dx^2}\right)\right]_{\Gamma_1}$$
$$+ \left[u\, g_1 + \frac{du}{dx} g_2\right]_{\Gamma_2}$$

where

$$\mathbf{f} = \{f_1, f_2\}\,; \qquad \mathbf{g} = \{g_1, g_2\}$$

## 0.4 Weighted Residual Scheme

Let us now consider the set of equations represented by (0.36). Our strategy is to propose an approximate solution $\tilde{u}$ containing unknown parameters and to establish appropriate values for these parameters. To avoid proliferation of notation, we shall drop the superscript tilde on $\tilde{u}$, and one must keep in mind that $u$ is now only an approximation. The errors for the respective equations are

$$\begin{aligned} \varepsilon_I &= \mathscr{L}(u) + b \neq 0 && \text{in } \Omega \\ \boldsymbol{\varepsilon}_1 &= \mathbf{F}(u) - \mathbf{f} \neq \mathbf{0} && \text{on } \Gamma_1 \\ \boldsymbol{\varepsilon}_2 &= \mathbf{G}(u) - \mathbf{g} \neq \mathbf{0} && \text{on } \Gamma_2 \end{aligned} \tag{0.37}$$

These errors are reduced by distributing them with prescribed weighting functions in such a way that the product of the residuals by the weighting functions is set to zero over $\Omega$ and $\Gamma$, i.e.

$$\int_\Omega \varepsilon_I\, w\, d\Omega + \int_{\Gamma_1} \boldsymbol{\varepsilon}_1 \cdot \mathbf{w}_1\, d\Gamma + \int_{\Gamma_2} \boldsymbol{\varepsilon}_2 \cdot \mathbf{w}_2\, d\Gamma = 0 \tag{0.38}$$

where $w$, $\mathbf{w}_1$ and $\mathbf{w}_2$ are weighting functions (the last two are vectors). By applying (0.38) one can determine the unknown parameters in the approximate function chosen for $u$.

The key issue is the choice of weighting functions and here our approach can be guided by the result obtained with the variational method. Referring back to section 3, we established a functional whose stationary requirement was equivalent to the satisfaction of the system given by equation (0.36). Examination of equation (0.34) shows that the right hand side is similar in form to (0.38) and suggests that we take the weighting function for the *self-adjoint* case to be

$$w = \delta u$$
$$\mathbf{w}_1 = \mathbf{G}(w)$$
$$\mathbf{w}_2 = -\mathbf{F}(w)$$

With this choice, the weighted residual expression is identical to the stationary requirement. Then, for the self-adjoint case, we require,

$$\int [\mathscr{L}(u) + b]\, w\, d\Omega + \int_{\Gamma_1} (\mathbf{F}(u) - \mathbf{f}) \cdot \mathbf{G}(w)\, d\Gamma + \int_{\Gamma_2} (\mathbf{g} - \mathbf{G}(u)) \cdot \mathbf{F}(w)\, d\Gamma = 0 \tag{0.40}$$

If we require $u$ to satisfy the boundary conditions, the boundary integrals vanish. If, in addition, we select $w$ from the basis functions contained in $u$, i.e.

$$u = u_p + \sum_{i=1}^{N} \alpha_i \varphi_i \quad \text{and} \quad w = \varphi_i \tag{0.41}$$

then we obtain the conventional Galerkin method. We do not constrain $w$ in this manner when working with boundary elements.

One can generate alternate forms of (0.40) by integrating $w\, L(u)$ by parts. In the conventional finite element method, one integrates until the order of differentiation for $u$ and $w$ is *balanced*, i.e. equalized. In the boundary element approach, we transform (0.40) to a form where differentiation of $u$ in the interior, $\Omega$, is eliminated. Using Green's formula, we can write (0.40) as,

$$\int_{\Omega} [u\mathscr{L}(w) + b\, w]\, d\Omega + \int_{\Gamma_1} [-\mathbf{f} \cdot \mathbf{G}(w) + \mathbf{F}(w) \cdot \mathbf{G}(u)]\, d\Gamma + \int_{\Gamma_2} [\mathbf{g} \cdot \mathbf{F}(w) - \mathbf{F}(u) \cdot \mathbf{G}(w)]\, d\Gamma = 0 \tag{0.42}$$

The attractiveness of (0.42) is related to the choice of $w$. Suppose one uses, as weighting functions, solutions of the homogeneous equation

$$\mathscr{L}(w) = 0 \tag{0.43}$$

for this choice, the interior integral involving $u$ vanishes and one has only to work with an expansion for $u$ on the *boundary*. Interior node points are not required. If, as usual in boundary elements, we choose a function such that

$$\mathscr{L}(w) = -\delta(\boldsymbol{r} - \boldsymbol{r}_p) \tag{0.44}$$

where $\delta$ is a delta function applied at a point $P$ and $r$ is the distance from the origin, the domain integral becomes the value of the function $u$ at the point $P$, i.e.

$$\int_{\Omega} (u \mathscr{L}(w))\, d\Omega \Rightarrow -u_p \tag{0.45}$$

Again, interior node points are not required.

The above procedure for defining the weighting functions is only valid for *self-adjoint* operators. One of the significant advantages of weighted residual formulations is that they can be used for non-self-adjoint problems as well as for self-adjoint ones. Because of this feature, it is important to establish a more general procedure for identifying appropriate weighting functions.

Let $u$ represent the exact solution of (0.36), where $\mathscr{L}$ is now treated as a non-self-adjoint operator. By definition, the domain integral vanishes for arbitrary $w$.

$$\langle w, \mathscr{L}(u) + b \rangle = 0 \tag{0.46}$$

Noting Green's formula, (0.14), and the boundary condition satisfied by $u$, (0.46) can be transformed to

$$\begin{aligned} &\langle b, w \rangle + \langle u, \mathscr{L}^*(w) \rangle + \langle \mathbf{F}^*(w)\, \mathbf{G}(u) - \mathbf{f} \cdot \mathbf{G}^*(w) \rangle_{\Gamma_1} \\ &\qquad + \langle \mathbf{F}^*(w) \cdot \mathbf{g} - \mathbf{F}(u) \cdot \mathbf{G}^*(w) \rangle_{\Gamma_2} = 0 \end{aligned} \tag{0.47}$$

Now, in the weighted residual approach, we form the inner product of the residuals with appropriate weighting functions. Our starting point is

$$\langle w, \mathscr{L}(\tilde{u}) + b \rangle + \langle (\mathbf{F}(\tilde{u}) - \mathbf{f}), \mathbf{w}_1 \rangle_{\Gamma_1} + \langle (\mathbf{G}(\tilde{u}) - \mathbf{g}), \mathbf{w}_2 \rangle_{\Gamma_2} = 0 \tag{0.48}$$

where $\tilde{u}$ denotes the approximate solution. Applying Green's formula to $\langle w\, L(\tilde{u}) \rangle$,

$$\langle w, \mathscr{L}(\tilde{u}) \rangle = \langle \tilde{u}, \mathscr{L}^*(w) \rangle + \langle \mathbf{F}^*(w), \mathbf{G}(\tilde{u}) - \mathbf{F}(\tilde{u}), \mathbf{G}^*(w) \rangle_{\Gamma} \tag{0.49}$$

and substituting in (0.48) leads to

$$\begin{aligned} &\langle b, w \rangle + \langle \tilde{u}, \mathscr{L}^*(w) \rangle \\ &\quad + \langle (\mathbf{F}(\tilde{u}) - \mathbf{f}), \mathbf{w}_1 + \mathbf{F}^*(w), \mathbf{G}(\tilde{u}) - \mathbf{F}(\tilde{u}), \mathbf{G}^*(w) \rangle_{\Gamma_1} \\ &\quad + \langle (\mathbf{G}(\tilde{u}) - \mathbf{g}), \mathbf{w}_2 + \mathbf{F}^*(w), \mathbf{G}(\tilde{u}) - \mathbf{F}(\tilde{u}), \mathbf{G}^*(w) \rangle_{\Gamma_2} \end{aligned} \tag{0.50}$$

Comparing (0.47) with (0.50) shows that both forms are identical when we select $\mathbf{w}$, and $\mathbf{w}_2$ as follows:

$$\begin{aligned} \mathbf{w}_1 &= \mathbf{G}^*(w) \\ \mathbf{w}_2 &= -\mathbf{F}^*(w) \end{aligned} \tag{0.51}$$

Then, the appropriate weighted residual expression, (0.48), is

$$\langle \mathscr{L}(u) + b, w \rangle + \langle (\mathbf{F}(u) - \mathbf{f}, \mathbf{G}^*(w) \rangle_{\Gamma_1} - \langle (\mathbf{G}(u) - \mathbf{g}), \mathbf{F}^*(w) \rangle_{\Gamma_2} = 0 \tag{0.52}$$

or equivalently

$$\int_{\Omega} \varepsilon\, w\, d\Omega + \int_{\Gamma_1} \varepsilon_1 \cdot \mathbf{G}^*(w)\, d\Gamma - \int_{\Gamma_2} \varepsilon_2 \cdot \mathbf{F}^*(w)\, d\Gamma = 0 \tag{0.53}$$

Equation (0.53) reduces to (0.40) for the self-adjoint case.

## 0.5 Boundary Element Formulation of Poisson Equation

We consider first the case where the governing equation is Poisson's equation. The weighted residuals, are

$$\begin{aligned} \varepsilon_I &= \nabla^2 u + b && \text{in } \Omega \\ \varepsilon_1 &= u - \bar{u} && \text{on } \Gamma_1 \\ \varepsilon_2 &= q - \bar{q} && \text{on } \Gamma_2 \end{aligned} \tag{0.54}$$

where $q = \partial u/\partial n$. Bar superscripts indicate these values are prescribed. Notice that here

$$\begin{aligned} \mathbf{F}(u) &= u, \quad & \mathbf{f} &= \bar{u} \\ \mathbf{G}(u) &= q, \quad & \mathbf{q} &= \bar{q} \end{aligned} \tag{0.55}$$

Substituting in the weighted residual expression, (0.40), results in

$$\int_\Omega (\nabla^2 u + b)\, w\, d\Omega + \int_{\Gamma_1} (u - \bar{u}) \frac{\partial w}{\partial n} d\Gamma + \int_{\Gamma_2} (\bar{q} - q)\, w\, d\Gamma = 0 \tag{0.56}$$

Applying Green's formula to $w\,\nabla^2 u$ transforms (0.56) to

$$\int_\Omega (b\,w + u \nabla^2 w)\, d\Omega + \int_{\Gamma_1} \left( w\,q - \bar{u} \frac{\partial w}{\partial n} \right) d\Gamma + \int_{\Gamma_2} \left( w\,\bar{q} - u \frac{\partial w}{\partial n} \right) d\Gamma = 0 \tag{0.57}$$

This latter form is the one we employ in the boundary element method, along with the fundamental solution

$$\nabla^2 w + \delta(\boldsymbol{x} - \boldsymbol{x}_p) = 0 \tag{0.58}$$

where

$$\begin{aligned} w &= \frac{1}{4\pi r} = \frac{1}{4\pi\,|\boldsymbol{x} - \boldsymbol{x}_p|} && \text{for 3–D} \\ w &= -\frac{1}{2\pi} \ln r = -\frac{1}{2\pi} \ln |\boldsymbol{x} - \boldsymbol{x}_p| && \text{for 2–D} \end{aligned} \tag{0.59}$$

Substituting (0.58) into (0.57) gives for an interior point $P$ the well known result,

$$\begin{aligned} - u_p + \int_\Omega b\,w\,d\Omega + \int_{\Gamma_1} \left( \frac{\partial u}{\partial n} w - \bar{u} \frac{\partial w}{\partial n} \right) d\Gamma \\ + \int_{\Gamma_2} \left( \bar{q}\, w - u \frac{\partial w}{\partial n} \right) d\Gamma = 0 \end{aligned} \tag{0.60}$$

## 0.6 Boundary Element Formulation of Navier Equations

Next we consider the equilibrium equations for an elastic solid which can be written as

$$\boldsymbol{\varepsilon} = \operatorname{div} \sigma + \boldsymbol{b} = \mu + \boldsymbol{u} + (\lambda + \mu) \nabla \operatorname{div} \boldsymbol{u} + \boldsymbol{b} \tag{0.61}$$

$\mu$, $\lambda$ are the Lame constants; div is the divergence operator; $\boldsymbol{u}$ is the displacement vector; $\sigma$ the stress tensor; $\nabla$ the gradient operator; $\boldsymbol{b}$ is the body force vector; and $\boldsymbol{\varepsilon}$

is the error vector. The boundary residuals are

$$\begin{aligned} \varepsilon_1 &= \boldsymbol{u} - \boldsymbol{u} \quad \text{on } \Gamma_1 \\ \varepsilon_2 &= \boldsymbol{p} - \boldsymbol{p} \quad \text{on } \Gamma_2 \end{aligned} \tag{0.62}$$

where $\boldsymbol{p}$ are the tractions on the boundary.

The weighted residual expression in this case becomes

$$\int_{\Omega} \boldsymbol{\varepsilon} \cdot \boldsymbol{w} \, d\Omega + \int_{\Gamma_1} \boldsymbol{\varepsilon}_1 \cdot \boldsymbol{q} \, d\Gamma - \int_{\Gamma_2} \boldsymbol{\varepsilon}_2 \cdot \boldsymbol{w} \, d\Gamma \tag{0.63}$$

where $\boldsymbol{q}$ represents the boundary tractions corresponding to the weighting function displacements $\boldsymbol{w}$.

Integrating experession (0.63) by parts twice we obtain,

$$\begin{aligned} &\int_{\Omega} (\mathbf{b} \cdot \mathbf{w} + \{\mu \Delta \mathbf{w} + (\lambda + \mu) \nabla \operatorname{div} \mathbf{w}\} \cdot \mathbf{u}) \, d\Omega \\ &\qquad + \int_{\Gamma_1} (\boldsymbol{p} \cdot \boldsymbol{w} - \boldsymbol{u} \cdot \boldsymbol{q}) \, d\Gamma + \int_{\Gamma_2} (\boldsymbol{p} \cdot \boldsymbol{w} - \boldsymbol{u} \cdot \boldsymbol{p}) \, d\Gamma \end{aligned} \tag{0.64}$$

We can apply this statement in boundary elements together with the fundamental (Kelvin) solution,

$$\mu \Delta \mathbf{w} + (\lambda + \mu) \nabla \operatorname{div} \boldsymbol{w} + \boldsymbol{\delta} = 0 \tag{0.65}$$

where $\boldsymbol{\delta}$ indicates a vector of dirac delta functions, each component corresponding to a particular direction. Hence equation (0.64) becomes for an internal point $\mathbf{p}$

$$-\boldsymbol{u}_p + \int_{\Gamma} \boldsymbol{b} \cdot \boldsymbol{w} \, d\Omega + \int_{\Gamma_1} (\boldsymbol{p} \cdot \boldsymbol{w} - \boldsymbol{u} \cdot \boldsymbol{q}) \, d\Gamma + \int_{\Gamma_2} (\boldsymbol{p} \cdot \boldsymbol{w} - \boldsymbol{u} \cdot \boldsymbol{q}) \, d\Gamma = 0 \tag{0.66}$$

## 0.7 Final Remarks

This chapter demonstrated how one can establish the correct boundary conditions for a given problem using the concepts of functional analysis.

In addition, the importance and convenience of weighted residual techniques for formulating the governing integral statements are discussed. In particular these techniques are useful when confronted with new problems for which an integral statement is not available. Such is the case with many time dependent and non-linear problems. Weighted residuals provide a simple and very convenient way to initiate the boundary element method.

Some examples are presented in the chapter to illustrate how the governing equations for the boundary element technique can be obtained and the meaning of the different terms.

## Bibliography

Brebbia, C. A., *The Boundary Element Method for Engineers.* Pentech Press, London, Halstead Press, NY, 1978, Second Edition 1980

Brebbia, C. A., S. Walker, *Boundary Element Techniques in Engineering.* Butterworths, London, Boston, 1980

Brebbia, C. A., J. Telles, L. Wrobel, *Boundary Element Techniques – Theory and Applications in Engineering*. Springer-Verlag, Berlin, NY, 1984

Brebbia, C. A. (Ed.), *Progress in Boundary Element Methods, Vol. 1*. Pentech Press, London and Halstead Press, NY, 1981

Brebbia, C. A. (Ed.), *Progress in Boundary Element Methods, Vol. 2*. Pentech Press, London and Springer-Verlag, NY, 1983

Brebbia, C. A. (Ed.), *Recent Advances in Boundary Elements*. Proceedings of the 1st Int. Conf. on BEM. Pentech Press, London, 1978

Brebbia, C. A. (Ed.), *New Developments in Boundary Element Methods*. Proceedings of the 2nd Int. Conf. on BEM, Southampton 1980. CML Publications, Southampton 1980. Second Edition 1983

Brebbia, C. A. (Ed.), *Boundary Element Methods*. Proceedings of the 3rd Int. Conf. on BEM, California 1981, Springer-Verlag, Berlin, NY, 1981

Brebbia, C. A. (Ed.), *Boundary Element Methods in Engineering*. Proceedings of the 4th Int. Conf. on BEM, Southampton 1982. Springer-Verlag, Berlin, NY, 1982

Brebbia, C. A. (Ed.), *Boundary Element Methods*. Proceedings of the 5th Int. Conf. on BEM, Hiroshima, 1983, Springer-Verlag, Berlin, NY, 1983

# Chapter 1

# A Review of the Theory

*by M. A. Jaswon*

## 1.1 Historical Introduction

The modern theory of boundary integral equations began with Fredholm [1], who established the existence of solutions on the basis of his limiting discretisation procedure. It was not envisaged by Fredholm or his immediate successors that solutions could actually be constructed in this way. However the advent of fast digital computers, some 50 years later, opened up the possibility of implementing the discretisation process arithmetically and so enabled numerical solutions of tolerable accuracy to be attempted. This possibility in turn gave a considerable impetus to the development of new and improved boundary integral formulations. In 1962, Hess and Smith [2, 3] formulated a Fredholm integral equation of the second kind for the distribution of simple sources over a surface of revolution. By solving this equation numerically, they were able to compute the perturbation of a uniform potential flow by the surface. In 1963, Jaswon and Ponter [4] threw the torsion problem on to the boundary by formulating an integral equation of the second kind for the warping function, which was solved numerically as a means of computing the torsional rigidity and boundary shear stress for cross-sections inaccessible to other methods of attack. This was one of the first published papers which effectively exploited Green's formula on the boundary, by emphasising its role as a functional relation between the boundary values and normal derivatives of an arbitrary harmonic function. Also in 1963, Jaswon [5] formulated the electrostatic capacitance problem in terms of a Fredholm integral equation of the first kind for the charge distribution, a formulation which had been noted and discarded by Volterra [6] because of apparent difficulties with the two-dimensional theory. This equation was solved numerically by Symm [7, 8 a, 8 b, 9] and adapted by him [10, 11, 12] and others [13, 14, 15] to attack the problem of numerical conformal mapping.

In 1967, Jaswon *et al.* [16, 17] formulated the biharmonic problems of plane elastostatics by means of two coupled integral equations, based upon Almansi's representation of a biharmonic function in terms of two harmonic functions. However, a more far-reaching development was Rizzo's [18] exploitation of Somigliana's formula on the boundary. This provided a functional relation between the boundary displacements and tractions, which immediately led to integral equation formulations covering all the main boundary-value problems of linear elastostatics. Although these are non-Fredholm equations of a vector character, their theory closely parallels that of the corresponding scalar integral equations derived from

Green's formula on the boundary [19]. Much of the mathematical foundation for Rizzo's formulation had already been laid down by Kupradze [20]. However his indirect approach, i.e. employing hypothetical source densities on the boundary, did not make the same impact as Rizzo's direct approach which involved only the quantities of engineering interest, i.e., the boundary tractions and displacements. The essential equivalence between the direct and indirect formulations was demonstrated by Jaswon and Symm [21]. Rizzo's equations were solved numerically by Rizzo *et al.* [22, 23] and vigorously applied by Cruse [24, 25, 26] to problems of technological interest.

The engineering viewpoint has been fostered particularly by Brebbia in his books [27, 28] and reports of conference proceedings [29, 30, 31] under the title Boundary Element Method (BEM). This terminology of course brings out the fact that useful common ground exists between BEM and the longer established Finite Element Method (FEM). Ambitious attempts at combining BEM with FEM have been proposed by Zienkiewicz [32] and by Brebbia and Georgiou [33]. Although these developments are encouraging, it must be said that a practically effective numerical and error analysis of the discretisation procedure over curved boundaries remains to be completed. Some important theoretical contributions by Wendland [34] point the way forward.

The basic scalar and vector formulations are reviewed in this chapter together with the necessary background of theory. Some previously unpublished material is included.

## 1.2 Potential Theory: Green's Formula

*The Simple-Layer Potential*

This provides a specialised representation for harmonic functions. which proves to be particularly convenient for solving certain boundary-value problems. Physically speaking it models the properties of continuous electrostatic charge distributions over closed conductors, so providing an easy entry into the theory. Thus, if charges are introduced on a smooth, closed, conducting surface $\partial B$, we posit a continuous charge density $\sigma(\mathbf{q})$ at every $\mathbf{q} \subset \partial B$. It is convenient to write $dq$ for the area element at $\mathbf{q}$, in which case $\sigma(\mathbf{q})\,dq$ defines the charge strength associated with $dq$. This generates an electrostatic potential $g(\mathbf{p}, \mathbf{q})\,\sigma(\mathbf{q})\,dq$ at any point $\mathbf{p}$ of space, where

$$g(\mathbf{p}, \mathbf{q}) = g(\mathbf{q}, \mathbf{p}) = |\,\mathbf{p} - \mathbf{q}\,|^{-1} \tag{1.1}$$

Superposing the contributions from all over $\partial B$, we construct the simple-layer potential

$$V(\mathbf{p}) = \int_{\partial B} g(\mathbf{p}, \mathbf{q})\,\sigma(\mathbf{q})\,dq\,; \qquad \mathbf{p} \subset B_i, B_e, \partial B \tag{1.2}$$

where $B_i$ denotes the interior domain enclosed by $\partial B$ and $B_e$ denotes the infinite domain exterior to $\partial B$. Although the main properties of $V$ are intuitively clear on physical grounds, its precise mathematical properties depend upon the smoothness of $\partial B$ and of $\sigma$ as will now be discussed.

To ensure that $V$ has the mathematical properties detailed below, we first require that $\sigma$ be Hölder-continuous on $\partial B$. Roughly speaking, Hölder-continuity is stronger than ordinary continuity but weaker than differentiability. More precisely, it means that

$$|\sigma(\mathbf{p}) - \sigma(\mathbf{q})| < D\,|\mathbf{p} - \mathbf{q}|^{\alpha}; \qquad 0 < \alpha \leqq 1,\ D > 1 \tag{1.3}$$

for any $\mathbf{p}, \mathbf{q} \subset \partial B$. The strongest class of Hölder-continuous functions is defined by $\alpha = 1$, since the condition (i.e. the Lipschitz condition)

$$|\sigma(\mathbf{p}) - \sigma(\mathbf{q})| < D\,|\mathbf{p} - \mathbf{q}|; \qquad D > 1 \tag{1.4}$$

implies (1.3) as $|\mathbf{p} - \mathbf{q}| \to 0$ but not vice versa. Hölder-continuity is stronger than ordinary continuity, because any admissible choice of $\alpha$, $D$ in (1.3) guarantees ordinary continuity but not vice-versa. However, differentiability is stronger than (1.4), since the existence of

$$\lim \frac{|\sigma(\mathbf{p}) - \sigma(\mathbf{q})|}{|\mathbf{p} - \mathbf{q}|} \quad \text{as } |\mathbf{p} - \mathbf{q}| \to 0 \tag{1.5}$$

implies (1.4) but not vice-versa. Some simple illustrations appear in [35].

Similar considerations hold for the smoothness properties of $\partial B$, i.e. we require

$$\cos^{-1}(\mathbf{n}_p \cdot \mathbf{n}_q) < D\,|\mathbf{p} - \mathbf{q}|^{\alpha}; \qquad 0 < \alpha \leqq 1,\ D > 0 \tag{1.6}$$

where $\mathbf{n}_p$, $\mathbf{n}_q$ are the unit normal vectors at any $\mathbf{p}, \mathbf{q} \subset \partial B$. This is a Liapunov surface. If $\partial B$ has the local equation $z = z(x, y)$ at any point $\mathbf{p} \subset \partial B$, with the $z$-axis pointing in the normal direction at $\mathbf{p}$, then (1.6) implies that that the partial derivatives $z_x$, $z_y$ are Hölder-continuous at $\mathbf{p}$. However $z_{xx}$, etc., do not necessarily exist at $\mathbf{p}$, i.e. the Liapunov surface has a continuously varying normal direction – but not necessarily a curvature – at every point. A simple example is provided by the circular cylinder capped by hemispheres at each end, since the curvature jumps on passing from the cylinder to the hemisphere whilst the normal direction remains continuous.

Subject to (1.3), (1.6) above, the main properties of $V$ may be summarised as follows [35]:

1) $V$ is continuous and differentiable to any order in $B_i$, $B_e$. Also

$$\frac{\partial V(\mathbf{p})}{\partial p_1} = \int_{\partial B} \frac{\partial g}{\partial p_1}(\mathbf{p}, \mathbf{q})\,\sigma(\mathbf{q})\,dq; \ \ \mathbf{p} \subset B_i, B_e \tag{1.7}$$

etc., where $\mathbf{p} = (p_1, p_2, p_3)$.

2) $V$ satisfies Laplace's equation in $B_i$, $B_e$ i.e.

$$\nabla^2 V(\mathbf{p}) = \int_{\partial B} \nabla^2 g(\mathbf{p}, \mathbf{q})\,\sigma(\mathbf{q})\,dq = 0; \ \ \mathbf{p} \subset B_i, B_e \tag{1.8}$$

since $\nabla^2 g = 0$, $\mathbf{p} \neq \mathbf{q}$. Accordingly, $V$ is a harmonic function everywhere except at $\partial B$.

3)
$$V(\mathbf{p}) = |\mathbf{p}|^{-1} \int_{\partial B} \sigma(\mathbf{q})\,dq + |\mathbf{p}|^{-3} \int_{\partial B} (\mathbf{p} \cdot \mathbf{q})\,\sigma(\mathbf{q})\,dq + 0\,|\mathbf{p}|^{-3} \tag{1.9}$$

as $|\mathbf{p}| \to \infty$. Accordingly, $V$ is a regular harmonic function in $B_e$.

4) $V(\mathbf{p})$ exists at every $\mathbf{p} \subset \partial B$, and it is continuous at $\mathbf{p}$ with respect to its neighbouring values in $B_i$, $B_e$ i.e.

$$\left.\begin{aligned} V(\mathbf{p}_i) &\to V(\mathbf{p}) \quad \text{as} \quad \mathbf{p}_i \to \mathbf{p};\ \mathbf{p}_i \subset B_i \\ V(\mathbf{p}_e) &\to V(\mathbf{p}) \quad \text{as} \quad \mathbf{p}_e \to \mathbf{p};\ \mathbf{p}_e \subset B_e \end{aligned}\right\} \tag{1.10}$$

5) $V$ is continuous and differentiable on $\partial B$. Also, in line with (1.7),

$$\frac{\partial V(\mathbf{p})}{\partial t} = \int_{\partial B} \frac{\partial g}{\partial t}(\mathbf{p}, \mathbf{q})\, \sigma(\mathbf{q})\, dq; \qquad \mathbf{p} \subset \partial B\,, \tag{1.11}$$

where $\partial/\partial t$ denotes differentiation along any tangential direction to $\partial B$ at $\mathbf{p}$.

6) $V$ has two formally distinct normal derivatives $V_e'$, $V_i'$ at $\mathbf{p} \subset \partial B$ pointing into $B_e$, $B_i$ respectively. These may be constructed by writing

$$V_e'(\mathbf{p}) = \int_{\partial B} g_e'(\mathbf{p}, \mathbf{q})\, \sigma(\mathbf{q})\, dq - 2\pi\, \sigma(\mathbf{p}); \qquad \mathbf{p} \subset \partial B \tag{1.12}$$

$$V_i'(\mathbf{p}) = \int_{\partial B} g_i'(\mathbf{p}, \mathbf{q})\, \sigma(\mathbf{q})\, dq - 2\pi\, \sigma(\mathbf{p}); \qquad \mathbf{p} \subset \partial B \tag{1.13}$$

where $g_e'(\mathbf{p}, \mathbf{q})$ denotes the exterior normal derivative of $g$ at $\mathbf{p}$ keeping $\mathbf{q}$ fixed, and similarly for $g_i'$. Since

$$g_i'(\mathbf{p}, \mathbf{q}) + g_e'(\mathbf{p}, \mathbf{q}) = 0\,, \tag{1.14}$$

it follows that

$$V_i'(\mathbf{p}) + V_e'(\mathbf{p}) = -4\pi\, \sigma(\mathbf{p}) \tag{1.15}$$

The integrands in (1.12), (1.13) become singular when $\mathbf{q} = \mathbf{p}$. For example, if $\partial B$ is a sphere of radius $a$, with

$$\mathbf{p} = (0, 0, r)_{r=a}, \qquad \mathbf{q} = a\,(\sin\theta \cos\psi, \sin\theta \sin\psi, \cos\theta)\,,$$

then

$$g(\mathbf{p}, \mathbf{q}( = (a^2 + r^2 - 2\,\mathrm{arcos}\,\theta)_{r=a}^{-1/2} = \left(2a \sin\frac{\theta}{2}\right)^{-1},$$

$$\left.\begin{aligned} g_e'(\mathbf{p}, \mathbf{q}) \\ -g_i'(\mathbf{p}, \mathbf{q}) \end{aligned}\right\} = \left(\frac{\partial g}{\partial r}\right)_{r=a} = \left\{\frac{r - \alpha\cos\theta}{(a^2 + r^2 - 2\,\mathrm{arcos}\,\theta)^{3/2}}\right\}_{r=a} = \left(-4a^2 \sin\frac{\theta}{2}\right)^{-1},$$

showing that

$$\lim g_e'(\mathbf{p}, \mathbf{q}) \quad \text{as} \quad \mathbf{q} \to \mathbf{p} = \lim \left(-4a^2 \sin\frac{\theta}{2}\right)^{-1} \quad \text{as} \quad \theta \to 0, \quad \text{etc.} \tag{1.16}$$

However, the integrals exist and may be readily evaluated to give

$$\int_{\partial B} g_e'(\mathbf{p}, \mathbf{q})\, dq = -\int_{\partial B} g_i'(\mathbf{p}, \mathbf{q})\, dq = -2\pi; \quad \mathbf{p} \subset \partial B \tag{1.17}$$

on putting $dq = a^2 \sin d\theta\, d\psi$. It is straightforward, though tedious, to perform the integrations for $\mathbf{p} \subset B_e, B_i$ i.e. $r > a$, $r < a$ respectively. We find

$$\int_{\partial B} \frac{\partial g_e}{\partial r}\, dq = -\frac{4\pi a^2}{r^2}, \qquad \int_{\partial B} \frac{\partial g_i}{\partial r}\, dq = 0 \tag{1.18}$$

and clearly these integrals jump by $2\pi$ when $r=a$ in consequence of the singularity. A useful exercise is to construct the $V$ generated by $\sigma=\sigma_0$ ($a$ cons.) on $r=a$. Simple analysis gives

$$V_e = \frac{4\pi a^2 \sigma_0}{r}, \qquad V_i = 4\pi a \sigma_0, \tag{1.19}$$

from which

$$\frac{\partial V_e}{\partial r} = -\frac{4\pi a^2 \sigma_0}{r^2}, \qquad \frac{\partial V_i}{\partial r} = 0 \tag{1.20}$$

in accordance with (1.18) since

$$\frac{\partial V_e}{\partial r} = \frac{\partial}{\partial r} \int_{\partial B} g_e(\mathbf{p}, \mathbf{q})\, \sigma_0\, dq = \sigma_0 \int_{\partial B} \frac{\partial g_e}{\partial r}\, dq, \quad \text{etc.} \tag{1.21}$$

At $r=a$, the expressions (1.20) become

$$\left(\frac{\partial V_e}{\partial r}\right)_{r=a} = -4\pi\sigma_0 \equiv V_e'(\mathbf{p}), \qquad \left(\frac{\partial V_i}{\partial r}\right)_{r=a} = 0 \equiv V_i'(\mathbf{p}) \tag{1.22}$$

These may alternatively be obtained by substituting (1.17) into (1.12), (1.13):

$$V_e'(\mathbf{p}) = -2\pi\sigma_0 - 2\pi\sigma_0 = -4\pi\sigma_0, \qquad V_i'(\mathbf{p}) = 2\pi\sigma_0 - 2\pi\sigma_0 = 0 \tag{1.23}$$

i.e. the contribution $-2\pi\sigma$ in (1.12), (1.13) cancels the jump in the integrals (1.18) at $\partial B$. This inclusion may be generalised to any Liapunov surface by an appropriate limiting analysis.

*The Double-Layer Potential*

This provides a specialised representation for harmonic functions of equal importance to that of the simple-layer potential. Physically speaking it models the properties of continuous dipole distributions, e.g. elementary magnets, over open or closed smooth surfaces. A unit dipole at $\mathbf{q} \subset \partial B$ generates the dipole potential $g(\mathbf{p}, \mathbf{q})'$, i.e. the normal derivative of $g$ at $\mathbf{q}$ keeping $\mathbf{p}$ fixed, at any point $p$ of space. As before, we introduce two formally distinct normal derivatives $g(\mathbf{p}, \mathbf{q})_e'$, $g(\mathbf{p}, \mathbf{q})_i'$ at $\mathbf{q}$ whether for closed or open boundaries, and which have equal status since $g_e' = -g_i'$. Utilising $g(\mathbf{p}, \mathbf{q})_i'$, a dipole of strength-density $\mu(\mathbf{q})$ per unit area at $\mathbf{q}$ generates the potential $g(\mathbf{p}, \mathbf{q})_i' \mu(\mathbf{q})\, dq$ at $\mathbf{p}$. Superposing the contributions from all over $\partial B$, we construct the double-layer potential

$$W(\mathbf{p}) = \int_{\partial B} g(\mathbf{p}, \mathbf{q})_i' \mu(\mathbf{q})\, dq; \qquad \mathbf{p} \not\subset \partial B \tag{1.24}$$

at any point $\mathbf{p}$ outside $\partial B$.

Subject to $\mu$ being Hölder-continuous and $\partial B$ a Liapunov surface, the main properties of $W$ may be summarised as follows [35]:

1) $W$ is continuous and differentiable to any order outside $\partial B$. Also

$$\frac{\partial W(\mathbf{p})}{\partial p_1} = \int_{\partial B} \frac{\partial g}{\partial p_1} (\mathbf{p}, \mathbf{q})_i' \mu(\mathbf{q})\, dq; \qquad \mathbf{p} \not\subset \partial B \tag{1.25}$$

etc., where $\mathbf{p} = (p_1, p_2, p_3)$.

2) $W$ satisfies Laplace's equation everywhere outside $\partial B$, i.e.

$$\nabla^2 W(\mathbf{p}) = \int_{\partial B} \nabla^2 g(\mathbf{p}, \mathbf{q})'_i \, \mu(\mathbf{q}) \, dq = 0; \qquad \mathbf{p} \not\subset \partial B \tag{1.26}$$

since $\nabla^2 g(\mathbf{p}, \mathbf{q})'_i = 0$, $\mathbf{p} \neq \mathbf{q}$. Accordingly, $W$ is a harmonic function everywhere except at $\partial B$.

3)
$$W(\mathbf{p}) = 0 \, | \, \mathbf{p} \, |^{-2} \quad \text{as} \quad | \, \mathbf{p} \, | \to \infty, \tag{1.27}$$

so that $W$ can only represent a restricted class of regular harmonic functions as $|\mathbf{p}| \to \infty$.

4) $W(\mathbf{p})$ exists at every $\mathbf{p} \subset \partial B$, but it jumps on passing from $\partial B$ to neighbouring points outside $\partial B$. More precisely, if $\mathbf{p}_i$, $\mathbf{p}_e$ are points on each side of $\partial B$ near $\mathbf{p}$, then

$$\left.\begin{aligned} \lim_{\mathbf{p}_i \to \mathbf{p}} W(\mathbf{p}_i) &= W(\mathbf{p}) + 2\pi \mu(\mathbf{p}) \\ \lim_{\mathbf{p}_e \to \mathbf{p}} W(\mathbf{p}_e) &= W(\mathbf{p}) - 2\pi \mu(\mathbf{p}) \end{aligned}\right\}; \quad \mathbf{p} \subset \partial B \tag{1.28}$$

assuming that $W(\mathbf{p})$ is defined everywhere by (1.24).

5) Even if $\mu$ is Hölder-continuous, the normal derivatives $W'_i$, $W'_e$ are not necessarily finite though they satisfy the relation

$$\lim_{\mathbf{p}_i \to \mathbf{p}} W'_i(\mathbf{p}_i) + \lim_{\mathbf{p}_e \to \mathbf{p}} W'_e(\mathbf{p}_e) = 0; \quad \mathbf{p} \subset \partial B \tag{1.29}$$

6) The separate limits in (1.29) exist if $\mu$ is Hölder-continuously differentiable, in which case the tangential derivative of $W$ also exists:

$$\frac{\partial W(\mathbf{p})}{\partial t} = \int_{\partial B} \frac{\partial g}{\partial t}(\mathbf{p}, \mathbf{q})'_i \, \mu(\mathbf{q}) \, dq; \quad \mathbf{p} \subset \partial B \tag{1.30}$$

as in (1.11). Also $\partial W / \partial t$ satisfies the limiting relations

$$\left.\begin{aligned} \lim_{\mathbf{p}_i \to \mathbf{p}} \frac{\partial W(\mathbf{p}_i)}{\partial t_i} &= \frac{\partial W(\mathbf{p})}{\partial t} + 2\pi \frac{\partial \mu(\mathbf{p})}{\partial t} \\ \lim_{\mathbf{p}_e \to \mathbf{p}} \frac{\partial W(\mathbf{p}_e)}{\partial t_e} &= \frac{\partial W(\mathbf{p})}{\partial t} - 2\pi \frac{\partial \mu(\mathbf{p})}{\partial t} \end{aligned}\right\} \tag{1.31}$$

where $\partial/\partial t_i$, $\partial/\partial t_e$ denote differentiations parallel to $\partial/\partial t$ at $\mathbf{p}_i$, $\mathbf{p}_e$ respectively.

Given $\mu = \mu_0$ (a cons.) over an open $\partial B$, we obtain

$$W(\mathbf{p}) = \mu_0 \int_{\partial B} g(\mathbf{p}, \mathbf{q})'_i \, dq = \mu_0 \Omega \tag{1.32}$$

where $\Omega$ denotes the solid angle subtended at $\mathbf{p}$ by $\partial B$. This interpretation immediately yields the result

$$\int_{\partial B} g(\mathbf{p}, \mathbf{q})'_i \, dq = 4\pi; \quad \mathbf{p} \subset B_i \tag{1.33}$$

for $\partial B$ closed, which may be viewed as the Gauss flux theorem applied to a unit simple source at $\mathbf{p}$. It follows from (1.28) that

$$\int_{\partial B} g(\mathbf{p}, \mathbf{q})'_i \, dq = 2\pi; \quad \mathbf{p} \subset \partial B \tag{1.34}$$

$$= 0; \quad \mathbf{p} \subset B_e \tag{1.35}$$

As will be seen later, these simple topological results play an important role in the theory of Fredholm integral equations. For ease of reference, we note the corresponding results:

$$\int_{\partial B} g(\mathbf{p}, \mathbf{q})'_e \, dq = -4\pi; \quad \mathbf{p} \subset B_i \tag{1.36}$$

$$= -2\pi; \quad \mathbf{p} \subset \partial B \tag{1.37}$$

$$= \quad 0; \quad \mathbf{p} \subset B_e \tag{1.38}$$

These may be verified for the sphere of radius $a$ by choosing

$$\mathbf{p} = (0, 0, z), \quad \mathbf{q} = (r \sin\theta \cos\psi, r \sin\theta \sin\psi, r \cos\theta)_{r=a},$$

so that

$$dq = a^2 \sin\theta \, d\theta \, d\psi,$$

$$g(\mathbf{p}, \mathbf{q}) = (z^2 + r^2 - 2zr\cos\theta)^{-1/2}_{r=a},$$

$$g(\mathbf{p}, \mathbf{q})'_e = \left(\frac{\partial g}{\partial r}\right)_{r=a} = -\left\{\frac{r - z\cos\theta}{(z^2 + r^2 - 2zr\cos\theta)^{3/2}}\right\}_{r=a},$$

and performing the integration for the cases $z < a$, $z = a$, $z > a$ respectively.

*Green's Formula*

It has been noted above that the simple-layer and double-layer potentials are harmonic functions under broad conditions. However, an arbitrary harmonic function may not always be representable by such potentials. For instance, the harmonic function $\varphi = k$ (*a cons.*) can not be represented by a simple-layer potential inside the unit circle. Again, the harmonic function $\varphi = r^{-1}$ can not be represented by a double-layer potential in the infinite domain exterior to a closed surface. To construct a more powerful potential representation for harmonic functions, we posit a harmonic function $\varphi$ in $B_i$, which assumes a continuous set of boundary values $\varphi(\mathbf{q})$, and boundary normal derivatives $\varphi_i(\mathbf{q})$, as $\mathbf{q}$ runs over $\partial B$. These boundary data, regarded as source densities, generate the double-layer potential

$$\int_{\partial B} g(\mathbf{p}, \mathbf{q})'_i \, \varphi(\mathbf{q}) \, dq, \tag{1.39}$$

and the simple-layer potential

$$-\int_{\partial B} g(\mathbf{p}, \mathbf{q}) \, \varphi'_e(\mathbf{q}) \, dq, \tag{1.40}$$

which have the properties associated with $W$, $V$ respectively for Hölder-continuous boundary data on a Liapunov surface $\partial B$. Superposing (1.39) and (1.40) yields the identity

$$\int_{\partial B} g(\mathbf{p}, \mathbf{q})'_i \, \varphi(\mathbf{q}) \, dq - \int_{\partial B} g(\mathbf{p}, \mathbf{q}) \, \varphi'_i(\mathbf{q}) \, dq = 4\pi \varphi(\mathbf{p}); \quad \mathbf{p} \subset B_i \tag{1.41}$$

valid for any harmonic $\varphi$ in $B_i$. This is Green's formula [35].

Formula (1.41) can be readily extended from $\mathbf{p} \subset B_i$ to $\mathbf{p} \subset \partial B$. Thus (1.40) remains continuous, but (1.39) jumps by $-2\pi\varphi(\mathbf{p})$, at $\partial B$ so providing the boundary formula

$$\int_{\partial B} g(\mathbf{p}, \mathbf{q})'_i \, \varphi(\mathbf{q}) \, dq - \int_{\partial B} g(\mathbf{p}, \mathbf{q}) \, \varphi'_i(\mathbf{q}) \, dq = 2\pi \varphi(\mathbf{p}); \quad \mathbf{p} \subset \partial B \tag{1.42}$$

This differs essentially from (1.41) in that $\varphi$ on the right-hand side is now a boundary value of $\varphi$, i.e. from the same set as enters into the first integral, i.e. (1.42) is a functional relation between $\varphi$, $\varphi_i'$ on $\partial B$ which ensures their compatibility as boundary data. As $\mathbf{p}$ crosses from $\partial B$ to $B_e$, the double-layer integral suffers a second jump $-2\pi\varphi(\mathbf{p})$ yielding Green's reciprocal theorem

$$\int_{\partial B} g(\mathbf{p},\mathbf{q})_i'\varphi(\mathbf{q})\,dq - \int_{\partial B} g(\mathbf{p},\mathbf{q})\varphi_i'(\mathbf{q})\,dq = 0; \quad \mathbf{p}\subset B_e \tag{1.43}$$

This may be proved directly by noting that $g(\mathbf{p},\mathbf{q})$ is a harmonic function of $\mathbf{q}$ in $B_i$ for any fixed $\mathbf{p}\subset B_e$: also $\varphi(\mathbf{q})$ is a harmonic function of $\mathbf{q}$ in $B_i$, and any two harmonic functions $\varphi$, $\psi$ in $B_i$ satisfy the reciprocal relation

$$\int_{\partial B}(\varphi\psi_i' - \psi\varphi_i')\,dq = \int_{B_i}(\varphi\nabla^2\psi - \psi\nabla^2\varphi)\,dq = 0 \tag{1.44}$$

as follows directly from the Gauss divergence theorem. Starting from (1.43), and reversing our steps, we successively recover (1.42), (1.41). Of course, this procedure hinges upon the validity of the boundary jumps (1.28), which can only be justified by the same kind of limiting analysis as would be involved in the direct proof of (1.41).

Green's formula may be readily adapted to a regular harmonic function $f$ in $B_e$, which assumes boundary values $f(\mathbf{q})$, and boundary normal derivatives $f_e'(\mathbf{q})$, as $\mathbf{q}$ runs over $\partial B$. Corresponding with (1.41), (1.42), (1.43) we have

$$\int_{\partial B} g(\mathbf{p},\mathbf{q})_e' f(\mathbf{q})\,dq - \int_{\partial B} g(\mathbf{p},\mathbf{q}) f_e'(\mathbf{q})\,dq = 4\pi f(\mathbf{p}); \quad \mathbf{p}\subset B_e \tag{1.45}$$

$$= 2\pi f(\mathbf{p}); \quad \mathbf{p}\subset \partial B \tag{1.46}$$

$$= 0; \quad \mathbf{p}\subset B_i \tag{1.47}$$

respectively, where

$$f = -\frac{1}{4\pi}|\mathbf{p}|^{-1}\int_{\partial B} f_e'(\mathbf{q})\,dq + 0\,|\mathbf{p}|^{-2} \quad \text{as} \quad |\mathbf{p}|\to\infty \tag{1.48}$$

All these exterior formulae are important in their own right, and (1.47) may be exploited to justify the single-potential representations (1.2), (1.24). Thus superposing (1.41) and (1.47), assuming $\varphi$ given and $f$ arbitrary, we obtain the generalised representation

$$\int_{\partial B} g(\mathbf{p},\mathbf{q})_i'[\varphi(\mathbf{q}) - f(\mathbf{q})]\,dq - \int_{\partial B} g(\mathbf{p},\mathbf{q})[\varphi_i'(\mathbf{q}) + f_e'(\mathbf{q})]\,dq = 4\pi\varphi(\mathbf{p}); \quad \mathbf{p}\subset B_i. \tag{1.49}$$

Two natural possibilities for $f$ now arise:

1) $f = \varphi$ on $\partial B$, giving the representation

$$-\int_{\partial B} g(\mathbf{p},\mathbf{q})[\varphi_i'(\mathbf{q}) + f_e'(\mathbf{q})]\,dq = 4\pi\varphi(\mathbf{p}); \quad \mathbf{p}\subset B_i, \tag{1.50}$$

which may be identified as a simple-layer potential with source density

$$\sigma = -\frac{1}{4\pi}[\varphi_i' + f_e'] \tag{1.51}$$

This possibility hinges upon the existence of a unique regular $f$ in $B_e$ satisfying $f = \varphi$ on $\partial B$, as is guaranteed by the exterior Dirichlet existence theorem in three dimensions. However, a breakdown may occur in two dimensions as will be discussed later.

2) $f'_e = -\varphi'_i$ on $\partial B$, giving the representation

$$\int_{\partial B} g(\mathbf{p}, \mathbf{q})'_i [\varphi(\mathbf{q}) - f(\mathbf{q})]\, dq = 4\pi\varphi(\mathbf{p}); \quad \mathbf{p} \subset B_i \tag{1.52}$$

which may be identified as a double-layer potential with source density

$$\mu = \frac{1}{4\pi}[\varphi - f] \tag{1.53}$$

This possibility hinges upon the existence of a unique regular $f$ in $B_e$ satisfying $f'_e = -\varphi'_i$ on $\partial B$, as is always guaranteed by the exterior Neumann existence theorem.

Accordingly we have recovered the simple-layer potential and the double-layer potential as specialised versions of Green's formula.

## 1.3 Boundary Integral Equations

*Simple-Layer Formulations*

According to the Dirichlet existence-uniqueness theorem, there exists a unique harmonic function $\varphi$ in $B_i$ which assumes prescribed continuous boundary values on a Liapunov surface $\partial B$. To construct $\varphi$ in $B_i$, we write

$$\varphi(\mathbf{p}) = \int_{\partial B} g(\mathbf{p}, \mathbf{q})\, \sigma(\mathbf{q})\, dq; \quad \mathbf{p} \subset B_i \tag{1.54}$$

where $\sigma$ appears as a hypothetical Hölder-continuous source density to be determined. In principle $\sigma = -\frac{1}{4\pi}(\varphi'_i + f'_e)$, but neither $\varphi'_i$ nor $f'_e$ are known *ab initio.* An effective way forward is to note that both sides of (1.54) remain continuous at $\partial B$, i.e. both $\varphi$ and its representation remain continuous at $\partial B$, so yielding the boundary relation

$$\int_{\partial B} g(\mathbf{p}, \mathbf{q})\, \sigma(\mathbf{q})\, dq = \varphi(\mathbf{p}); \quad \mathbf{p} \subset \partial B \tag{1.55}$$

This may be viewed as a Fredholm integral equation of the first kind for $\sigma$ in terms of $\varphi$ on $\partial B$, with a unique solution enabling us to generate $\varphi$ throughout $B_i$ from (1.54).

According to the exterior Dirichlet existence-uniqueness theorem, there exists a unique regular harmonic function $\varphi$ in $B_e$ which assumes prescribed continuous boundary values on $\partial B$. Clearly $\varphi$ may be constructed by solving (1.55) as before and utilising $\sigma$ to generate the simple-layer potential

$$\varphi(\mathbf{p}) = \int_{\partial B} g(\mathbf{p}, \mathbf{q})\, \sigma(\mathbf{q})\, dq; \quad \mathbf{p} \subset B_e \tag{1.56}$$

Despite its power and simplicity, equation (1.55) has not yet been widely exploited in the solution of Dirichlet problems. However, an exceptional case is $\varphi = 1$ i.e. the capacitance problem of electrostatics. In this case the charge density $\lambda$ satisfies the equation

$$\int_{\partial B} g(\mathbf{p}, \mathbf{q})\, \lambda(\mathbf{q})\, dq = 1; \quad \mathbf{p} \subset \partial B \tag{1.57}$$

and we note that

1) $\lambda > 0$ on $\partial B$, so providing the capacitance

$$\varkappa = \int_{\partial B} \lambda(\mathbf{q})\, dq > 0; \tag{1.58}$$

2) $\lambda$ generates the simple-layer potential

$$\int_{\partial B} g(\mathbf{p}, \mathbf{q})\, \lambda(\mathbf{q})\, dq = 1, \quad \mathbf{p} \subset B_i; \tag{1.59}$$

3) $\lambda$ satisfies, in addition to (1.57), the normal derivative equation (1.65) presented below.

Equation (1.57) has been solved numerically for various closed surfaces as a means of computing their electrostatic capacitance.

According to the interior Neumann existence theorem, there exists a unique (up to an arbitrary constant) harmonic function $\varphi$ in $B_i$, which assumes prescribed continuous normal-derivative values $\varphi'_i$ on $\partial B$ subject to the Gauss condition

$$\int_{\partial B} \varphi'_i(\mathbf{q})\, dq = 0 \tag{1.60}$$

To construct $\varphi$ in $B_i$, we introduce the representation (1.54) as before, with $\varphi'_i$ assumed Hölder-continuous on a Liapunov surface $\partial B$. This yields the boundary relation

$$\int_{\partial B} g'_i(\mathbf{p}, \mathbf{q})\, \sigma(\mathbf{q})\, dq - 2\pi\, \sigma(\mathbf{p}) = \varphi'_i(\mathbf{p}); \quad \mathbf{p} \subset \partial B \tag{1.61}$$

which may be viewed as a Fredholm integral equation of the second kind for $\sigma$ in terms of $\varphi'_i$. Despite the singularity in the kernel, see (1.16), classical Fredholm theory applies: the homogeneous equation

$$\int_{\partial B} g'_i(\mathbf{p}, \mathbf{q})\, \sigma(\mathbf{q})\, dq - 2\pi\, \sigma(\mathbf{p}) = 0; \quad \mathbf{p} \subset \partial B \tag{1.62}$$

has a non-trivial solution $\sigma = \Lambda$, corresponding with the non-trivial solution $\mu = 1$ of the adjoint homogeneous equation

$$\int_{\partial B} g(\mathbf{p}, \mathbf{q})'_i\, \mu(\mathbf{q})\, dq - 2\pi\, \mu(\mathbf{p}) = 0; \quad \mathbf{p} \subset \partial B \tag{1.63}$$

as may be seen from (1.34), so that a solution exists subject to the orthogonality condition

$$\int_{\partial B} \mu(\mathbf{p})\, \varphi'_i(\mathbf{p})\, dp = \int_{\partial B} \varphi'_i(\mathbf{p})\, dp = 0 \tag{1.64}$$

i.e. the Gauss condition (1.60). By suitable normalisation, $\Lambda$ may be identified with the unique solution, $\lambda$, of equation (1.57), since this yields the normal derivative equation

$$\int_{\partial B} g'_i(\mathbf{p}, \mathbf{q})\, \lambda(\mathbf{q})\, dq - 2\pi\, \lambda(\mathbf{p}) = 0; \quad \mathbf{p} \subset \partial B \tag{1.65}$$

Accordingly, equation (1.61) has the general solution

$$\sigma = \sigma_0 + k\,\lambda \tag{1.66}$$

within the space of Hölder-continuous functions, where $\sigma_0$ is a particular solution and $k$ is an arbitrary constant. This solution generates the class of potentials

$$\int_{\partial B} g(\mathbf{p}, \mathbf{q})\, \sigma_0(\mathbf{q})\, dq + k \int_{\partial B} g(\mathbf{p}, \mathbf{q})\, \lambda(\mathbf{q})\, dq \tag{1.67}$$

i.e.

$$\int_{\partial B} g(\mathbf{p}, \mathbf{q})\, \sigma_0(\mathbf{q})\, dq + k; \quad \mathbf{p} \subset B_i + \partial B \tag{1.68}$$

each characterised by the same normal derivative $\varphi_i'$ on $\partial B$.

According to the exterior Neumann existence-uniqueness theorem, there exists a unique regular harmonic function $\varphi$ in $B_e$, which assumes prescribed continuous normal-derivative values $\varphi_e'$ on $\partial B$. The Gauss condition (1.60) is not necessary. Also, $\varphi$ may not be unique if $\partial B$ has a sharp edge, so providing the fundamental loophole for aerofoil theory. Utilising the representation (1.56), we obtain the integral equation

$$\int_{\partial B} g_e'(\mathbf{p}, \mathbf{q})\, \sigma(\mathbf{q})\, dq - 2\pi\, \sigma(\mathbf{p}) = \varphi_e'(\mathbf{p}); \quad \mathbf{p} \subset \partial B \tag{1.69}$$

for $\sigma$ in terms of $\varphi_e'$. In this case, however, the homogeneous equation

$$\int_{\partial B} g_e'(\mathbf{p}, \mathbf{q})\, \sigma(\mathbf{q})\, dq - 2\pi\, \sigma(\mathbf{p}) = 0; \quad \mathbf{p} \subset \partial B \tag{1.70}$$

and therefore also its associated adjoint equation

$$\int_{\partial B} g(\mathbf{p}, \mathbf{q})_e'\, \tau(\mathbf{q})\, dq - 2\pi\, \tau(\mathbf{p}) = 0; \quad \mathbf{p} \subset \partial B \tag{1.71}$$

has no non-trivial solution. Accordingly, equation (1.69) has a unique solution $\sigma_0$, which generates the unique exterior potential

$$\int_{\partial B} g(\mathbf{p}, \mathbf{q})\, \sigma_0(\mathbf{q})\, dq; \quad \mathbf{p} \subset B_e + \partial B \tag{1.72}$$

characterised by $\varphi_e'$ on $\partial B$.

Operating with $\int_{\partial B} \ldots dp$ upon both sides of (1.69), we find

$$\int_{\partial B} \varphi_e'(\mathbf{p})\, dp = \int_{\partial B} dp \left\{ \int_{\partial B} g_e'(\mathbf{p}, \mathbf{q})\, \sigma(\mathbf{q})\, dq \right\} - 2\pi \int_{\partial B} \sigma(\mathbf{p})\, dp$$

$$= \int_{\partial B} \sigma(\mathbf{q})\, dq \int_{\partial B} g_e'(\mathbf{p}, \mathbf{q})\, dp - 2\pi \int_{\partial B} \sigma(\mathbf{p})\, dp$$

where the order of the double integral has been interchanged by invoking Fubini's theorem [36]. Now

$$\int_{\partial B} g_e'(\mathbf{p}, \mathbf{q})\, dp = \int_{\partial B} g(\mathbf{p}, \mathbf{q})_e'\, dq = -2\pi; \quad \mathbf{q}, \mathbf{p} \subset \partial B$$

from (1.37), whence

$$\int_{\partial B} \varphi_e'(\mathbf{p})\, dp = -2\pi \int_{\partial B} \sigma(\mathbf{q})\, dq - 2\pi \int_{\partial B} \sigma(\mathbf{p})\, dp = -4\pi \int \sigma(\mathbf{p})\, dp \tag{1.73}$$

This provides a useful check on $\sigma$, and it also shows – from (1.9) – that

$$\varphi = 0\,(r^{-2}) \quad \text{as} \quad r \to \infty \quad \text{if} \quad \int_{\partial B} \varphi_e'(\mathbf{p})\, dp = 0$$

The same procedure applied to (1.61) gives the expected result $\int_{\partial B} \varphi_i'(\mathbf{p})\, dp = 0$. An important companion result is

$$\int_{\partial B} \varphi(\mathbf{p})\, \lambda(\mathbf{p})\, dp = \int_{\partial B} \sigma(\mathbf{p})\, dp \tag{1.74}$$

as may be proved by operating with $\int_{\partial B} \ldots \lambda(\mathbf{p})\, dp$ upon both sides if (1.55) and again invoking Fubini's theorem:

$$\int_{\partial B} \varphi(\mathbf{p})\, \lambda(\mathbf{p})\, dp = \int_{\partial B} \lambda(\mathbf{p}) \left\{ \int_{\partial B} g(\mathbf{p}, \mathbf{q})\, \sigma(\mathbf{q})\, dq \right\} dp$$

$$= \int_{\partial B} \sigma(\mathbf{q})\, dq \int_{\partial B} g(\mathbf{p}, \mathbf{q})\, \lambda(\mathbf{p})\, dp = \int_{\partial B} \sigma(\mathbf{q})\, dq$$

since

$$\int_{\partial B} g(\mathbf{p}, \mathbf{q})\, \lambda(\mathbf{p})\, dp = \int_{\partial B} g(\mathbf{q}, \mathbf{p})\, \lambda(\mathbf{p})\, dp = 1; \quad \mathbf{q}, \mathbf{p} \subset \partial B$$

*Double-Layer Formulations*

The representation

$$\varphi(\mathbf{p}) = \int_{\partial B} g(\mathbf{p}, \mathbf{q})_i'\, \mu(\mathbf{q})\, dq; \quad \mathbf{p} \subset B_i \tag{1.75}$$

provides a classically preferred alternative to (1.54) for Dirichlet problems, where $\mu$ appears as a hypothetical Hölder-continuous source density to be determined. In principle $\mu = \frac{1}{4\pi}(\varphi - f)$, but of course $f$ is not available *ab initio*. An effective way forward is to note that $\varphi$ remains continuous at $\partial B$, whereas the integral jumps by $-2\pi\mu(\mathbf{p})$ at $\mathbf{p} \subset \partial B$, so yielding the boundary relation

$$\int_{\partial B} g(\mathbf{p}, \mathbf{q})_i'\, \mu(\mathbf{q})\, dq + 2\pi\mu(\mathbf{p}) = \varphi(\mathbf{p}); \quad \mathbf{p} \subset \partial B \tag{1.76}$$

This may be viewed as a Fredholm integral equation of the second kind for $\mu$ in terms of $\varphi$, to which classical Fredholm theory applies: the homogeneous equation

$$\int_{\partial B} g(\mathbf{p}, \mathbf{q})_i'\, \mu(\mathbf{q})\, dq + 2\pi\mu(\mathbf{p}) = 0; \quad \mathbf{p} \subset \partial B \tag{1.77}$$

is mathematically equivalent to (1.71) and therefore has no non-trivial solution; consequently equation (1.76) has a unique solution, which generates $\varphi$ in $B_i$ via (1.75).

The representation

$$\varphi(\mathbf{p}) = \int_{\partial B} g(\mathbf{p}, \mathbf{q})_e'\, \mu(\mathbf{q})\, dq; \quad \mathbf{p} \subset B_e \tag{1.78}$$

does not, in general, provide an alternative to (1.56) for exterior Dirichlet problems. Thus, replacing equation (1.76) by its exterior version

$$\int_{\partial B} g(\mathbf{p}, \mathbf{q})_e'\, \mu(\mathbf{q})\, dq + 2\pi\mu(\mathbf{p}) = \varphi(\mathbf{p}); \quad \mathbf{p} \subset \partial B \tag{1.79}$$

we obtain the homogeneous equation

$$\int_{\partial B} g(\mathbf{p}, \mathbf{q})_e'\, \mu(\mathbf{q})\, dq + 2\pi\mu(\mathbf{p}) = 0; \quad \mathbf{p} \subset \partial B \tag{1.80}$$

and associated adjoint equation

$$\int_{\partial B} g'_e(\mathbf{p}, \mathbf{q})\, \Lambda(\mathbf{q})\, dq + 2\pi \Lambda(\mathbf{p}) = 0; \quad \mathbf{p} \subset \partial B \tag{1.81}$$

which are mathematically equivalent to (1.63), (1.62) respectively. Clearly, therefore, equation (1.79) only has a solution subject to the orthogonality condition $\int_{\partial B} \varphi(\mathbf{p})\, \lambda(\mathbf{p})\, dp = 0$, i.e. provided $\varphi = 0(r^{-2})$ as $r \to \infty$ as follows from (1.74) and (1.9). This limitation stems fundamentally from the fact that $\int_{\partial B} g(\mathbf{p}, \mathbf{q})'_e\, \mu(\mathbf{q})\, dq = 0(r^{-2})$ as $r \to \infty$, as noted in (1.27), whereas in general, $\varphi = 0(r^{-1})$ as $r \to \infty$. The difficulty may be overcome by writing

$$\varphi(\mathbf{p}) = \int_{\partial B} g(\mathbf{p}, \mathbf{q})'_e\, \mu(\mathbf{q})\, dq + \frac{c}{|\mathbf{p}|}; \quad \mathbf{p} \subset B_e \tag{1.82}$$

i.e.

$$\int_{\partial B} g(\mathbf{p}, \mathbf{q})'_e\, \mu(\mathbf{q})\, dq + 2\pi \mu(\mathbf{p}) = \varphi(\mathbf{p}) - \frac{c}{|\mathbf{p}|}; \quad \mathbf{p} \subset \partial B \tag{1.83}$$

and choosing the constant $c$ so that

$$\int_{\partial B} \lambda(\mathbf{p}) \left\{\varphi(\mathbf{p}) - \frac{c}{|\mathbf{p}|}\right\} dp = \int_{\partial B} \lambda(\mathbf{p})\, \varphi(\mathbf{p})\, dp - c \int_{\partial B} \frac{\lambda(\mathbf{p})}{|\mathbf{p}|}\, dp = 0 \tag{1.84}$$

This value of $c$ ensures that equation (1.83) has a general solution $\mu = \mu_0 + k$, where $\mu_0$ is a particular solution and $k$ is an arbitrary constant, which generates the unique potential

$$\int_{\partial B} g(\mathbf{p}, \mathbf{q})'_e\, \mu_0(\mathbf{q})\, dq + k \int_{\partial B} g(\mathbf{p}, \mathbf{q})'_e\, dq; \quad \mathbf{p} \subset B_e \tag{1.85}$$

i.e.

$$\int_{\partial B} g(\mathbf{p}, \mathbf{q})'_e\, \mu_0(\mathbf{q})\, dq; \quad \mathbf{p} \subset B_e \tag{1.86}$$

by virtue of (1.38), for insertion into the representation (1.82).

*Direct Formulations*

Both the Dirichlet and Neumann problems may be formulated directly through Green's boundary formula (1.42). Thus, given $\varphi$ on $\partial B$ (interior Dirichlet problem), it becomes a Fredholm integral equation of the first kind for $\varphi'_i$ in terms of $\varphi$, i.e.

$$\int_{\partial B} g(\mathbf{p}, \mathbf{q})\, \varphi'_i(\mathbf{q})\, dq = \int_{\partial B} g(\mathbf{p}, \mathbf{q})'_i\, \varphi(\mathbf{q})\, dq - 2\pi \varphi(\mathbf{p}); \quad \mathbf{p} \subset \partial B \tag{1.87}$$

This has a unique solution as already noted in connection with equation (1.55). With $\varphi$, $\varphi'_i$ both known on $\partial B$, we may generate $\varphi$ throughout $B_i$ via Green's formula (1.41). It requires proof, however, that $\varphi'_i$ satisfies the Gauss condition (1.60): operating with $\int_{\partial B} \ldots \lambda(\mathbf{p})\, dp$ upon both sides of (1.87), and interchanging the order of integration where appropriate, we obtain

$$\int_{\partial B} \varphi'_i(\mathbf{q})\, dq \int_{\partial B} g(\mathbf{p}, \mathbf{q})\, \lambda(\mathbf{p})\, dp = \int_{\partial B} \varphi(\mathbf{q})\, dq \int_{\partial B} g(\mathbf{p}, \mathbf{q})'_i\, \lambda(\mathbf{p})\, dp - 2\pi \int_{\partial B} \varphi(\mathbf{p})\, \lambda(\mathbf{p})\, dp$$

i.e.,

$$\int_{\partial B} \varphi_i'(\mathbf{q})\, dq \int_{\partial B} g(\mathbf{q}, \mathbf{p})\, \lambda(\mathbf{p})\, dp = \int_{\partial B} \varphi(\mathbf{q})\, dq \int_{\partial B} g_i'(\mathbf{q}, \mathbf{p})\, \lambda(\mathbf{p})\, dp - 2\pi \int_{\partial B} \varphi(\mathbf{p})\, \lambda(\mathbf{p})\, dp$$

i.e.,

$$\int_{\partial B} \varphi_i'(\mathbf{q})\, dq = 2\pi \int_{\partial B} \varphi(\mathbf{q})\, \lambda(\mathbf{q})\, dq - 2\pi \int_{\partial B} \varphi(\mathbf{p})\, \lambda(\mathbf{p})\, dp = 0 \tag{1.88}$$

on bearing in mind (1.57), (1.65) with **p**, **q** interchanged.

Given $\varphi_i'$ on $\partial B$ (interior Neumann problem), equation (1.87) now reads as

$$\int_{\partial B} g(\mathbf{p}, \mathbf{q})_i'\, \varphi(\mathbf{q})\, dq - 2\pi\, \varphi(\mathbf{p}) = \int_{\partial B} g(\mathbf{p}, \mathbf{q})\, \varphi_i'(\mathbf{q})\, dq; \quad \mathbf{p} \subset \partial B \tag{1.89}$$

i.e. a Fredholm integral equation of the second kind for $\varphi$ in terms of $\varphi_i'$. This has an associated homogeneous equation

$$\int_{\partial B} g(\mathbf{p}, \mathbf{q})_i'\, \varphi(\mathbf{q})\, dq - 2\pi\, \varphi(\mathbf{p}) = 0; \quad \mathbf{p} \subset \partial B \tag{1.90}$$

and adjoint equation

$$\int_{\partial B} g_i'(\mathbf{p}, \mathbf{q})\, \lambda(\mathbf{q})\, dq - 2\pi\, \lambda(\mathbf{p}) = 0; \quad \mathbf{p} \subset \partial B \tag{1.91}$$

which are mathematically equivalent to (1.80), (1.81). Consequently, equation (1.89) only has a solution subject to the orthogonality condition

$$\int_{\partial B} \lambda(\mathbf{p}) \left\{ \int_{\partial B} g(\mathbf{p}, \mathbf{q})\, \varphi_i'(\mathbf{q}\, dq \right\} dp = 0$$

i.e.

$$\int_{\partial B} \varphi_i'(\mathbf{q})\, dq \int_{\partial B} g(\mathbf{p}, \mathbf{q})\, \lambda(\mathbf{p})\, dp = \int_{\partial B} \varphi_i'(\mathbf{q})\, dq \int_{\partial B} g(\mathbf{p}, \mathbf{q})\, \lambda(\mathbf{p})\, dp = \int_{\partial B} \varphi_i'(\mathbf{q})\, dq = 0 \tag{1.92}$$

which is the expected Gauss condition. In this case, equation (1.89) has the general solution

$$\varphi = \varphi_0 + k \tag{1.93}$$

where $\varphi_0$ is a particular solution and $k$ is an arbitrary constant, in complete agreement with the family of simple-source solutions (1.66). With $\varphi$, $\varphi_i'$ both known on $\partial B$, we may generate $\varphi$ throughout $B_i$ via Green's formula (1.41).

The exterior Dirichlet problem is formulated by the equation

$$\int_{\partial B} g(\mathbf{p}, \mathbf{q})\, \varphi_e'(\mathbf{q})\, dq = \int_{\partial B} g(\mathbf{p}, \mathbf{q})_e'\, \varphi(\mathbf{q})\, dq - 2\pi\, \varphi(\mathbf{p}); \quad \mathbf{p} \subset \partial B \tag{1.94}$$

i.e. a Fredholm integral equation of the first kind for $\varphi_e'$ in terms of $\varphi$. This has a unique solution, which satisfies the relation

$$\int_{\partial B} \varphi_e'(\mathbf{q})\, dq = -4\pi \int_{\partial B} \lambda(\mathbf{p})\, \varphi(\mathbf{p})\, dp \tag{1.95}$$

as also follows by eliminating $\sigma$ between (1.73) and (1.74).

The exterior Neumann problem is formulated by the equation

$$\int_{\partial B} g(\mathbf{p}, \mathbf{q})_e'\, \varphi(\mathbf{q})\, dq - 2\pi\, \varphi(\mathbf{p}) = \int_{\partial B} g(\mathbf{p}, \mathbf{q})\, \varphi_e'(\mathbf{q})\, dq; \quad \mathbf{p} \subset \partial B \tag{1.96}$$

i.e. a Fredholm equation of the second kind for $\Phi$ in terms of $\varphi_e'$. This always has a unique solution as follows by reference to (1.71).

The Dirichlet and Neumann conditions are particular cases of a prescribed linear relation

$$\alpha\,\varphi + \beta\,\varphi' = \gamma; \quad \varphi' = \varphi_i' \quad \text{or} \quad \varphi_e' \tag{1.97}$$

at each point of $\partial B$. Thus,

$$\alpha = 1, \quad \beta = 0, \quad \gamma \text{ given (continuous) on } \partial B \tag{1.98}$$

defines the Dirichlet problem, and

$$\alpha = 0, \quad \beta = 1, \quad \gamma \text{ given (continuous) on } \partial B \tag{1.99}$$

defines the Neumann problem. An existence-uniqueness theorem is available for the Robin problem of heat conduction, defined by

$$\alpha < 0, \quad \beta = 1, \quad \gamma \text{ given (continuous) on } \partial B \tag{1.100}$$

Finally, an existence-uniqueness theorem is available for the difficult mixed problem defined by

$$\left.\begin{array}{ll} \alpha = 1, \quad \beta = 0 \quad \text{on } \partial B_1; & \alpha = 0, \quad \beta = 1 \text{ on } \partial B_2 \\ \partial B = \partial B_1 + \partial B_2, & \gamma \text{ given (continuous) on } \partial B \end{array}\right\} \tag{1.101}$$

Essentially, in all these cases, we couple the local condition (1.97) with the global condition (1.42) or (1.46) to determine a compatible $\varphi$, $\varphi'$ on $\partial B$, so allowing us to generate $\varphi$ in $B_i$ or $B_e$ via (1.41) or (1.45) respectively. Generally speaking, an analytical solution is out of the question even for specialised boundaries. However, we may attempt a numerical solution by discreting $\partial B$ into $n$ surface elements centred about the $n$ pivotal (collocation) points $q_1, \ldots, q_n$. Condition (1.97) provides $n$ linear equations between the $2n$ unknowns $\varphi_1, \ldots, \varphi_n$; $\varphi_1', \ldots, \varphi_n'$ at these points. Also condition (1.42) or (1.46) may be discretised to provide another set of $n$ linear equations between these $2n$ unknowns. Accordingly, we have sufficient equations to compute all the unknowns, so enabling $\varphi$ to be computed through any part of $B_i$ or $B_e$ via a discretised version of (1.41) or (1.45).

## 1.4 Vector Potential Theory: Somigliana's Formula

*Identification of Scalar and Vector Symbolism*

Classical linear elastostatics may be formulated by a vector potential theory which closely parallels scalar potential theory. It would, indeed, be advantageous to employ the same symbolism in each theory, its interpretation depending on the context. Thus the scalar potential $\varphi$ becomes the elastostatic displacement vector. The normal derivative $\varphi'$ becomes the traction vector associated with $\varphi$. The Newtonian unit-source potential $g(\mathbf{p}, \mathbf{q})$ becomes the fundamental displacement dyadic of the medium. More precisely, in this context we mean that

$$g(\mathbf{p}, \mathbf{q}) = \begin{bmatrix} g(p_1 q_1) & g(p_1 q_2) & g(p_1 q_3) \\ g(p_2 q_1) & g(p_2 q_2) & g(p_2 q_3) \\ g(p_3 q_1) & g(p_3 q_2) & g(p_3 q_3) \end{bmatrix} \tag{1.102}$$

where $g(p_\alpha q_\eta)$ signifies the displacement component in the $\alpha$-direction at **p** generated by a unit point-force in the $\gamma$-direction at **q**. Clearly column 1 defines the displacement vector at **p** generated by a unit force acting in the 1-direction at **q**, etc. By virtue of $g(p_\alpha q_\eta) = g(q_\eta p_\alpha)$, we see that row 1 defines the displacement vector at **q** generated by a unit point-force acting in the 1-direction at **p**, etc. Corresponding with $g'(\mathbf{p}, \mathbf{q})$ we construct the fundamental traction dyadic of the medium, i.e.

$$g'(\mathbf{p}, \mathbf{q}) = \begin{bmatrix} g'(p_1 q_1) & g'(p_1 q_2) & g'(p_1 q_3) \\ g'(p_2 q_1) & g'(p_2 q_2) & g'(p_2 q_3) \\ g'(p_3 q_1) & g'(p_3 q_2) & g'(p_3 q_3) \end{bmatrix} \tag{1.103}$$

where $g'(p_\alpha q_\eta)$ signifies the traction component in the $\alpha$-direction at **p** generated by a unit point-force acting in the $\eta$-direction at **q**. Clearly column 1 defines the traction vector at **p** generated by a unit point-force acting in the 1-direction at **q**, etc. Finally, corresponding with $g(\mathbf{p}, \mathbf{q})'$ we construct the traction dyadic

$$g(\mathbf{p}, \mathbf{q})' = \begin{bmatrix} g(p_1 q_1)' & g(p_1 q_2)' & g(p_1 q_3)' \\ g(p_2 q_1)' & g(p_2 q_2)' & g(p_2 q_3)' \\ g(p_3 q_1)' & g(p_3 q_2)' & g(p_3 q_3)' \end{bmatrix} \tag{1.104}$$

where row 1 defines the traction vector at **q** generated by a unit point-force acting in the 1-direction at **p**, etc. It may be readily proved that column 1 defines a singular displacement vector at **p**, i.e., that generated by a unit traction-source associated with the 1-direction at **q**, etc., in line with the fact that $g(\mathbf{p}, \mathbf{q})'$ may function as a unit dipole-potential generated at **q**. By the same token, row 1 of (1.103) defines the singular displacement vector at **q** generated by a unit traction-source associated with the 1-direction at **p**. Any individual component of (1.103) or (1.104) carries two possible interpretations: either the traction component generated by a unit point-force or the displacement component generated by a unit traction-source. The interpretation will always be clear from the context. We note that $g'(\mathbf{p}, \mathbf{q})$ stands for $g'_e(\mathbf{p}, \mathbf{q})$ or $g'_i(\mathbf{p}, \mathbf{q})$ as the case may be, and similarly for $g(\mathbf{p}, \mathbf{q})'$.

The simple-source density $\sigma$ now becomes a vector simple-source density $\sigma = \langle \sigma_1, \sigma_2, \sigma_3 ) \rangle$, so allowing us to regard (1.2) as a vector simple-layer potential with components

$$V_\alpha(\mathbf{p}) = \int_{\partial B} g(p_\alpha q_\eta)\, \sigma_\eta(\mathbf{q})\, dq; \quad \alpha, \eta = 1, 2, 3 \tag{1.105}$$

This has properties at $\partial B$ entirely analogous to those of the scalar simple-source potential, e.g. formulae (1.12), (1.13) may be read as traction formulae, and it defines an elastostatic displacement field for any choice of **p**. These properties have been proved by Kupradze for the linear isotropic elastic continuum, but we may conjecture that they also hold for the general linear anisotropic elastic continuum. Similarly, the double-source density $\mu$ becomes a vector double-source density $\mu = \langle \mu_1, \mu_2, \mu_3 \rangle$, so allowing us to regard (1.24) as a vector double-layer potential with components

$$W_\alpha(\mathbf{p}) = \int_{\partial B} g(p_\alpha q_\eta)'\, \mu_\eta(\mathbf{q})\, dq; \quad \alpha, \eta = 1, 2, 3 \tag{1.106}$$

This has properties at $\partial B$ entirely analogous to those of the scalar double-source potential, e.g. formulae (1.28) may be read as vector formulae, and it defines an elastostatic displacement field everywhere except at $\mathbf{p} \subset \partial B$.

*Somigliana's Formula*

Green's formula (1.41) now reads as Somigliana's formula [37], i.e. it represents an arbitrary displacement vector $\boldsymbol{\varphi}$ as the superposition of a vector simple-layer potential and a vector double-layer potential, generated respectively by the boundary tractions and boundary displacements associated with $\boldsymbol{\varphi}$. Green's boundary formula (1.42) now reads as Somigliana's boundary formula, which provides a vector functional relation between tractions and displacements on $\partial B$. Green's reciprocal theorem (1.43) now reads at Betti's reciprocal theorem. Corresponding exterior formulae hold for a displacement field which remains regular at infinity, so allowing us to introduce the generalised Somigliana formula (1.49). The boundary-conditions of linear elastostatics are all covered by (1.97), reading $\boldsymbol{\varphi}$, $\boldsymbol{\varphi}'$, $\gamma$ as vectors and $\alpha$, $\beta$ as scalars. Assuming the validity of the fundamental existence-uniqueness theorems of linear elastostatics, we may readily prove the validity of the vector representations $V$, $W$ with $\sigma$, $\mu$ identified by the vector equations (1.51), (1.53) respectively.

*Rigid-Body Displacement Field*

Corresponding to the scalar harmonic function $\varphi = 1$, we introduce the rigid-body displacement field

$$\varphi(\mathbf{p}) = \mathbf{a} + \mathbf{b} \wedge \mathbf{p} \tag{1.107}$$

where $\mathbf{a}$, $\mathbf{b}$ are given constant vectors. This has the following properties in parallel with those of $\varphi = 1$:

1) $\varphi = \mathbf{a} + \mathbf{b} \wedge \mathbf{p}$ on $\partial B$ implies $\varphi = \mathbf{a} + \mathbf{b} \wedge \mathbf{p}$ in $B_i + \partial B$;
2) $\varphi_i' = \mathbf{0}$ on $\partial B$ i.e. no tractions are associated with the rigid-body displacement field;
3) given $\varphi_i' = \mathbf{0}$ on $\partial B$, we may infer that $\varphi = \mathbf{a} + \mathbf{b} \wedge \mathbf{p}$ on $B_i + \partial B$ where $\mathbf{a}$, $\mathbf{b}$ are arbitrary constant vectors.

It is convenient to break down (1.107) into the six independent vectors

$$\left.\begin{array}{lll} \mu_1 = \langle 1, 0, 0\rangle, & \mu_2 = \langle 0, 1, 0\rangle, & \mu_3 = \langle 0, 0, 1\rangle \\ \mu_4 = \langle 1, 0, 0\rangle \wedge \mathbf{p}, & \mu_5 = \langle 0, 1, 0\rangle \wedge \mathbf{p}, & \mu_6 = \langle 0, 0, 1\rangle \wedge \mathbf{p} \end{array}\right\} \tag{1.108}$$

which provide the six vector double-layer identities

$$\int_{\partial B} g(\mathbf{p}, \mathbf{q})_i' \mu_s(\mathbf{q})\, dq = 4\pi\, \mu_s(\mathbf{p}); \quad s = 1, 2, \ldots, 6, \quad \mathbf{p} \subset B_i \tag{1.109}$$

as may be proved by substituting $\varphi = \mu_s$, $\varphi_i' = \mu_s' = \mathbf{0}$ into Somigliana's formula (1.41). Essentially these are vector generalisations of the Gauss flux theorem as expressed by (1.33). Their physical significance is most readily understood by envisaging a unit point-force acting in the 1-direction at $\mathbf{p}$, so generating the traction vector $g(p_1 q_\eta)_i'$; $\eta = 1, 2, 3$ over $\partial B$. In this case the identities (1.109) yield component identities showing that (1) the resultant force of the tractions balances the point-force at **p**, and that (2) their resultant moment about any axis balances its moment about this axis. Easy deductions from (1.109) are

$$\int_{\partial B} g(\mathbf{p}, \mathbf{q})'_i \, \mu_s(\mathbf{q}) \, dq = 2\pi \, \mu_s(\mathbf{p}); \quad s = 1, 2, \ldots, 6, \quad \mathbf{p} \subset \partial B \qquad (1.110)$$

$$= 0; \qquad \mathbf{p} \subset B_e \qquad (1.111)$$

in parallel with (1.34), (1.35) respectively.

## 1.5 Indirect Vector Formulations

*Introduction*

Corresponding with the electrostatic integral equation (1.57), we introduce the six vector integral equations

$$\int_{\partial B} g(\mathbf{p}, \mathbf{q}) \, \lambda_s(\mathbf{q}) \, dq = \mu_s(\mathbf{p}); \quad s = 1, 2, \ldots, 6, \quad \mathbf{p} \subset \partial B \qquad (1.112)$$

These are not Fredholm equations since $g(p_\alpha q_\eta) = 0 \, |\mathbf{p} - \mathbf{q}|^{-2}$ as $|\mathbf{p} - \mathbf{q}| \Rightarrow 0$. However, they have solutions in principle given by

$$\lambda_s = -\frac{1}{4\pi} (f_s)'_e; \quad s = 1, 2, \ldots, 6 \qquad (1.113)$$

where $f_s$ is the unique regular displacement field in $B_e$ which satisfies $f_s = \mu_s$ on $\partial B$. Since $\mu'_s = 0$ on $\partial B$, it follows from (1.112) that $\lambda_s$ also satisfy the traction equations

$$\int_{\partial B} g'_i(\mathbf{p}, \mathbf{q}) \, \lambda_s(\mathbf{q}) \, dq - 2\pi \, \lambda_s(\mathbf{p}) = 0; \quad s = 1, 2, \ldots, 6, \quad \mathbf{p} \subset \partial B, \qquad (1.114)$$

i.e. the adjoint system of equations to the system (1.110). Although these are not Fredholm systems, they play an entirely parallel role to that played by (1.63), (1.65) in the scalar theory.

Vector Dirichlet problems may be formulated by the vector equation (1.55), following a parallel analysis to that of the scalar theory. Thus, operating upon both sides of (1.55) by $\int \ldots \lambda_s(\mathbf{p}) \, dp$, and interchanging the order of integration at appropriate stages, we find

$$\int_{\partial B} \varphi(\mathbf{p}) \, \lambda_s(\mathbf{p}) \, dp = \int_{\partial B} \sigma(\mathbf{q}) \, \mu_s(\mathbf{q}) \, dq; \quad s = 1, 2, \ldots, 6 \qquad (1.115)$$

in parallel with (1.74).

*Kupradze Formulations*

Kupradze was the first to propose that the vector interior Neumann problem could be formulated by the vector integral equation (1.61), with the associated homogeneous systems (1.62), (1.63). Naively applying classical Fredholm theory to this system, it follows that a solution only exists if

$$\int_{\partial B} \varphi'_e(\mathbf{p}) \, \mu_s(\mathbf{p}) \, dp = 0; \quad s = 1, 2, \ldots, 6 \qquad (1.116)$$

i.e. if the prescribed tractions form a self-equilibrated distribution of forces and moments over $\partial B$, as expected on physical grounds. Subject to (1.116), the general solution of equation (1.61) now appears as

$$\sigma = \sigma_0 + \sum_1^6 k_s \mu_s \tag{1.117}$$

where $\sigma_0$ is a particular solution and $k_s$ are arbitrary scalar coefficients. This generates the class of displacement fields

$$\int_{\partial B} g(\mathbf{p}, \mathbf{q})\, \sigma_0(\mathbf{q})\, dq + k_s \int_{\partial B} g(\mathbf{p}, \mathbf{q})\, \lambda_s(\mathbf{q})\, dq; \quad \mathbf{p} \subset B_i + \partial B$$

i.e.

$$\int_{\partial B} g(\mathbf{p}, \mathbf{q})\, \sigma_0(q)\, dq + \sum_1^6 k_s \mu_s; \quad \mathbf{p} \subset B_i + \partial B_i \tag{1.118}$$

each characterised by the same traction vector $\varphi_e'$ on $\partial B$.

The vector exterior Neumann problem is formulated by (1.69), with the associated homogeneous systems (1.70), (1.71). As for the scalar case, this has a unique solution which generates the regular exterior displacement field (1.72). Operating with $\int_{\partial B} \ldots \mu_s(\mathbf{p})\, dp$ upon both sides of (1.69), we find

$$\int_{\partial B} \varphi_e'(\mathbf{p})\, \mu_s(\mathbf{p})\, dp = -4\pi \int_{\partial B} \sigma(\mathbf{p})\, \mu_s(\mathbf{p})\, dp; \quad s = 1, 2, \ldots, 6 \tag{1.119}$$

i.e.

$$\int_{\partial B} \varphi_e'(\mathbf{p})\, \mu_s(\mathbf{p})\, dp = -4\pi \int_{\partial B} \varphi(\mathbf{p})\, \lambda_s(\mathbf{p})\, dp \tag{1.120}$$

from (1.115), i.e. the resultant force and moment of the exterior tractions on $\partial B$ are directly known in terms of $\varphi$, $\lambda_s$ on $\partial B$.

Vector Dirichlet problems may also be formulated by the vector double-layer equations (1.76), (1.79). The interior equation (1.76) has a unique solution. However, the exterior equation (1.79) only has a solution subject to the orthogonality condition

$$\int_{\partial B} \varphi(\mathbf{p})\, \lambda_s(\mathbf{p})\, dp = 0; \quad s = 1, 2, \ldots, 6 \tag{1.121}$$

which means – as expected – that the tractions associated with $\varphi$ on $\partial B$ must form a self-equilibrated system. If so, the general solution of this equation appears as

$$\mu = \mu_0 + \sum_1^6 k_s \mu_s \tag{1.122}$$

where $\mu_0$ is a particular solution and $k_s$ are arbitrary scalar coefficients. This generates the unique displacement field

$$\int_{\partial B} g(\mathbf{p}, \mathbf{q})_e'\, \mu_0(\mathbf{q})\, dq + \sum_1^6 k_s \int_{\partial B} g(\mathbf{p}, \mathbf{q})_e'\, \mu_s(\mathbf{q})\, dq; \quad \mathbf{p} \subset B_e \tag{1.123}$$

i.e.

$$\int_{\partial B} g(\mathbf{p}, \mathbf{q})_e'\, \mu_0(\mathbf{q})\, dq; \quad \mathbf{p} \subset B_e + \partial B \tag{1.124}$$

on bearing in mind (1.109).

Condition (1.121) stems essentially from the fact that the vector representation (1.78) has only $O(r^{-2})$ behaviour at infinity, whereas a regular exterior $\varphi$ has in general $O(r^{-1})$ behaviour at infinity.

Following (1.82), we extend the vector representation (1.78) by writing

$$\varphi(\mathbf{p}) = \int_{\partial B} g(\mathbf{p}, \mathbf{q})'_e \mu(\mathbf{q})\, dq + a \cdot g(\mathbf{q}, \mathbf{p})_{\mathbf{q}=\mathbf{0}} + b \wedge \nabla \cdot g(\mathbf{q}, \mathbf{p})_{\mathbf{q}=\mathbf{0}}; \quad \mathbf{p} \subset B_e \tag{1.125}$$

where $a$, $b$ are constant vectors to be determined:

$$a = \langle a_1, a_2, a_3 \rangle, \quad b = \langle b_1, b_2, b_3 \rangle \equiv \langle b_4, b_5, b_6 \rangle \tag{1.126}$$

It will be noted that both additional terms in (1.125) have $0(r^{-1})$ behaviour as $r \to \infty$. Written out in component form these appear as

$$a \cdot g(\mathbf{q}, \mathbf{p}) = \sum_1^3 a_\alpha g(q_\alpha p_\eta); \quad \eta = 1, 2, 3$$

$$b \wedge \nabla = \left\langle b_2 \frac{\partial}{\partial q_3} - b_3 \frac{\partial}{\partial q_2},\; b_3 \frac{\partial}{\partial q_1} - b_1 \frac{\partial}{\partial q_3},\; b_1 \frac{\partial}{\partial q_2} - b_2 \frac{\partial}{\partial q_1} \right\rangle$$

$$b \wedge \nabla \cdot g(\mathbf{q}, \mathbf{p}) = \sum_1^3 (b \wedge \nabla)_\beta g(q_\beta p_\eta); \quad \eta = 1, 2, 3$$

Operating upon both sides of (1.123) by $\int_{\partial B} \ldots \lambda_s(\mathbf{p})\, dp$, we find

$$\int_{\partial B} \varphi(\mathbf{p}) \lambda_s(\mathbf{p})\, dp = a \cdot \mu_s(\mathbf{q}) + b \wedge \nabla \cdot \mu_s(\mathbf{q}) = a \cdot \mu_s(\mathbf{q}) = a_s; \quad s = 1, 2, 3 \tag{1.127}$$

$$= a \cdot \mu_s(\mathbf{q})_{\mathbf{q}=0} + b \wedge \nabla \cdot \mu_s(\mathbf{q})_{\mathbf{q}=0} = b \wedge \nabla \cdot \mu_s(\mathbf{q})_{\mathbf{q}=0} = 2b_s; \quad s = 4, 5, 6 \tag{1.128}$$

With these values of $a$, $b$ the integral equation

$$\int_{\partial B} g(\mathbf{p}, \mathbf{q})'_e \mu(\mathbf{q})\, dq + 2\pi \mu(\mathbf{p}) = \varphi(\mathbf{p}) - a \cdot g(\mathbf{q}, \mathbf{p})_{\mathbf{q}=0} - b \wedge \nabla \cdot g(\mathbf{q}, \mathbf{p})_{\mathbf{q}=0} \tag{1.129}$$

always has a solution, of the form (1.122), which generates $\varphi$ in $B_e$ via (1.125).

*Rizzo Formulations*

As first proposed by Rizzo, Somigliana's boundary formula (1.42) may be exploited to attack vector Dirichlet and Neumann problems. Thus, given the displacement vector $\boldsymbol{\varphi}$ on $\partial B$ (interior Dirichlet problem), equation (1.87) now reads as a vector integral equation of the first kind for $\varphi'_i$ in terms of $\boldsymbol{\varphi}$. This has a unique solution $\varphi'_i$, as already noted for the vector equation (1.55). Operating with $\int_{\partial B} \ldots \lambda_s(p)\, dp$ upon both sides of (1.87), we see that $\varphi'_i$ satisfies the relations

$$\int_{\partial B} \varphi'_i(\mathbf{p}) \mu_s(\mathbf{p})\, dp = 0; \quad s = 1, 2, \ldots, 6 \tag{1.130}$$

as expected on physical grounds.

Given the traction vector $\varphi'_i$ on $\partial B$ (interior Neumann problem), equation (1.89) now reads as a vector integral equation of the second kind for $\boldsymbol{\varphi}$ in terms of $\varphi'_i$.

Subject to condition (1.130), this has a solution which appears as

$$\varphi = \varphi_0 + \sum_1^6 k_s \mu_s \tag{1.131}$$

where $\varphi_0$ is a particular solution and $k_s$ are arbitrary coefficients.

The exterior Dirichlet problem is formulated by equation (1.94), read as a vector integral equation of the first kind for $\varphi_e'$ in terms of $\varphi$. This has a unique solution, which satisfies the relations

$$\int_{\partial B} \varphi_e'(\mathbf{q})\, \mu_s(\mathbf{q})\, dq = -4\pi \int_{\partial B} \lambda_s(\mathbf{p})\, \varphi(\mathbf{p})\, dp; \quad s = 1, 2, \ldots, 6 \tag{1.132}$$

so confirming the result (1.120) already obtained by earlier arguments.

Given $\varphi_e'$ on $\partial B$ (exterior Neumann problem), equation (1.96) now reads as a vector integral equation of the second kind for $\varphi$ in terms of $\varphi_e'$. This always has a unique solution.

As already mentioned, the vector Dirichlet and Neumann conditions are particular cases of the vector boundary relations (1.97) connecting $\varphi$, $\varphi'$ on $\partial B$. Coupling this local relation with the global relation (1.42) or (1.46), we have sufficient vector equations to determine $\varphi$, $\varphi'$ on $\partial B$.

## 1.6 Two-Dimensional Potential Theory

*The Logarithmic Potential*

The preceding scalar theory may be adapted to two dimensions by writing $g(\mathbf{p}, \mathbf{q}) = \log|\mathbf{p}-\mathbf{q}|$ in place of $|\mathbf{p}-\mathbf{q}|^{-1}$ and $-\frac{\pi}{2}$ in place of $\pi$. Thus a continuous distribution of simple sources on a closed Liapunov contour $\partial B$, of line density $\sigma(\mathbf{q})$ at $\mathbf{q} \subset \partial B$, generates the simple-layer potentials

$$V(\mathbf{p}) = \int_{\partial B} \log|\mathbf{p}-\mathbf{q}|\, \sigma(\mathbf{q})\, dq; \quad \mathbf{p} \subset B_i + \partial B \tag{1.133}$$

$$V(\mathbf{p}) = \int_{\partial B} \log|\mathbf{p}-\mathbf{q}|\, \sigma(\mathbf{q})\, dq; \quad \mathbf{p} \subset B_e + \partial B \tag{1.134}$$

These satisfy the relations (1.7)–(1.15) adapted as appropriate, except that (1.9) gets replaced by

$$V(\mathbf{p}) = \log|\mathbf{p}| \int_{\partial B} \sigma(\mathbf{q})\, dq - |\mathbf{p}|^{-2} \int_{\partial B} (\mathbf{p}\cdot\mathbf{q})\, \sigma(\mathbf{q})\, dq + 0\,|\mathbf{p}|^{-2} \text{ as } |\mathbf{p}| \to \infty \tag{1.135}$$

Similarly, a continuous distribution of double sources on $\partial B$ of line density $\mu(\mathbf{q})$ at $\mathbf{q} \subset \partial B$, generates the double-layer potentials

$$W(\mathbf{p}) = \int_{\partial B} \log|\mathbf{p}-\mathbf{q}|_i'\, \mu(\mathbf{q}); \quad \mathbf{p} \subset B_i \tag{1.136}$$

$$W(\mathbf{p}) = \int_{\partial B} \log|\mathbf{p}-\mathbf{q}|_e'\, \mu(\mathbf{q})\, dq; \quad \mathbf{p} \subset B_e \tag{1.137}$$

These satisfy the relations (1.25)–(1.31) adapted as appropriate, except that (1.27) gets replaced by

$$W(\mathbf{p}) = 0\,|\mathbf{p}|^{-2} \quad \text{as} \quad |\mathbf{p}| \to \infty \tag{1.138}$$

A convenient feature of two-dimensional theory is the Cauchy-Riemann relation

$$\log|\mathbf{p}-\mathbf{q}|_i' = \log_i'|\mathbf{q}-\mathbf{p}| = -\frac{\partial\theta}{\partial q}(\mathbf{q}-\mathbf{p}); \quad \mathbf{p} \subset B_i \tag{1.139}$$

where $\theta$ denotes the angle (counter-clockwise sense) between a fixed direction through $\mathbf{p}$ and the vetor joining $\mathbf{p}$ to $\mathbf{q}$. This relation immediately yields

$$\int_{\partial B} \log|\mathbf{p}-\mathbf{q}|'_i \, dq = -\int_{\partial B} \frac{\partial\theta}{\partial q}(\mathbf{q}-\mathbf{p})\, dq = -2\pi; \quad \mathbf{p} \subset B_i \tag{1.140}$$

i.e. the two-dimensional Gauss flux theorem, and also the boundary relation

$$\int_{\partial B} \log|\mathbf{p}-\mathbf{q}|'_e \, dq = -\pi; \quad \mathbf{p} \subset \partial B \tag{1.141}$$

More generally, relaxing the restriction to Liapunov contours, it readily follows from (1.39) that

$$\int_{\partial B} \log|\mathbf{p}-\mathbf{q}|'_i \, dq = -\Omega(\mathbf{p}); \quad \mathbf{p} \subset \partial B \tag{1.142}$$

if $\mathbf{p}$ is located at a corner of $\partial B$ having the interior angle $\Omega(\mathbf{p})$.

The two-dimensional existence-uniqueness theorems are identical with the corresponding three-dimensional theorems for interior problems. However, complications arise for exterior problems because of possible logarithmic behaviour at infinity, i.e. the behaviour

$$\varphi = \varkappa \log r + 0(r^{-1}) \quad \text{as} \quad r \to \infty \tag{1.143}$$

appertaining to simple-layer potentials. For instance, given $\varphi = k$ ($a$ cons.) on the circle $r = a$, there exists a unique harmonic function

$$\varphi(r) = \frac{k \log r}{\log a}; \quad r \geqq a \tag{1.144}$$

conforming to (1.143). However this breaks down when $a = 1$, so ruling out the possibility of a general existence theorem embodying (1.135). On the other hand, there exists a unique harmonic function

$$\varphi = k; \quad r \geqq a \tag{1.145}$$

subject to bounded behaviour at infinity, which clearly holds for all contours. Classical theory therefore emphasises bounded behaviour at infinity in preference to logarithmic behaviour.

Complications of a different kind arise with the exterior Neumann problem. For instance, given $\varphi'_e = K$ ($a$ cons.) on $r = a$ i.e. $\left(\frac{\partial\varphi}{\partial r}\right)_{r=a} = K$, there exists a unique harmonic function

$$\varphi = K a \log r; \quad r \geqq a \tag{1.146}$$

conforming to (1.143). By contrast, however, no solution exists subject to bounded behaviour at infinity. The difficulty is surmounted in classical theory by imposing the exterior Gauss condition $\int_{\partial B} \varphi'_e(\mathbf{q})\, dq = 0$. This forces the choice $K = 0$, which is consistent with the class of solutions

$$\varphi(r) = c \text{ (any constant including zero)}; \quad r \geqq a \tag{1.147}$$

One of these solutions, i.e. $\varphi = 0$, is uniquely determined in the class (1.146) by putting $K = 0$.

To summarise, logarithmic behaviour at infinity offers advantages over bounded behaviour for the exterior Neumann problem. It is competitive with bounded behaviour for the exterior Dirichlet problem, apart from breakdown cases examplified by the unit circle. These will now be investigated.

*$\Gamma$-Contours*

A distinguishing feature of two-dimensional theory is the existence of contours, termed $\Gamma$-contours, for which the equation

$$\int_{\partial B} \log|\mathbf{p}-\mathbf{q}|\,\lambda(\mathbf{q})\,dq = 1; \quad \mathbf{p} \subset \partial B \tag{1.148}$$

does not have a solution. To identify these contours, start with a contour $\partial B$ defined by the Cartesian equation $f(x, y) = 0$, and construct the family of similar contours $f(x/m, y/m) = 0$ where $m > 0$ is a continuously varying parameter. Then, according to theorems proved by Jaswon & Symm:

(i) one, and only one, member of this family is a $\Gamma$-contour;
(ii) on $\Gamma$, and only on $\Gamma$, the homogeneous equation

$$\int_{\partial B} \log|\mathbf{p}-\mathbf{q}|\,\lambda(\mathbf{q})\,dq = 0 \tag{1.149}$$

has a non-trivial solution.

To locate $\Gamma$, we label the $m^{th}$ contour $\partial B_m$ and introduce the normal derivative equation

$$\int_{\partial B_1} \log_i'|\mathbf{p}-\mathbf{q}|\,\lambda(\mathbf{q})\,dq + \pi\,\lambda(\mathbf{p}) = 0; \quad \mathbf{p} \subset \partial B \tag{1.150}$$

on $\partial B_1\,(\equiv \partial B)$. This has a non-trivial solution $\Lambda$ corresponding with the non-trivial solution $\mu = 1$ of the adjoint equation

$$\int_{\partial B_1} \log|\mathbf{p}-\mathbf{q}|_i'\,\mu(\mathbf{q})\,dq + \pi\,\mu(\mathbf{p}) = 0; \quad \mathbf{p} \subset \partial B_1 \tag{1.151}$$

Clearly $\Lambda$ generates the potential

$$\int_{\partial B_1} \log|\mathbf{p}-\mathbf{q}|\,\Lambda(\mathbf{q})\,dq = k\ (a\ \text{cons.}); \quad \mathbf{p} \subset \partial B \tag{1.152}$$

and two distinct possibilities arise for $k$:

(i) $k \neq 0$, in which case equation (1.148) has the solution $\lambda = \Lambda/k$. If so,

$$\int_{\partial B_m} \log|\mathbf{p}-\mathbf{q}|\,\lambda(\mathbf{q})\,dq = \varkappa\, m \log m + m; \quad \varkappa = \int_{\partial B_1} \lambda(\mathbf{q})\,dq \tag{1.153}$$

showing that $\partial B_m = \Gamma$ when $m = \exp(-1/\varkappa)$.
(ii) $k = 0$, in which case $\Lambda$ provides a non-trivial solution of equation (1.149), i.e. $\partial B_1 = \Gamma$. If so,

$$\int_{\partial B_m} \log|\mathbf{p}-\mathbf{q}|\,\lambda(\mathbf{q})\,dq = \varkappa\, m \log m \tag{1.154}$$

showing that $\partial B_m \neq \Gamma$ unless $m = 1$.

As already indicated in connection with (1.144), the unit circle is a $\Gamma$-contour for the family of concentric circles: by symmetry equation (1.148) has the solution

$\sigma(\mathbf{q}) = \sigma_0$ which gives

$$\sigma_0 \int_{\partial B_1} \log |\mathbf{p}-\mathbf{q}|\, dq = \sigma_0\, 2\pi a \log a = 0 \quad \text{when} \quad a = 1 \tag{1.155}$$

The more general equation

$$\int_{\partial B} \log |\mathbf{p}-\mathbf{q}|\, \sigma(\mathbf{q})\, dq = \varphi(\mathbf{p}); \quad \mathbf{p} \subset \partial B (\neq \Gamma) \tag{1.156}$$

has a unique solution if $\partial B \neq \Gamma$, which solves both the interior and exterior Dirichlet problems for $\partial B$. However, it generates logarithmic behaviour at infinity. To ensure bounded behaviour, we must write

$$c + \int_{\partial B} \log |\mathbf{p}-\mathbf{q}|\, \sigma(\mathbf{q})\, dq = \varphi(\mathbf{p}); \quad \mathbf{p} \subset \partial B \tag{1.157}$$

coupled with the complementary side conditions

$$\int_{\partial B} \sigma(\mathbf{q})\, dq = 0, \quad c = \int_{\partial B} \varphi(\mathbf{p})\, \lambda(\mathbf{p})\, dp \Big| \int_{\partial B} \lambda(\mathbf{p})\, dp \tag{1.158}$$

This formulation is entirely equivalent to (1.156) for interior problems, and it holds for both interior and exterior problems even if $\partial B = \Gamma$.

Subject to $\int_{\partial B} \varphi_i'(\mathbf{p})\, dp = 0$, the interior normal derivative equation

$$\int_{\partial B} \log_i' |\mathbf{p}-\mathbf{q}|\, \sigma(\mathbf{q})\, dq + \pi\, \sigma(\mathbf{p}) = \varphi_i'(\mathbf{p}); \quad \mathbf{p} \subset \partial B \tag{1.159}$$

always has a non-unique solution. The corresponding exterior equation

$$\int_{\partial B} \log_e' |\mathbf{p}-\mathbf{q}|\, \sigma(\mathbf{q})\, dq + \pi\, \sigma(\mathbf{p}) = \varphi_e'(\mathbf{p}); \quad \mathbf{p} \subset \partial B \tag{1.160}$$

always has a unique solution subject to:

either (i) $\int_{\partial B} \varphi_e'(\mathbf{p})\, dp = 0$, which implies $\int_{\partial B} \sigma(\mathbf{q})\, dq = 0$ and therefore the representation (1.157);

or (ii) $\int_{\partial B} \varphi_e'(\mathbf{p})\, dp \neq 0$, which implies $\int_{\partial B} \sigma(\mathbf{q})\, dq \neq 0$ and therefore the representation (1.156).

*Two-Dimensional Green's Formula*

Formulae (1.41), (1.42) now become

$$\int_{\partial B} \log |\mathbf{p}-\mathbf{q}|_i'\, \varphi(\mathbf{q})\, dq - \int_{\partial B} \log |\mathbf{p}-\mathbf{q}|\, \varphi_i'(\mathbf{q})\, dq = -2\pi\, \varphi(\mathbf{p}); \quad \mathbf{p} \subset B_i \tag{1.161}$$

$$= -\pi\, \varphi(\mathbf{p}); \quad \mathbf{p} \subset \partial B \tag{1.162}$$

and (1.162) will be recognised as the real-variable analogue of Plemelj's complex-variable formula. Changing *i* into *e* yields the exterior formulae

$$\int_{\partial B} \log |\mathbf{p}-\mathbf{q}|_e'\, \varphi(\mathbf{q})\, dq - \int_{\partial B} \log |\mathbf{p}-\mathbf{q}|\, \varphi_e'(\mathbf{q})\, dq = -2\pi\, \varphi(\mathbf{p}); \quad \mathbf{p} \subset B_e \tag{1.163}$$

$$= -\pi\, \varphi(\mathbf{p}); \quad \mathbf{p} \subset \partial B (\neq \Gamma) \tag{1.164}$$

and it will be noted that

$$\varphi(\mathbf{p}) = \frac{1}{2\pi} \log r \int_{\partial B} \varphi'_e(\mathbf{q})\, dq + 0(r^{-1}) \quad \text{as} \quad r = |\mathbf{p}| \to \infty \tag{1.165}$$

We may generate integral equations from (1.162), (1.164) which have the usual properties if $\partial B \neq \Gamma$. However, to ensure bounded behaviour at infinity, we must replace (1.163), (1.164) by

$$-2\pi c + \int_{\partial B} \log |\mathbf{p}-\mathbf{q}|'_e \varphi(\mathbf{q}) - \int_{\partial B} \log |\mathbf{p}-\mathbf{q}|\, \varphi'_e(\mathbf{q})\, dq = -2\pi\, \varphi(\mathbf{p}); \quad \mathbf{p} \subset B_e \tag{1.166}$$

$$= -\pi\, \varphi(\mathbf{p}); \quad \mathbf{p} \subset \partial B \tag{1.167}$$

coupled with the complementary side conditions

$$\int_{\partial B} \varphi'_e(\mathbf{q})\, dq = 0, \quad c = \int_{\partial B} \varphi(\mathbf{p})\, \lambda(\mathbf{p})\, dp \Bigg| \int_{\partial B} \lambda(\mathbf{p})\, dp \tag{1.168}$$

The formulation (1.162) for $\varphi'_i$ holds even if $\partial B = \Gamma$, with uniqueness ensured by $\int_{\partial B} \varphi'_i(\mathbf{q})\, dq = 0$. This does not apply to the formulation (1.164) for $\varphi'_e$, which must be replaced by (1.166) if $\partial B = \Gamma$.

*Plane Elastostatics*

Vector potential theory may be adapted to two dimensions by using appropriate expansions for the dyadics $g(\mathbf{p}, \mathbf{q})$ or $g'(\mathbf{p}, \mathbf{q})$ or $g(\mathbf{p}, \mathbf{q})'$. A simpler alternative approach is to work with a scalar biharmonic function $\chi(\nabla^2 \nabla^2 \chi = 0)$, which signifies either the Airy stress function (plane stress or plane strain) or the transverse deflection of a thin plate. According to a theorem of Hadamard, $\chi$ is uniquely determined in $B_i$ if $\chi$, $\chi'_i$ are given on $\partial B$ – data readily available from the boundary tractions or boundary deflections as the case may be. To construct $\chi$ in $B_i$, we employ the Almansi representation

$$\chi = r^2 \varphi + \psi; \quad r^2 = x^2 + y^2, \quad \nabla^2 \varphi = 0, \quad \nabla^2 \psi = 0 \tag{1.169}$$

with $\varphi$, $\psi$ represented as simple-layer potentials. Thus, writing

$$\varphi(\mathbf{p}) = \int_{\partial B} \log |\mathbf{p}-\mathbf{q}|\, \sigma(\mathbf{q})\, dq, \quad \psi(\mathbf{p}) = \int_{\partial B} \log |\mathbf{p}-\mathbf{q}|\, \eta(\mathbf{q})\, dq; \quad \mathbf{p} \subset B_i + \partial B \tag{1.170}$$

we obtain two coupled integral equations

$$\left.\begin{aligned} \chi(\mathbf{p}) &= r^2 \varphi(\mathbf{p}) + \psi(\mathbf{p}) \\ \chi'_i(\mathbf{p}) &= 2r\, r'_i\, \varphi(\mathbf{p}) + r^2 \varphi'_i(\mathbf{p}) + \varphi'_i(\mathbf{p}) \end{aligned}\right\} \quad \begin{aligned} &r = |\mathbf{p}| \\ &\mathbf{p} \subset \partial B \neq \Gamma \end{aligned} \tag{1.171}$$

for the hypothetical source densities $\sigma$, $\eta$. Once computed, these enable $\varphi$, $\psi$ and hence $\chi$ to be generated throughout $B_i$. Some problems have been attacked by this method [16, 17].

It is straightforward to determine $\chi$, $\chi'_i$ from given boundary tractions but not from given boundary displacements. In the latter case we employ the alternative Almansi representation

$$\chi = x\, \varphi + \psi; \quad \nabla^2 \varphi = 0, \quad \nabla^2 \psi = 0 \tag{1.172}$$

and note * that

$$2\mu\, u_1 = 2(1-\nu)\,\varphi - \frac{\partial\chi}{\partial x}, \quad 2\mu\, u_2 = 2(1-\nu)\,\bar{\varphi} - \frac{\partial\chi}{\partial\eta} \tag{1.173}$$

where $\langle u_1, u_2, 0\rangle$ signifies the displacement vector and $\bar{\varphi}$ is the harmonic conjugate of $\varphi$. The formulae (1.173) may be proved directly, but they also follow from the Papkovich-Neuber representation

$$\left.\begin{aligned} &\mu\langle u_1, u_2, u_3\rangle = \langle h_1, h_2, h_3\rangle - \varkappa\nabla(x\,h_1 + y\,h_2 + z\,h_3)\\ &\nabla^2\langle h_1, h_2, h_3\rangle = 0, \quad \varkappa^{-1} = 4(1-\nu)\end{aligned}\right\} \tag{1.174}$$

on identifying

$$h_1 = (1-\nu)\,\varphi, \quad h_2 = (1-\nu)\,\bar{\varphi}, \quad h_3 = 0$$

and noting that

$$x\,\varphi + y\,\bar{\varphi} = 2x\,\varphi + \psi; \quad \nabla^2\psi = 0$$

Given $u_1$, $u_2$ on $\partial B$, formulae (1.173) provide a pair of coupled integral equations for $\varphi$ (and therefore also $\bar{\varphi}$) and $\psi$ on $\partial B$, so enabling $\varphi$, $\bar{\varphi}$, $\psi$ and therefore also $u_1$, $u_2$ to be generated in $B_i$. Details are given in a recent paper by Symm and Bhattacharyya [38] building upon earlier work by Jaswon and Symm. Very recently [39] the representation (1.172) has been employed to attack traction problems as an alternative to (1.168).

Uniqueness for exterior problems requires $\chi = 0(r)$ as $r \to \infty$. There is no difficulty in adapting the representation (1.169) to this requirement: we write

$$\chi = r^2\,\varphi + \psi + c \tag{1.175}$$

where $\varphi$, $\psi$ are provided by (1.170) and $c$ is a constant balanced by $\int_{\partial B} \sigma(\mathbf{q})\, dq = 0$ (which ensures $\varphi = 0(r^{-1})$ and therefore $\chi = 0(r)$ as $r \to \infty$). However, the adaption of (1.172) to exterior domains presents considerable difficulties which go beyond the scope of this review.

## References


1 Fredholm, I., Sur une classe d'equations fonctionelles. Acta Math. **27,** 365–390 (1903)

2 Hess, J. L., Smith, A. M. O., Calculation of non-lifting potential flow about arbitrary three-dimensional bodies. Report No. E. S. 40622, Douglas Aircraft Co., Long Beach 1962

3 Hess, J. L., Smith, A. M. O., Calculation of potential flow about arbitrary bodies. In: *Progress in Aeronautical Sciences, Vol. 8.* (D. Kuchemann, Ed.). Pergamon Press, London 1967

4 Jaswon, M. A., Ponter, A. R. S., An integral equation solution of the torsion problem. Proc. Roy. Soc. **A, 273,** 237–246 (1963)

5 Jaswon, M. A., Integral equation methods in potential theory, I, Proc. Roy. Soc., **A, 275,** 23–32 (1963)

6 Volterra, V., *Theory of Functionals and of Integral and Integro-Differential Equations.* Dover, New York 1959


* These formulae refer to plane strain, but they also hold for plane stress on replacing $\nu$ by $\nu/(1+\nu)$

7 Symm, G. T., Integral equation methods in potential theory, II. Proc. Roy. Soc., **A, 275,** 33–46 (1963)

8a Symm, G. T., External thermal resistance of buried cables and troughs. Proc. I.E.E.E., **116** (10), 1695–1698 (1969)

8b Symm, G. T., Computation of potential in a multiwise proportional counter of arbitrary cross-section. Nucl. Instrum. Methods, **118,** 605–607 (1974)

9 Symm, G. T., Capacitance of coaxial lines with steps and tapers. In: *Recent Advances in B.E.M., Proc. 1st Int. Conf. B.E.M.,* Southampton Univ., (C. A. Brebbia, Ed.). Pentech Press, London 1978

10 Symm, G. T., An integral equation method in conformal mapping. Num. Math., **9,** 250–258 (1966)

11 Symm, G. T., Numerical mapping of exterior domains. Num. Math., **10,** 437–445 (1967)

12 Symm, G. T., Conformal mapping of doubly-connected domains. Num. Math., **13,** 448–457 (1969)

13 Papamichael, N., Whiteman, J. R., A numerical conformal transformation method for harmonic mixed boundary value problems in polynomial domains. J. App. Math. Phys. (ZAMP), **24,** 304–316 (1973)

14 Gaier, D., Integralgleichungen erster Art und konforme Abbildung. Math. Zeit., **147,** 113–129 (1976)

15 Hough, D. M., Papamichael, N., An integral equation method for the numerical conformal mapping of interior, exterior and doubly connected domains. Num. Math. (in the press)

16 Jaswon, M. A., Maitit, M., Symm, G. T., Numerical biharmonic analysis and some applications. Int. J. Solids Structures, **3,** 309–332 (1967)

17 Jaswon, M. A., Maiti, M., An integral equation formulation of plate bending problems. J. Eng. Math., **2** (1), 83–93 (1968)

18 Rizzo, F. J., An integral equation approach to boundary value problems of classical elastostatics. Quart App. Math., **25** (1), 83–95 (1967)

19 Jaswon, M. A., Some theoretical aspects of boundary integral equations. Appl. Math. Modell., **5,** 409–413 (1981)

20 Kupradze, V. D., *Potential Methods in the Theory of Elasticity.* Israel Program for Scientific Translations. Jerusalem 1965

21 Jaswon, M. A., Symm, G. T., *Integral Equation Methods in Potential Theory and Elastostatics.* Academic Press, London 1977

22 Rizzo, F. J., Shippy, D. J., A Method for Stress Determination in Plane Anisotropic Elastic Bodies. J. Composite Materials, **4,** 36–60 (1970)

23 Vogel, S. M., Rizzo, F. J., An integral equation formulation of three dimensional anisotropic elastostatic boundary value problems. J. Elast. **3** (3), 203–216 (1973)

24 Cruse, T. A., Numerical solutions in three dimensional elastostatics. Int. J. of Solids Structures **5,** 1259–1274 (1969)

25 Cruse, T. A., An application of the boundary-integral equation method to three-dimensional stress analysis. Computers & Structures **3,** 509–527 (1973)

26 Cruse, T. A., An improved boundary-integral equation method for three-dimensional elastic stress analysis. Computers & Structures **4,** 741–754 (1974)

27 Brebbia, C. A., *The Boundary Element Method for Engineers.* Pentech Press, London 1978

28 Brebbia, C. A., Walker, S., Boundary Element Techniques in Engineering. Butterworth 1979

29 Brebbia, C. A. (Ed.), *Recent Advances in B.E.M. Proc. 1st Int. Conf. B.E.M.* Southampton Univ., Pentech Press, London 1978

30 Brebbia, C. A. (Ed.), *New Developments in B.E.M., Proc. 2nd Int. Conf. B.E.M.* C.M.L Publications, Southampton 1980

31 Brebbia, S. A. (Ed.), *B.E.M., Proc. 3rd Int. Seminar. Irving, California.* C.M.L. Publications, Springer 1981

32 Zinkiewicz, O. K., Marriage à la Mode-Finite Element and Boundary Integral Method. In: *International Symposium on Innovative Numerical Analysis in Applied Engineering Science, Versailles.* CETIM Publications, Paris 1977

33 Brebbia, C. A., Georgiou, P., Combination of boundary and finite elements in elastostatics. App. Math. Modell. **3**, 213–220 (1978)
34 Wendland, W. L., On applications and the convergence of boundary integral methods. In: *Treatment of Integral Equations by Numerical Methods*, (C. T. Baker & G. F. Miller, Eds.). Proc. Durham Symposium, Academic Press 1982
35 Kellogg, O. D., *Foundations of Potential Theory*. Springer 1929
36 Petrovsky, I. G., *Lectures on Theory of Integral Equations*. M.I.R., Moscow 1971
37 Love, A. E. H., *A Treatise on the Mathematical Theory of Elasticity*. Cambridge 1927
38 Bhattacharyya, P. K., Symm, G. T., A novel formulation and solution of the plane elastostatic displacement problem. J. Comp. Math. App. **6**, 443–448 (1982)
39 Bhattacharyya, P. K., Symm, G. T., A new formulation and solution of the plane elastostatic traction problem. Appl. Math. Modell (in the press)

# Chapter 2

# Applications in Transient Heat Conduction

*by H. L. G. Pina and J. L. M. Fernandes*

## 2.1 Introduction

The use of singularities to represent instantaneous sources or sinks of heat for solving time dependent heat conduction problems is described in reference [1, ch.X] where Kelvin is credited with having made systematic use of this method to obtain analytical solutions. The integral representation to be derived below in Section 2 appears in reference [2], but without any numerical treatment.

Reference [3] appears to be the first paper where the time dependent heat equation is solved by numerical methods based on integral equations. The differential equation is transformed via Laplace transformation for the time variable and the resulting integral equation is solved numerically. The subsequent inversion of the Laplace transform is also done numerically.

In references [4, 5] the integral representation is employed to set an integral equation which is discretized in both space and time using finite elements type functions. Reference [6] of these series presents a comprehensive review of the solution of parabolic problems using boundary integral equations. In reference [7] this has been extended to solve the difficult problem of axisymmetric time-dependent heat conduction also via a space and time discretization.

In this chapter we describe how to solve heat conduction problems in homogeneous isotropic bodies by the Boundary Element Method (BEM). The problem is to find the solution of the following heat equation

$$\frac{\partial u}{\partial t} = k\,\nabla^2 u \quad \text{in} \quad \Omega \times T \tag{2.1}$$

satisfying the initial condition

$$u(\mathbf{x}, t_0) = u_0(\mathbf{x}) \quad \text{in} \quad \Omega \tag{2.2}$$

and appropriate boundary conditions. In heat conduction problems we assume the following forms for the boundary conditions:

– Dirichlet boundary condition:

$$u(\mathbf{x}, t) = \bar{u}(\mathbf{x}, t) \quad \text{on} \quad \Gamma_1 \times T \tag{2.3 a}$$

– Neumann boundary condition:

$$q(\mathbf{x}, t) = \bar{q}(\mathbf{x}, t) \quad \text{on} \quad \Gamma_2 \times T \tag{2.3 b}$$

– Robin boundary condition:

$$q(\mathbf{x}, t) + \alpha(\mathbf{x}, t)\, u = \beta(\mathbf{x}, t) \quad \text{on} \quad \Gamma_3 \times T \tag{2.3c}$$

In the expressions above $\bar{u}$, $\bar{q}$, $\alpha$ and $\beta$ are given functions of space and time, with the restriction that $\alpha \geqq 0$. $\Gamma_1$, $\Gamma_2$ and $\Gamma_3$ are different parts of the boundary $\Gamma$ such that their total gives $\Gamma$.

## 2.2 Integral Formulation of Heat Conduction Problems

*Integral Representation of the Heat Equation Solutions*

The differential equation governing heat conduction in homogeneous isotropic bodies is, as we have already remarked,

$$\frac{\partial u}{\partial t} = k\,\nabla^2 u \quad \text{in} \quad \Omega \times T \tag{2.4}$$

It is known [1, ch.X] that the function

$$u^*(\mathbf{x}, t; \mathbf{x}', t') = \frac{1}{[4\pi\varkappa(t-t')]^{d/2}} \exp\left[-\frac{r^2}{4\pi\varkappa(t-t')}\right] \tag{2.5}$$

is a particular solution of (2.4). Notice that $\mathbf{x} \in R^d$ and all $t > t'$ and $r$ represents the euclidean distance between points $\mathbf{x}$ and $\mathbf{x}'$, i.e.,

$$r = (\mathbf{r} \cdot \mathbf{r})^{1/2}, \quad \mathbf{r} = \mathbf{x} - \mathbf{x}' \tag{2.6}$$

It can be shown (see reference [1]) that this function has the following properties:

$$\lim_{t' \to t} u^*(\mathbf{x}, t; \mathbf{x}', t') = \delta(\mathbf{x} - \mathbf{x}') \tag{2.7a}$$

where $\delta(\cdot)$ stands for the Dirac distribution centered at the origin;

$$\int_{R^d} u^*(\mathbf{x}, t; \mathbf{x}', t')\, d\Omega(\mathbf{x}) = \int_{R^d} u^*(\mathbf{x}, t; \mathbf{x}', t')\, d\Omega(\mathbf{x}') = 1 \tag{2.7b}$$

The function $u(\mathbf{x}, t; \mathbf{x}'; t')$ is called the fundamental solution of equation (2.4). It represents the field of temperature produced by an instantaneous point source of heat placed at point $\mathbf{x}'$ and instant $t'$. Next we demonstrate how this fundamental solution can be used to obtain an integral representation for the solution of the heat equation. Consider the following

$$u = u(x, t'), \qquad \frac{\partial u}{\partial t'} = k\,\nabla^2 u, \tag{2.8a}$$

$$u^* = u^*(\mathbf{x}, t; \mathbf{x}', t'), \qquad \frac{\partial u^*}{\partial t'} = -k\,\nabla^2 u^*, \tag{2.8b}$$

from which we see that $u$ and $u^*$ are solutions of two adjoint equations. We also known that:

$$\frac{\partial}{\partial t'}(u\,u^*) = u^* \frac{\partial u}{\partial t'} + u \frac{\partial u^*}{\partial t'} = k\,(u^* \nabla^2 u - u \nabla^2 u^*) \tag{2.9}$$

Integrating both sides of this equation over the cylinder $\Gamma\, x\, (t_0, t)$ we obtain

$$\int_{t_0}^{t} \int_{\Omega} \frac{\partial}{\partial t'} (u\, u^*)\, d\Omega\, (\mathbf{x}')\, dt' = k \int_{t_0}^{t} \int_{\Omega} (u^*\, \nabla^2 u - u\, \nabla^2 u^*)\, d\Omega\, (\mathbf{x}')\, dt' \tag{2.10}$$

The left hand side of this expression can be modified by interchanging the order of integration, i.e.,

$$\int_{t_0}^{t} \int_{\Omega} \frac{\partial}{\partial t'} (u\, u^*)\, d\Omega\, (\mathbf{x}')\, dt' = \int_{\Omega} \int_{t_0}^{t} \frac{\partial}{\partial t'} (u\, u^*)\, dt'\, d\Omega\, (\mathbf{x}') = \int_{\Omega} [u\, u^*]_{t_0}^{t}\, d\Omega\, (\mathbf{x}')$$

$$= \int_{\Omega} u\, (\mathbf{x}, t)\, u^*\, (\mathbf{x}, t; \mathbf{x}', t)\, d\Omega\, (\mathbf{x}') - \int_{\Omega} u\, (\mathbf{x}, t_0)\, u^*\, (\mathbf{x}, t; \mathbf{x}', t_0)\, d\Omega\, (\mathbf{x}')$$

We can now recall (2.7 a) and put for simplicity

$$u_0^*\, (\mathbf{x}, t; \mathbf{x}') = u^*\, (\mathbf{x}, t; \mathbf{x}', t_0)\,, \tag{2.11 a}$$

$$c\, (\mathbf{x}) = \int_{\Omega} \delta (\mathbf{x} - \mathbf{x}')\, d\Omega\, (\mathbf{x}') \tag{2.11 b}$$

Hence we find the following expression

$$\int_{t_0}^{t} \int_{\Omega} \frac{\partial}{\partial t'} (u\, u^*)\, d\Omega\, (\mathbf{x}')\, dt' = c\, (x)\, u\, (\mathbf{x}, t) - \int_{\Omega} u_0\, u_0^*\, d\Omega\, (\mathbf{x}') \tag{2.12}$$

If the point $\boldsymbol{x}$ lies in the interior of $\Omega$ then, by the well known properties of the Dirac distribution, $c(\boldsymbol{x}) = 1$. If $\boldsymbol{x}$ is on the boundary $\Gamma$ of $\Omega$ then $c(\boldsymbol{x})$ equals the fraction of solid angle subtended by $\Gamma$ at $\boldsymbol{x}$, relative to the solid angle of the sphere in $R^d$ [21]. For smooth points on $\Gamma$, $c(\boldsymbol{x}) = 1/2$.

Applying Green's second identity to the right hand side of (2.10) we find that

$$\int_{\Omega} (u^*\, \nabla^2 u - u\, \nabla^2 u^*)\, d\Omega\, (\mathbf{x}') = \int_{\Gamma} \left( u^*\, \frac{\partial u}{\partial n} - u\, \frac{\partial u^*}{\partial n} \right) d\Gamma\, (\mathbf{x}') \tag{2.13}$$

Collecting this result and (2.12) and abbreviating slightly the notation, we arrive at the expression

$$c\, u = \int_{\Omega} u_0\, u_0^*\, d\Omega + \int_{\Gamma} \int_{t_0}^{t} (u^*\, q - u\, q^*)\, dt'\, d\Gamma \tag{2.14}$$

This relation shows that the value of the function $u$ at interior points ($c = 1$) and for any instants $t > t_0$ can be explicitly evaluated by integration, once the initial temperature field and boundary values of temperature and flux are known. Or, to use other words, (2.14) gives an integral representation for the solutions of the heat equation.

We notice that in (2.14) the fundamental solution flux is

$$q^* \equiv k\, \frac{\partial u^*}{\partial n} = -\frac{1}{2\, (t - t')}\, \mathbf{r} \cdot \mathbf{n} \cdot u^* \tag{2.15}$$

We have two kinds of space integrals in the representation (2.14), one in the domain $\Omega$ and another on the boundary $\Gamma$. Now we can show that, if $u_0$ is a harmonic function in $\Omega$, the space integral can be transformed into a boundary integral. Indeed, applying Green's second identity to $u_0$ and another function $\varphi$ we have

$$\int_{\Omega} (u_0 \nabla^2 \varphi - \varphi \nabla^2 u_0)\, d\Omega = \int_{\Gamma} \left( u_0\, \frac{\partial \varphi}{\partial n} - \varphi\, \frac{\partial u_0}{\partial n} \right) d\Gamma \tag{2.16}$$

But, by definition,

$$\nabla^2 u_0 = 0 \tag{2.17}$$

If $\varphi$ is choosen such that it satisfies

$$\nabla^2 \varphi = u_0^* \tag{2.18}$$

we arrive at the following expression

$$\int_\Omega u_0 u_0^* \, d\Omega = \int_\Gamma \left( u_0 \frac{\partial \varphi}{\partial n} - \varphi \frac{\partial u_0}{\partial n} \right) d\Gamma \tag{2.19}$$

The integration of (2.18) for two and three dimensional problems brings no special difficulties. For example, for the two dimensional case employing polar co-ordinates we have

$$\frac{1}{r} \frac{d}{dr} \left( r \frac{d\varphi}{dr} \right) = \frac{1}{4\pi k (t - t_0)} \exp \left[ - \frac{r^2}{4k(t - t_0)} \right] \tag{2.20 a}$$

Integrating this second order ordinary differential equation we obtain

$$\varphi = - \frac{1}{4\pi} E_1 \left( \frac{r^2}{4k(t - t_0)} \right) + c_1 \ln \frac{r^2}{4\pi (t - t_0)} + c_2 \tag{2.20 b}$$

where $E_1(\cdot)$ is the exponential integral function [8]. We can choose the integration constants $c_1$ and $c_2$ in order to cancel the logarithmic singularity of $E_1(s)$ when $s \to 0$ and thus obtaining a smooth $\varphi$. The exponential integral $E_1$ can be expanded as shown in reference [8, ch.5], i.e.,

$$E_1(s) = -\gamma - \ln s + e_1(s) \tag{2.21}$$

where $e_1(s)$ is a smooth function. Therefore we can take

$$c_1 = -\frac{1}{4\pi}, \quad c_2 = -\frac{\gamma}{4\pi}$$

and arrive at the expression

$$\varphi = -\frac{1}{4\pi} E_1 \left( \frac{r^2}{4k(t - t_0)} \right) - \frac{1}{4\pi} \ln \left( \frac{r^2}{4k(t - t_0)} \right) - \frac{\gamma}{4\pi}$$

$$\varphi = -\frac{1}{4\pi} e_1 \left( \frac{r^2}{4k(t - t_0)} \right) \tag{2.22}$$

which yields a smooth function $\varphi$ for two-dimensional problems. For three-dimensional problems, we have already a smooth function when $r \to 0$, i.e.,

$$\varphi = -\frac{1}{4\pi r} \operatorname{erf} \left[ \frac{r}{(4k(t - t_0))^{1/2}} \right] \tag{2.23}$$

*Integral Equations for Heat Conduction Problems*

In this section we show how the integral representation (2.14) can be employed to transform the partial differential equation for the heat conduction problem into an integral formulation. To illustrate this we will consider the case of the heat equation, i.e.,

$$\frac{\partial u}{\partial t} = k\,\nabla^2 u \quad \text{in} \quad \Omega \times T \tag{2.24a}$$

subject to the following two types of boundary conditions,

$$u = \bar{u} \quad \text{on} \quad \Gamma_1 \times T \tag{2.24b}$$

$$q = \bar{q} \quad \text{on} \quad \Gamma_2 \times T \tag{2.24c}$$

where $\bar{u}$ is equal to the prescribed temperature values on $\Gamma_1$ and $\bar{q}$ represents the prescribed flux at points on the boundary $\Gamma_2$. As before $\Gamma_1$ and $\Gamma_2$ are complementary parts of the boundary $\Gamma$ of the domain $\Omega$. We recall expression (2.14) and, in view of (2.24 b, c), rewrite it as

$$c\,u = \int_\Omega u_0\, u_0^*\, d\Omega + \int_{\Gamma_1} \int_{t_0}^{t} (u^*\, q_1 - \bar{u}\, q^*)\, dt'\, d\Gamma + \int_{\Gamma_2} \int_{t_0}^{t} (u^*\, \bar{q} - u_2\, q^*)\, dt'\, d\Gamma \tag{2.25}$$

where $q_1$ represents the values of $q$ on $\Gamma_1$ and $u_2$ those of $u$ on $\Gamma_2$. Notice that, if we knew the $q_1$ and $u_2$ values, we could compute explicitly $u$ at any interior point ($c = 1$) by evaluating the integrals on the right hand side of the above expression.

In order to simplify the notation we can now introduce the operators $\hat{\mathscr{H}}$ and $\mathscr{G}$ and the function $\hat{b}$ given by

$$\hat{\mathscr{H}}\, u_2 = \int_{\Gamma_2} \int_{t_0}^{t} q^*\, u_2\, dt'\, d\Gamma \tag{2.26a}$$

$$\mathscr{G}\, q_1 = \int_{\Gamma} \int_{t_0}^{t} u^*\, q_1\, dt'\, d\Gamma \tag{2.26b}$$

$$\hat{b} = -\int_{\Gamma_1} \int_{t_0}^{t} \bar{u}\, q^*\, dt'\, d\Gamma + \int_{\Gamma_2} \int_{t_0}^{t} u^*\, \bar{q}\, dt'\, d\Gamma + \int_\Omega u_0\, u_0^*\, d\Omega \tag{2.26c}$$

We note that the operator $\hat{\mathscr{H}}$ transforms functions over $\Gamma_2$ into functions over $\Gamma$ and the operator $\mathscr{G}$ transforms functions over $\Gamma_1$ into functions over $\Gamma$. Then, the expression (2.25) can be simply written as

$$c\,u = -\hat{\mathscr{H}}\, u_2 + \mathscr{G}\, q_1 = \hat{b} \tag{2.27}$$

Equation (2.25), or its equivalent (2.27), is also valid at points on the boundary. Making $\boldsymbol{x} \in \Gamma_1$ we obtain the following relation

$$\hat{\mathscr{H}}\, u_2 - \mathscr{G}\, q_1 = \hat{b} - c\,\bar{u} \tag{2.28a}$$

and making $\boldsymbol{x} \in \Gamma_2$ we similarly find the following expression

$$c\,u_2 + \hat{\mathscr{H}}\, u_2 - \mathscr{G}\, q_1 = \hat{b} \tag{2.28b}$$

Defining a new operator $\mathscr{H}$ and a new function $b$ such that,

$$\mathscr{H}\, u_2 = c_2\, u_2 + \hat{\mathscr{H}}\, u_2 \tag{2.29a}$$

and

$$b = \hat{b} - c_1\, \bar{u} \tag{2.29b}$$

with $c_1$ and $c_2$ functions on $\Gamma$ given by

$$\begin{aligned} c_1 = c, \quad c_2 = 0 \quad \text{for} \quad \mathbf{x} \in \Gamma_1 \\ c_1 = 0, \quad c_2 = c \quad \text{for} \quad \mathbf{x} \in \Gamma_2 \end{aligned} \tag{2.30}$$

we arrive at the boundary integral equation

$$\mathscr{H} u_2 - \mathscr{G} q_1 = b \tag{2.31}$$

which is valid on the boundary $\Gamma \times T$ of the domain $\Omega \times T$. The functions $u_2$ and $q_1$ are unknown.

## 2.3 Numerical Solution of the Integral Equations

*Discretization of the Integral Equations*

The boundary element method solution of equation (2.31) rests broadly on two basic features. One is the construction of finite element approximations $\tilde{u}_2$ and $\tilde{q}_1$ to the unknown functions $u_2$ and $q_1$ over $\Gamma \times T$. The other consists in introducing these approximations into equation (2.31) and by collocation at a sufficient number of points on $\Gamma \times T$ to set up a system of equations to determine the nodal values of $\tilde{u}_2$ and $\tilde{q}_1$.

The right hand side $b$ in (2.31) contains an integral over the domain $\Omega$ which in general cannot be evaluated analytically, therefore we must resort to numerical integrations over $\Omega$. For domains of arbitrary shape this makes it necessary to discretize $\Omega$ into internal finite elements. As this discretization does not involve additional unknowns, some authors prefer to refer to these regions in $\Omega$ as "cells".

Let us consider the whole process in more detail. The discretization of the domain $\Omega$ and of the boundary $\Gamma$ into "cells" leads to an approximate domain $\tilde{\Omega}$ and an approximate boundary $\tilde{\Gamma}$ due to the assemblage of elements (see Figure 2.1 for details). We denote by $\Omega_e$ the element labelled $e$ in $\tilde{\Omega}$, by $E(\tilde{\Omega})$ the total number of such elements and by $N(\tilde{\Omega})$ the total number of nodes. We proceed likewise for $\tilde{\Gamma}$. We assume also that the discretization of the space-time domain $\Gamma \times T$ is the direct product of the discretization of $\Gamma$ on the interval $T$. The boundary discretization $\tilde{\Gamma}$ of $\Gamma$ may be determined by the internal discretization $\tilde{\Omega}$ of $\Omega$, but this is not necessary, i.e., we do not have to force the mesh on $\tilde{\Gamma}$ to be equal to the boundary mesh of $\tilde{\Omega}$. If we denote by $\psi_i(\boldsymbol{x})$ $(i = 1, \ldots, N(\tilde{\Gamma}))$ the basis functions over $\tilde{\Gamma}$ and by $\varphi_\alpha(t)$, $\alpha = 1, 2, \ldots$ the basis functions over $T$ we end up

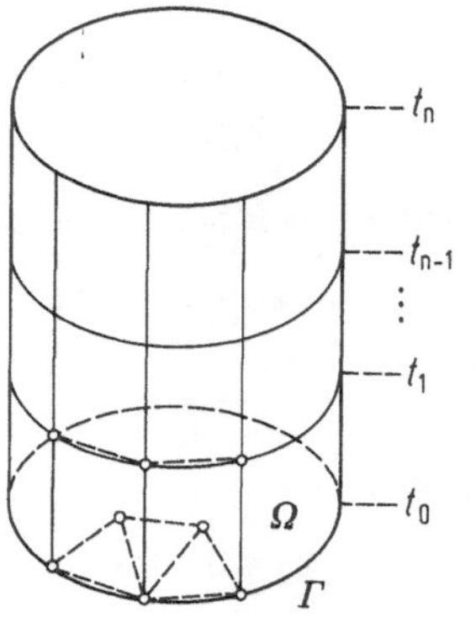


**Fig. 2.1** Geometry and discretization

with the approximations $\tilde{u}_2$ and $\tilde{q}_1$, i.e.,

$$u_2 \approx \tilde{u}_2 = \sum_i{}_2 \sum_\alpha \tilde{u}_{2i\alpha}\, \psi_i(\mathbf{x})\, \varphi_\alpha(t) \tag{2.32a}$$

$$q_1 \approx \tilde{q}_1 = \sum_i{}_1 \sum_\alpha \tilde{q}_{1i\alpha}\, \psi_i(\mathbf{x})\, \varphi_\alpha(t) \tag{2.32b}$$

where the symbols $\sum_i{}_1$ and $\sum_i{}_2$ stand for summation over the nodes of $\tilde{\Gamma}_1$ and of $\Gamma_2$. The parameters $\tilde{u}_{2i\alpha}$ and $\tilde{q}_{1i\alpha}$ represent the nodal values of $\tilde{u}_2$ and $\tilde{q}_1$ at node $i$ on $\Gamma$ and node $\alpha$ on $T$. Since by definition the $\psi_i$ and $\varphi_\alpha$ are interpolation functions, at the nodes $x_i$ and $t_\alpha$ we have that

$$\psi_i(\mathbf{x}_j) = \delta_{ij} \quad \text{and} \quad \varphi_\alpha(t_\beta) = \delta_{\alpha\beta} \tag{2.33}$$

Introducing relations (2.32) into equation (2.31) we obtain

$$\sum_i{}_2 \sum_\alpha (\mathscr{H}\, \psi_i\, \varphi_\alpha)\, \tilde{u}_{2i\alpha} + \sum_i{}_1 \sum_\alpha \mathscr{G}\, (\psi_i\, \varphi_\alpha)\, \tilde{q}_{1i\alpha} = \tilde{b} \tag{2.34}$$

To determine the nodal values $\tilde{u}_{2i\alpha}$ and $\tilde{q}_{1i\alpha}$ we collocate equations (2.34) at points $\mathbf{x}_j$ on $\Gamma$ and points $\tau_\beta$ in the interval $T$. Defining

$$(\mathscr{H}\, \psi_i\, \varphi_\alpha)\, (\mathbf{x}_j, \tau_\beta) = H_{j\beta i\alpha} \tag{2.35a}$$

$$(\mathscr{G}\, \psi_i\, \varphi_\alpha)\, (\mathbf{x}_j, \tau_\beta) = G_{j\beta i\alpha} \tag{2.35b}$$

$$\tilde{b}\, (\mathbf{x}_j, \tau_\beta) = b_{j\beta} \tag{2.35c}$$

results in the following system of linear algebraic equations,

$$\sum_i{}_2 \sum_\alpha H_{j\beta i\alpha}\, \tilde{u}_{2i\alpha} - \sum_i{}_1 \sum_\alpha G_{j\beta i\alpha}\, q_{1i\alpha} = b_{j\beta}\,, \tag{2.36}$$

which, once solved, yields the nodal values $\tilde{u}_{2i\alpha}$ and $\tilde{q}_{1i\alpha}$.

*Analytical and Numerical Integration Techniques*

The formation of system (2.36) requires the evaluation of the coefficients $H_{j\beta i\alpha}$, $G_{j\beta i\alpha}$ and $b_{j\beta}$, which in turn are due to the computation of integrals on $\tilde{\Gamma} \times T$ and $\tilde{\Omega}$. So does the computation of the temperature $u$ or flux $q$ at internal points. Therefore it is important to perform these integrations accurately and efficiently. To show how this can be done let us concentrate on the case of piecewise constant elements in time and continuous elements on space $R^2$. The interval $T$ is divided in subintervals $(t_0, t_1)$, $(t_1, t_2)$, $\ldots$, $(t_{\alpha-1}, t_\alpha)$, $\ldots$ such that

$$t_\alpha = t_0 + (\alpha - 1)\, \Delta t, \quad \alpha = 1, 2, \ldots \tag{2.37}$$

where $\Delta t$ is the constant time step (see Figure 2.2). Therefore the interpolation function $\varphi_\alpha$ is defined by

$$\varphi_\alpha(t) = \begin{cases} 1 & \text{for} \quad t_{\alpha-1} < t \leqq t_\alpha \\ 0 & \text{otherwise.} \end{cases} \tag{2.38}$$

Choosing the collocation points to be coincident with nodes $(\mathbf{x}_j, t_1)$ on $\tilde{\Gamma} \times (t_0, t_1)$ allows one to obtain from (2.36) the following system of linear equations in matrix notation

$$[H^{11}]\, \{\tilde{u}_2^1\} - [G^{11}]\, \{\tilde{q}_1^1\} = \{\, \tilde{b}^1\} \tag{2.39}$$

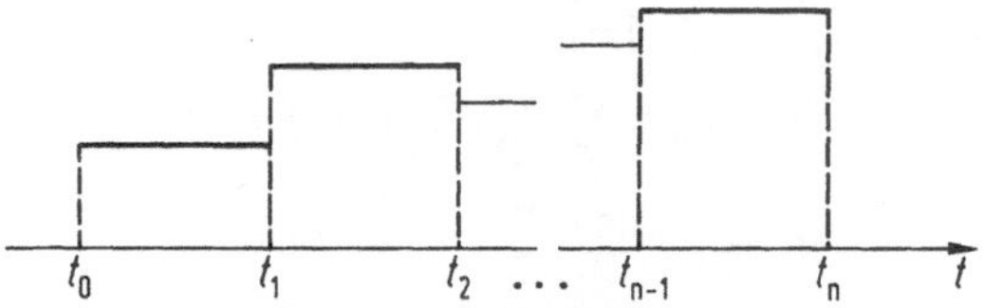

**Fig. 2.2.** Piecewise constant elements in time

where

$$[H^{11}] = H_{j1i1}, \quad [G_{ji}^{11}] = G_{j1i1} \quad \text{and} \quad \{\tilde{b}_j^1\} = \tilde{b}_{j1} \tag{2.40 a}$$

$[H^{11}]$ is a $N(\tilde{\Gamma}) \times N(\tilde{\Gamma}_2)$ matrix, $[G^{11}]$ is a $N(\tilde{\Gamma}) \times N(\tilde{\Gamma}_1)$ matrix and $\{\tilde{b}\}$ a $N(\tilde{\Gamma})$ vector. Since $N(\tilde{\Gamma}) = N(\tilde{\Gamma}_1) + N(\tilde{\Gamma}_2)$ we have the same number of equations as unknowns. We remember that at each point of $\Gamma$ either the temperature or the flux (but not both) is an unknown.

The integrals involved in matrix $[H^{11}]$ are

$$[\hat{H}_{ji}^{11}] \equiv \hat{H}_{j1i1} \equiv (\mathscr{H}\psi_i\,\varphi_\alpha)(\mathbf{x}_j, t_1) = \int_{\tilde{\Gamma}_2}\int_{t_0}^{t_1} q^*(\mathbf{x}_j, t_1; \mathbf{x}', t')\,\psi_i(\mathbf{x}')\,dt'\,d\Gamma(\mathbf{x}') \tag{2.40 b}$$

Similarly, matrix $[G^{11}]$ requires integrals of the form

$$[G_{ji}^{11}] \equiv G_{j1i1} \equiv (\mathscr{G}\psi_i\,\varphi_\alpha)(\mathbf{x}_j, t_1) = \int_{\tilde{\Gamma}_1}\int_{t_0}^{t} u^*(\mathbf{x}_j, t_1; \mathbf{x}', t')\,\psi_i(\mathbf{x}')\,dt'\,d\Gamma(\mathbf{x}') \tag{2.41}$$

To obtain $\{\tilde{b}^1\}$ we need to compute boundary integrals that have the same form as (2.40, 41) plus an integral over the domain.

Recalling expressions (2.5, 15) we can write that

$$[\hat{H}_{ji}^{11}] = -\int_{\tilde{\Gamma}_1}\int_{t_0}^{t_1} \frac{1}{2(t_1 - t')}\,\mathbf{r}\cdot\mathbf{n}\,\psi_i\,u^*\,dt'\,d\Gamma(\mathbf{x}') \tag{2.42 a}$$

and grouping all the terms depending on the variable $t'$ we obtain

$$[\hat{H}_{ji}^{11}] = -\frac{1}{8\pi k}\int_{\tilde{\Gamma}_1} \mathbf{r}\cdot\mathbf{n}\,\psi_i \int_{t_0}^{t_1} \frac{1}{(t_1 - t')^2}\exp\left[-\frac{r^2}{4k(t_1 - t')}\right] dt'\,d\Gamma(\mathbf{x}') \tag{2.42 b}$$

The integration on time can be done analytically. In fact making the change of variable

$$s = \frac{r^2}{4k(t_1 - t')}, \quad t' = t_1 - \frac{r^2}{4ks}, \quad s_0 = \frac{r}{4k\,\Delta t}, \quad s_1 = \infty, \tag{2.43}$$

we are able to write

$$\int_{t_0}^{t_1} \frac{1}{(t_1 - t')^2}\exp\left[-\frac{r^2}{4k(t_1 - t')}\right] dt' = \frac{4k}{r^2}\int_{s_0}^{\infty} \exp(-s)\,ds$$

$$= \frac{4k}{r^2}\exp(-s_0) = \frac{4k}{r^2}\exp\left(-\frac{r^2}{4k\,\Delta t}\right) \tag{2.44}$$

Introducing this result in (2.42 b) we arrive at

$$[H_{ji}^{11}] = -\frac{1}{2\pi}\int_{\tilde{\Gamma}_2} \frac{1}{r^2}\,\mathbf{r}\cdot\mathbf{n}\exp\left(-\frac{r^2}{4k\Delta t}\right)\psi_i\,d\Gamma(\mathbf{x}') \tag{2.45}$$

This integral has to be evaluated numerically, choosing an appropriate quadrature rule in each element of $\bar{\Gamma}_2$ and adding up the contributions from all the elements. It can be shown that the integrand is regular when $r \to 0$, therefore Gauss-Legendre quadrature rules are suitable. These rules however must be carefully selected. To choose too many integration points is time consuming but too few may yield poor accuracy. Also it is important to notice that for elements far away from the collocation point the integrand on the right hand-side of (2.45) varies little so we could use there fewer integration points than for elements near or containing the collocation point or singularity. This could lead to adaptive quadrature but so far nobody seems to have used them, due perhaps to the increased programming complexity. Experience gathered by several authors show the following heuristic rule to be a good compromise between accuracy and speed:

constant elements in space – 1 or 2 points Gauss-Legendre rule,
linear elements in space – 2 or 3 points Gauss-Legendre rule,
parabolic elements in space – 3 or 4 points Gauss-Legendre rule.

Coming back to expression (2.41) and repeating the procedure described above we have that

$$[G_{ji}^{11}] \equiv \int_{\bar{\Gamma}_1} \int_{t_0}^{t_1} \frac{1}{4\pi k (t_1 - t')} \exp\left[-\frac{r^2}{4k(t_1 - t')}\right] \psi_i \, dt' \, d\Gamma(\mathbf{x}')$$

$$= \frac{1}{4\pi k} \int_{\bar{\Gamma}_1} \psi_i \int_{t_0}^{t_1} \frac{1}{(t_1 - t')} \exp\left[-\frac{r^2}{4k(t_1 - t')}\right] dt' \, d\Gamma(\mathbf{x}') \qquad (2.46)$$

Performing the same change of variable the integral in time becomes

$$\int_{t_0}^{t_1} \frac{1}{(t_1 - t')} \exp\left[-\frac{r^2}{4k(t_1 - t')}\right] dt' = \int_{s_0}^{s_1} \frac{1}{s} \exp(-s) \, ds = E_1(s_0) - E_1(s_1)$$

$$= E_1(s_0) - E_1(\infty) = E_1(s_0) \qquad (2.47)$$

where

$$E_1(s) = \int_{s}^{\infty} \frac{1}{z} \exp(-z) \, dz \qquad (2.48)$$

is the exponential integral. Approximations suitable for computer applications can be found in reference [8, ch.5]. Introducing (2.47) into (2.46) we have that

$$[G_{ji}^{11}] = \frac{1}{4\pi k} \int_{\bar{\Gamma}_1} E_1\left(\frac{r^2}{4k\Delta t}\right) \psi_i \, d\Gamma(\mathbf{x}') \qquad (2.49)$$

To evaluate these integrals we proceed as before, except that now the integrand develops a singularity of the type $\ln r$ as $r \to 0$. This means that numerical integration over the elements containing the collocation point has to take this into consideration. According to (2.21) we can write

$$E_1(s) = -\ln s + e_1(s) \qquad (2.50)$$

where $e_1(s)$ is a smooth function. Then the integral in expression (2.49) can be split into a regular part and a singular part with a logarithmic singularity. We can apply the above considerations for the regular integral. For the singular one must employ special integration rules such as those due to Berthod-Zaborowski [9]. The whole procedure is analogous to that of steady potential problems so we do not need elaborate any further on this. For details see reference [10].

The integral in the domain $\tilde{\Omega}$, necessary to compute $\{\tilde{b}^1\}$ is

$$\int_{\tilde{\Omega}} u_0(\mathbf{x}')\, u_0^*(\mathbf{x}_j, t_1; \mathbf{x}', t_0)\, d\Omega(\mathbf{x}') = \frac{1}{4\pi k\, \Delta t} \int_{\tilde{\Omega}} u_0(\mathbf{x}') \exp\left(-\frac{r^2}{4k\,\Delta t}\right) d\Omega(\mathbf{x}') \tag{2.51}$$

where the integrand is easily recognized as regular. If the field $u_0(x')$ of initial temperature is harmonic this integral can be transformed into a boundary one, as was shown in 2.2, with the obvious advantages. Otherwise we have to evaluate it by using numerical integration over each cell in $\tilde{\Omega}$ and add up all the contributions. In reference [11] there are several quadrature formulae for standard triangles and quadrilaterals that have been used by Wrobel [12], Fernandes and Pina [13], namely:

- the seven point quintic rule of Hammer for linear and quadratic triangles,
- the $2 \times 2$ Gauss-Legendre rule for linear quadrilaterals,
- the seven point quintic rule or Radon, Albrecht and Collatz for quadratic quadrilaterals.

The evaluation of integrals to compute the temperature $u$ at interior points of $\tilde{\Omega}$ follows along the same lines as above. However, as the point comes close to the boundary $\tilde{\Gamma}$ the integrands very rapidly and to retain the accuracy an increasingly greater number of integration points have to be used. This means that some sort of adaptive integration scheme must be devised.

## 2.4 Time-Marching Procedures

In the previous section we have discussed how to evaluate the various integrals needed to form the system of equations (2.39). Leaving for the next section those aspects related to the actual solution of this system we admit that $\{\tilde{u}_2^1\}$ and $\{\tilde{q}_1^1\}$ have been found, and therefore, that we are able to calculate an approximation $\tilde{u}(x, t_1)$ to $u(x, t_1)$ over $\tilde{\Omega} \cup \tilde{\Gamma}$. The question is how we progress from instant $t_1$ to instant $t_2, t_3, \ldots$? There are basically two procedures, which we shall designate by I and II [6].

**Procedure I.** After obtaining the values of temperature and flux at time $t_1$, that is, $\tilde{u}_2(x, t_1)$ and $\tilde{q}_1(x, t_1)$ we can evaluate explicitly using (2.25) the value of the temperature at any interior point, which implies to construct an approximation $\tilde{u}(x, t)$ to $u(x, t)$. Procedure I consists in considering at instant $t_1$ as initial condition $u(x, t_1)$ and taking time $t_2 = t_1 + \Delta t$ as the new collocation point on the time domain. Once the solution is found at $t_2$ the procedure can be repeated with $t_2$ as the initial time. Retracing the analysis of section 3. we can easily arrive at the conclusion that the matrices $[H^{22}]$ and $[G^{22}]$ are identical to its counterparts $[H^{11}]$ and $[G^{11}]$ if $\Delta t$ is allways the same. The evolutionary character of the problem is carried out on the right hand side $\{\tilde{b}\}$ which is different for each time step.

The solution of system (2.39) can be conveniently done by LU factorization followed by forward and backward substitutions. The factorization needs only be done once at the first time-step, saved and used in the forward and backward substitutions required at all later time-steps.

In this procedure we require the computations of domain integrals $\int_{\tilde{\Omega}} u_\alpha u^*_\alpha \, d\tilde{\Omega}$ even if the inital temperature field $u_0$ is zero. To evaluate those integrals two methods can be enployed. One, which we designate by Method A, is to compute $u_\alpha$ at internal points coinciding with the integration points of the cells. The other, Method B, is to compute $u_\alpha$ at the nodes of the mesh and construct a finite element type approximation. As the number of integration points usually is greater than the number of nodes this second variant requires evaluation of the temperature of a smaller number of internal points, thus being more rapid and demanding less computer memory. The first method however tends to be more precise, but if the size of the cells is too small some integration points can become too close to the collocation points where the fundamental solutions changes rapidly and this may offset the greater theoretical accuracy.

**Procedure II.** This procedure consists in starting all time integrations from the initial instant $t_0$ up to the current instant $t_n$. Suppose the unknown nodal vectors $\{\tilde{u}_2^\alpha\}$ and $\{\tilde{q}_1^\alpha\}$ have been calculated at $t_\alpha$, $\alpha = 1, \ldots, n-1$. Equation (2.36) allows one to advance the solution to time $t_n$. Considering that the collocation time $t_\beta$ in (2.36) equals $t_n$, we obtain,

$$\sum_{\alpha=1}^{n} ([H^{n\alpha}]\{\tilde{u}_2^\alpha\} - [G^{n\alpha}]\{\tilde{q}_1^\alpha\}) = \{\tilde{b}^n\} \tag{2.52}$$

Isolating the unknowns $\{\tilde{u}_2^n\}$ and $\{\tilde{q}_1^n\}$ on the left hand-side we have the following system of linear equations

$$[H^{nn}]\{\tilde{u}_2^n\} - [G^{nn}]\{\tilde{q}_1^n\} = \{\tilde{b}^n\} - \sum_{\alpha=1}^{n-1} ([H^{n\alpha}]\{\tilde{u}_2^\alpha\} + [G^{n\alpha}]\{\tilde{q}_1^\alpha\}) \tag{2.53}$$

which, once solved, delivers $\{\tilde{u}_2\}$ and $\{\tilde{q}_1^n\}$. This procedure requires the formation of a triangular array of matrices that can be depicted as

| | $\alpha = 1$ | $\alpha = 2$ | $\alpha = 3$ | $\cdots$ | $\alpha = n$ |
|---|---|---|---|---|---|
| $\beta = 1$ | $[H^{11}]$ | | | | |
| $\beta = 2$ | $[H^{21}]$ | $[H^{22}]$ | | | |
| $\beta = 3$ | $[H^{31}]$ | $[H^{32}]$ | $[H^{33}]$ | | |
| $\vdots$ | $\vdots$ | $\vdots$ | $\vdots$ | | |
| $\beta = n$ | $[H^{n1}]$ | $[H^{n1}]$ | $[H^{n3}]$ | $\cdots$ | $[H^{nn}]$ |

and similarly for $[G^{\beta\alpha}]$. We recall that $t_\beta$ is the collocation time and the $t_\alpha$ are the nodes in the interval $T = (t_0, t_n)$. It can be easily shown that the matrices connected by the diagonal lines in this table are equal. Therefore at step $n$ only the matrices $[H^{n1}]$ and $[G^{n1}]$ and the right-hand side vector $\{\tilde{b}^n\}$ have to be calculated anew.

Albeit this procedure is more elaborate than Procedure I it presents some advantages. One is that computation of values of temperature at internal points is not required. The other rests on the fact that to evaluate the domain integral

$\int_{\Omega} u_0 u_0^* d\Omega$ in (2.26c) we need only to discretize the support of the function $u$ at time $t_0$. When $\Omega$ is an unbounded domain this can be a major benefit. Also if the initial temperature field $u_0(x)$ is harmonic in $\Omega$ the above integral can be transformed into a boundary one, as discussed in section 2.2. It should be remarked that the left-hand side of (2.53) is identical with that obtained with Procedure I. Therefore the LU factorization needs to be performed only once and saved. At each time step the solution of (2.53) requires only a forward and a backward substitution phases.

*Solution of Resulting Linear Algebraic System of Equations*

It was shown that no matter which Procedure, I or II, we selected we end up with a linear algebraic system of equations which can be written as

$$[H]\{u\} - [G]\{q\} = \{b\} \tag{2.54a}$$

or if we prefer, using partitioned matrices,

$$[H \,\vdots\, - G] \left\{ \begin{array}{c} u \\ \hdashline q \end{array} \right\} = \{b\} \tag{2.54b}$$

There is so far no general proof that the above system is solvable. However the special cases treated by [6] and [14] and the numerical experience gained from BEM practice validate the approach.

As we have remarked before, during the time marching process one system of the type (2.54) has to be solved for each time step using the same matrices but with different right-hand sides. Thus the LU factorization method seems the most adequate. As soon as the matrices $[H \,\vdots\, - G]$ are formed they are factorized and this factorization saved. At later steps only the forward and backward substitution phases need be performed. The matrix $[H \,\vdots\, - G]$ is a dense matrix, hence the number of operations (multiplications/divisions) that the factorization involves is $N(\tilde{\Gamma})^3/3$, the forward and backward substitution phases requiring $N(\tilde{\Gamma})^2/2$ each [15, ch.3].

We have been using subroutine SGECO of Linpack [16] to estimate the condition number of the resulting matrices for several test cases found that in all of them the condition number was typically less than $10^2$, indicating good stability for the system (2.54b).

## 2.5 Examples and Conclusions

**Example 1.** This example has been solved using finite elements and is taken from reference [17] and has been solved using linear boundary elements in reference [6]. It studies a two-dimensional infinite medium with unit thermal diffusivity cooled by a hole of unit radius as shown in Figure 2.3a. The boundary and initial conditions are:

$$\frac{\partial u}{\partial n} + h\,u = 0 \qquad \text{at} \qquad r = 1$$

$$u(r, 0) = 1 \qquad \text{for} \quad r > 1$$

The convection coefficient $h$ takes the values $h = 0.2, 1.0, 5.0$.

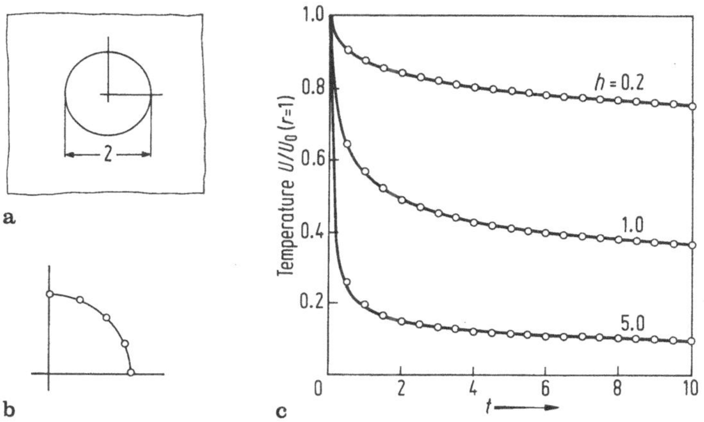


**Fig. 2.3.** **a** Geometry. **b** Discretization. **c** Results for $r = 1$

By a simple change of variable this problem can be transformed into one with zero initial condition, thus making Procedure II of section 3 specially suitable.

The mesh employed is depicted in Figure 2.3b. We see that because of symmetry one can discretize only one quarter of the hole circumference. Only 2 quadratic elements and 5 nodes were required to obtain the results shown above in Figure 2.3c.

The numerical integration were performed with 4 Gauss-Legendre points for the regular integrals and with 3 special points [9] for those with logarithmic singularities. The time step employed was $\Delta t = 0.5$.

The results obtained for the temperature at $r = 1$ are shown in Figure 2.3c and they agree well with those of references [6] and [17].

**Example 2.** This is one-dimensional test problem used in reference [18]. The region $\Omega$ is the interval $0 < x < 1$ subject to the initial condition $u_0(x) = \sin \pi x$ and to the Dirichlet boundary conditions $u(0, t) = u(1, t) = 0$ for all $t > 0$. The analytical solution is known to be $u(x, t) = \exp(-\pi^2 t) \sin \pi x$.

The mesh employed for this example is shown in Figure 2.4 Procedure I, Method A and Procedure I, Method B yield similar accuracy, Method A proving to be marginally better.

With linear elements the best accuracy was obtained with a time step $\Delta t = 0.01$ and with quadratic elements with $\Delta t = 0.005$. The accuracy of Procedure I, Method A remains almost unchanged for a large range of time steps. Procedure I, Method B is more sensitive to the choice of time step. Typical results are presented in Figure 2.4.

**Example 3.** The aim of this example is to show the applicability of the BEM to more complicated cases. The problem is to find the temperature field in a gas turbine rotor blade with cooling holes. The convection coefficient $h$ was obtained from [19, 20]. The geometry and mesh employed are depicted in Figure 2.5 as well as some isothermals for several times. A total of 45 quadratic elements and 90 nodes were used. In order to draw the isotherms the temperature was also computed at 30 interior points. Integrations were performed as in Example 1.

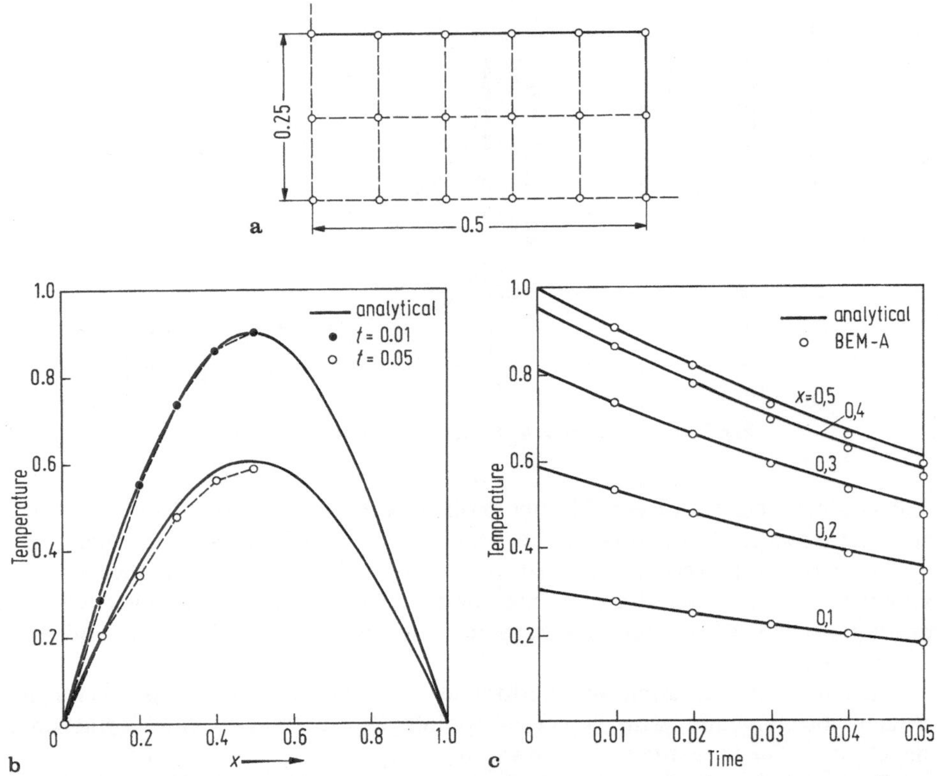


**Fig. 2.4.** **a** Geometry and discretization. **b** Temperature for two time instants. **c** Some internal point temperatures along time

**Example 4.** This example consits on a section of a structure with a hole filled with air. The geometry is shown in Figure 2.6. The boundary conditions are of mixed type, the top and bottom surfaces are radiating according to the following formula:

top surface: $\bar{q} = 900 \sin \pi \dfrac{(t-4)}{16}$ W/m$^2$ ($t$ in hours); absorptivity = 0.7

bottom surface: $\bar{q} = 83 \sin \pi \dfrac{(t-5)}{16}$ W/m$^2$ ($t$ in hours); absorptivity = 0.5

On the vertical surfaces the radiation intensities vary linearly on time, as follows.

| $t$ (hours) | Left surface | $t$ (hours) | Right surface |
|---|---|---|---|
| 4 | 0 W/m$^2$ | 0 | 0 W/m$^2$ |
| 10 | 400 W/m$^2$ | 10 | 100 W/m$^2$ |
| 14 | 200 W/m$^2$ | 15 | 600 W/m$^2$ |
| 20 | 0 W/m$^2$ | 20 | 0 W/m$^2$ |

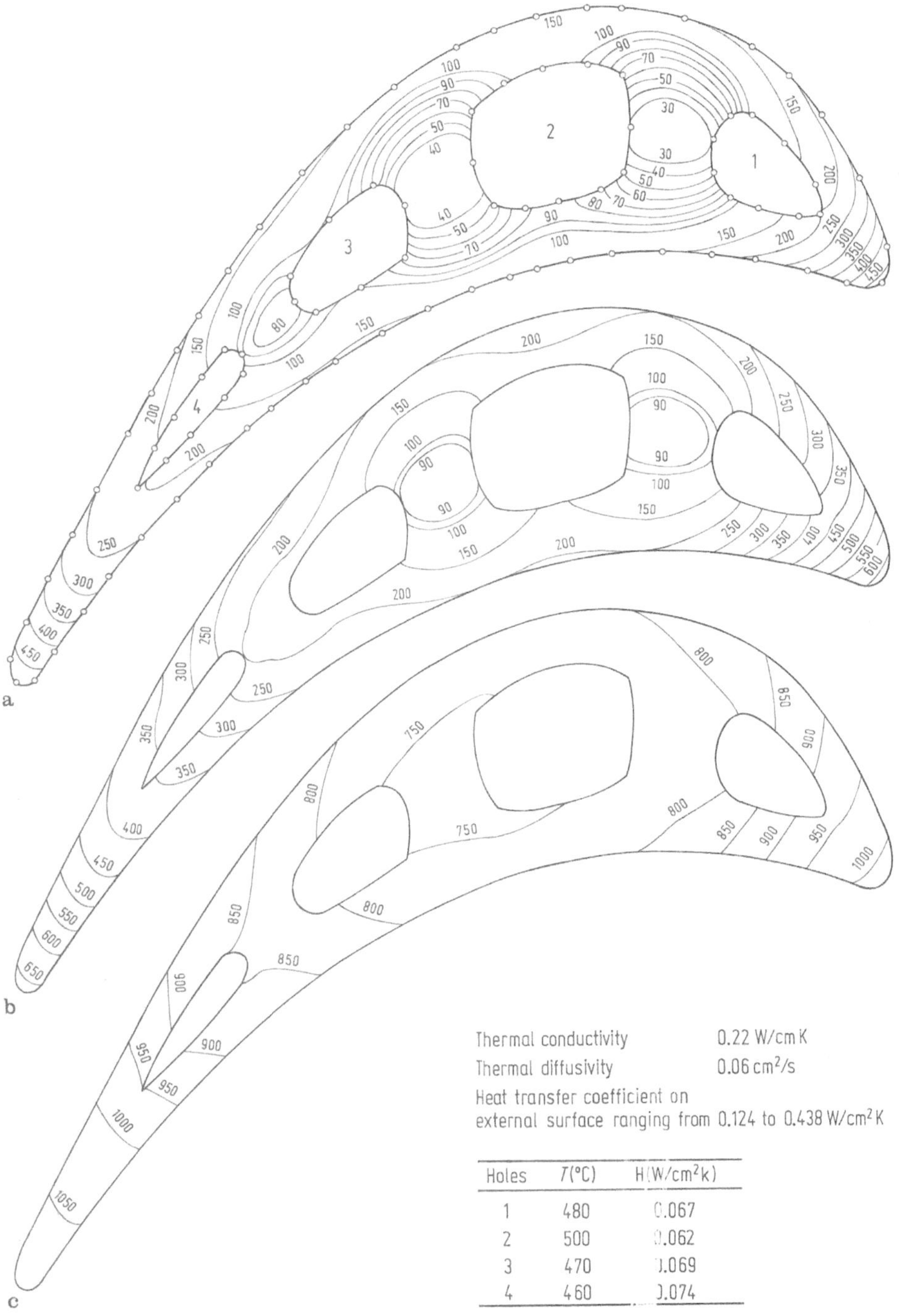


Thermal conductivity 0.22 W/cm K
Thermal diffusivity 0.06 cm²/s
Heat transfer coefficient on external surface ranging from 0.124 to 0.438 W/cm² K

| Holes | T(°C) | H (W/cm²k) |
|---|---|---|
| 1 | 480 | 0.067 |
| 2 | 500 | 0.062 |
| 3 | 470 | 0.069 |
| 4 | 460 | 0.074 |

**Fig. 2.5.** **a** Isotherms in °C for $t = 0.5$ s. **b** Isotherms for $t = 1$ s. **c** Isotherms for $t = 10$ s

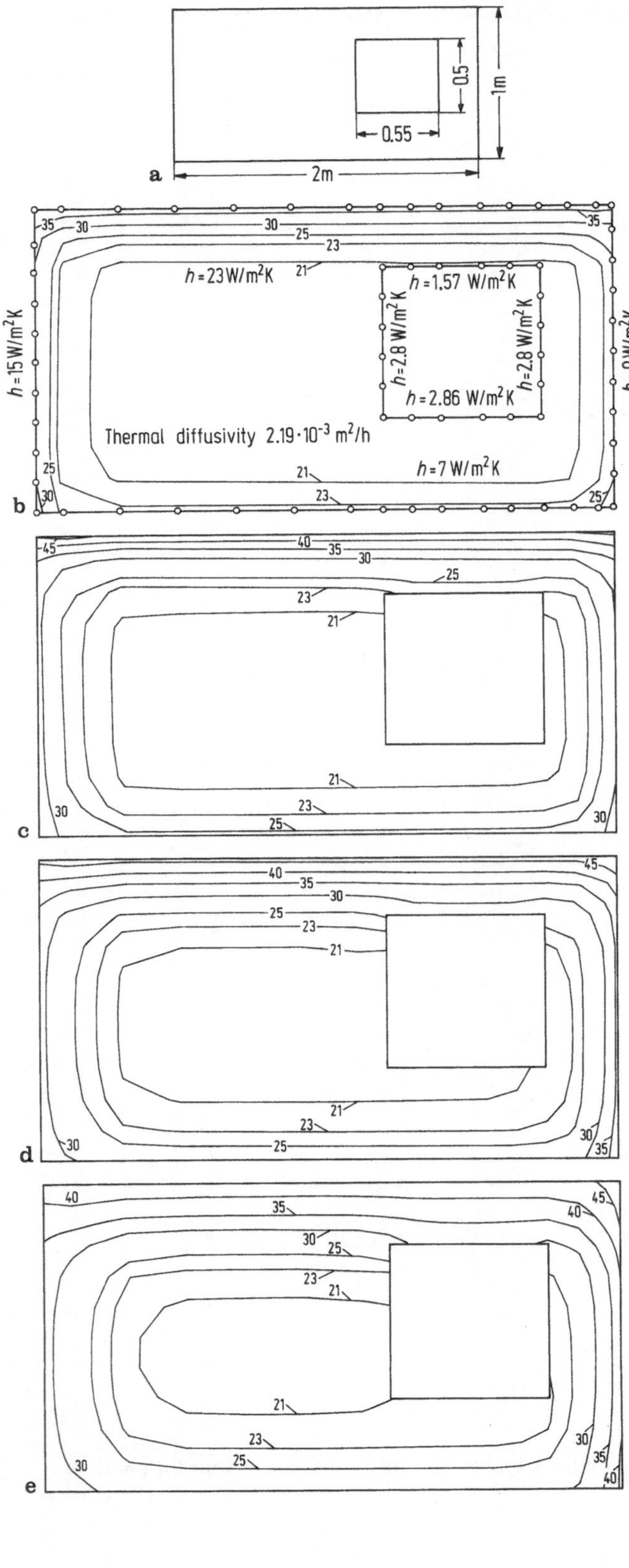


**Fig. 2.6.** **a** Geometry. **b** Discretization and isotherms at 10 o'clock. **c** Isotherms at 12 o'clock. **d** Isotherms at 14 o'clock. **e** Isotherms at 16 o'clock

On the lateral surfaces the absorptivity is 0.5. The thermal properties of the section are: *conductivity* = 1.4 W/m · K, *density* = 2400 kg/m$^3$, *specific heat* = 960 J/kg · K.

The ambient air temperature also has a linear variation on time,

| $t$ (hours) | |
|---|---|
| 4 | 22 °C |
| 15 | 34 °C |
| 20 | 26 °C |

The initial condition, assumed at $t = 8$ hours, is a uniform temperature $u = 20$ °C.

The mesh consists of 70 linear elements and 78 boundary nodes. The number of internal points required to draw the isotherms was 90. The time step was chosen $\Delta t = 1$ hour. The temperature of the air in the hole was computed via a energy balance equation, that is, the energy lost by convection through the sides of the hole was used to raise the internal energy of the air. The results obtained using Procedure II in this complex example are shown in Figure 2.6.

*Conclusions*

The examples discussed in this chapter and which range from simple tests to more complex practical applications demonstrate that the BEM can be easily applicable to the solution of linear heat conduction problems. The so-called Procedure II allows for the discretization only on the boundary thus leading to an effective reduction on the dimensionality of the problem. In the cases where comparisons where possible the BEM yielded good accuracy with time steps which are substantially larger than those employed in the finite element or finite difference methods. This conclusion has been confirmed by other authors, specially Wrobel and Brebbia in reference [6]

It is important to point out that the application of the BEM to the heat equation still requires more research work, namely with respect to the three dimensional case. Also it should be stressed that more effort is needed to develop computationally efficient codes.

**Acknowledgements.** The work reported in this paper was partially supported by Junta Nacional de Investigaçao Científica e Tecnológica.

## References

1 H. S. Carslaw, J. C. Jaeger, *Conduction of Heat in Solids.* Oxford U. Press 1959

2 B. A. Boley, J. H. Weiner, *Theory of Thermal Stresses.* J. Wiley 1960

3 F. J. Rizzo, D. H. Shippy, A method of solution for certain problems of transient heat conduction. AIAA J., **8, 11,** 2004–2009 (1970)

4 Y. P. Chang, C. S. Kang, D. J. Chen, The use of fundamental Green's functions for the solution of problems of heat conduction in anisotropic media. Int. J. Heat Mass Transfer, **16,** 1905–1918 (1973)

5 R. P. Shaw, An integral approach to diffusion. Int. J. Heat Mass Transfer, **17,** 693–699 (1974)

6 L. C. Wrobel, C. A. Brebbia, Time dependent potential problems. Chapter 6 in *Progress in Boundary Element Methods. Vol. 1.* (Ed. C. A. Brebbia), Pentech Press, London, Halstead Press, USA 1981

7 L. C. Wrobel, C. A. Brebbia, A formulation of the boundary element method for axisymmetric transient heat conduction. Int. J. Heat and Mass Transfer, **24,** 843–850 (1981)
8 M. Abramowitz, I. Stegun, *Handbook of Mathematical Functions.* Dover 1968
9 H. Mineur, *Techniques de Calcul Numerique.* Dunod 1966
10 C. A. Brebbia, H. L. G. Pina, J. L. M. Fernandes, The effect of mesh refinement in the boundary element solution of Laplace's equation with singularities. Third International Seminar on Recent Advances in Boundary Element methods, Irvine, Ca., USA 1981
11 A. H. Stroud, *Approximate Calculation of Multiple Integrals.* Prentice-Hall 1971
12 L. C. Wrobel, *Potential and Viscous Flow Problems Using the Boundary Element Methods* Ph.D. Thesis, University of Southampton 1981
13 J. L. M. Fernandes, H. L. G. Pina, Unsteady heat conduction using the boundary element method, Fourth International Seminar on BEM, Southampton 1982
14 K. Onishi, T. Kuroki, Y. Ohura, K. Obata, T. Ito, Boundary element method in transient heat transfer problems, Bull. Institute for Advanced Research of Fukuoka University, No. 55 (1981)
15 G. W. Steward, *Introduction to Matrix Computations.* Academic Press 1973
16 J. J. Dongarra, C. B. Moler, J. R. Bunch, G. W. Stewart, *Linpack User's Guide.* SIAM 1979
17 O. C. Zienkiewicz, C. J. Parekh, Transient field problems: two dimensional and three dimensional analysis by Isoparametric finite elements. Int. J. Num. Meth. Engng., **2,** 61–71 (1970)
18 D. Curran, M. Cross, B. A. Lewis, A preliminary analysis of boundary element methods applied to parabolic partial differential equations. Proceedings of the 2nd Int. Seminar on Recent Advances in Boundary Element Methods, Southampton 1980
19 K. Consigny, B. E. Richards, Short duration measurements of heat-transfer rate to a gas turbine rotor blade. Trans. of ASME, J. of Engineering for Power, **104,** 542–551 (1982)
20 D. K. Mukherjee, O. Frei, Experimental verification of cooled gas turbine stator blade design. Turboforum, **4,** 217–224 (1974)
21 O. D. Kellog, *Foundations of Potential Theory.* Dover 1954

# Notation

In this chapter the following general notation has been adopted:

| | |
|---|---|
| $R^d$ | the $d$-dimensional euclidean space. |
| $\Omega$ | an open connected set of $R^d$ representing the domain occupied by the body. |
| $\Gamma$ | the boundary of $\Omega$, which is supposed to be sufficiently smooth in the sense of Kellog [21]. |
| $\Gamma_1, \Gamma_2$ | complementary parts of $\Gamma$. |
| $T=(t_0, t_f)$ | an interval of time, were $t_0$ represents the initial instant and $t_f$ the final instant. |
| $\mathbf{x}, \mathbf{x}'$ | coordinates of points in $\Omega$ or on $\Gamma$. |
| $t, t'$ | instants in $T$. |
| $u(\mathbf{x}, t)$ | a real function over $\Omega \times T$ representing the temperature of the body. |
| $u_0(\mathbf{x})$ | a real function over $\Omega$ representing the (initial) temperature at instant $t_0$. |
| $\frac{\partial}{\partial t}(\cdot)$ | partial derivative with respect to time $t$. |
| $\nabla^2$ | the Laplacian operator. |
| $\mathbf{n}(\mathbf{x})$ | the unit exterior normal to $\Gamma$ at point $\mathbf{x}$. |
| $\frac{\partial}{\partial n}(\cdot)$ | the gradient in the direction of $\mathbf{n}$. |
| $k$ | a positive constant representing the thermal diffusity. |
| $q=k\frac{\partial u}{\partial n}$ | the flux associated to $u$. This flux differs from the real heat flux by a negative constant factor. |

## Chapter 3

# Fracture Mechanics Application in Thermoelastic States

*by M. Tanaka, H. Togoh and M. Kikuta*

## 3.1 Introduction

Thermal stress analysis is one of the most important subjects in engineering and technology. Several attempts have been so far made for analysis of thermoelastic problems in steady heat conduction states by the boundary element method (Rizzo and Shippy, 1978; Danson, 1981, for example). However, as far as the authors are aware, there are few investigations for thermal stress analysis in transient heat conduction states, although the transient heat conduction problems are often studied by means of the boundary element method (Wrobel and Brebbia, 1981). This is partly because the boundary integral formulation includes inevitably domain integrals which arise due to the temperature distributions in space and time, and in the transient heat conduction states they can not be transformed into the corresponding boundary integrals. Therefore, we have to innovate a more efficient numerical implementation for such problems.

In this paper, a relatively simple boundary element formulation and its numerical implementation are presented for the two-dimensional thermoelastic problems. The proposed method of solution makes full use of the numerical techniques developed in the finite element methods for evaluating the values of stresses and strains at internal points and also domain integrals.

A couple of sample problems in transient heat conduction as well as thermoelasticity are computed to demonstrate the potentiality of the proposed method. Then we apply the method to the stress intensity factor calculation of cracked structural components subjected to different thermal shocks. Finally, we discuss briefly the numerical accuracy of the proposed method.

## 3.2 Integral Equation Formulation

We shall first discuss the transient heat conduction problem of an isotropic medium which occupies domain $\Omega$ with boundary $\Gamma$ as shown in Fig. 3.1. Arbitrary points on $\Gamma$ and in $\Omega$ are denoted by $y$, $y'$ and $x$, $x'$, respectively. The governing equation of the heat conduction problem under consideration can be expressed by

$$\varkappa \nabla^2 \theta = \dot{\theta} \tag{3.1}$$

where $\theta$ is the excess temperature which has to be determined both in $\Omega$ and on $\Gamma$, and $\varkappa$ denotes the thermal diffusivity. The superposed dot stands for time deriva-

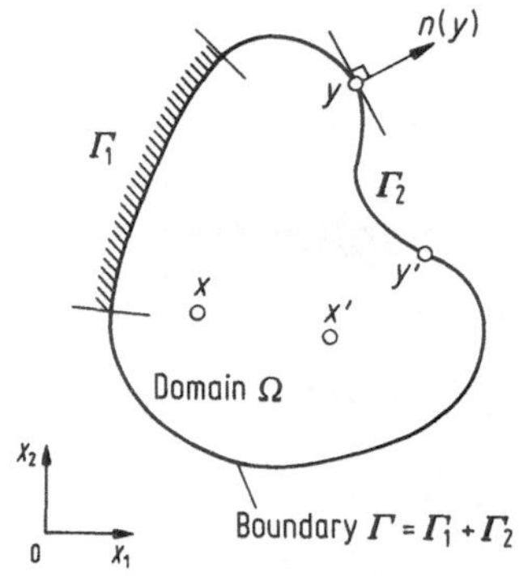


**Fig. 3.1.** Coordinate system and notation

tive. It is assumed throughout in this study that the material constants are invariant in space and time. The boundary conditions for the heat conduction equation (3.1) are such that

$$\begin{aligned} \theta &= \bar{\theta} && \text{on } \Gamma_1 \\ q &\equiv \partial\theta/\partial n = \bar{q} && \text{on } \Gamma_2 \\ q &= h\,(\theta_a - \theta) && \text{on } \Gamma_3 \end{aligned} \tag{3.2}$$

where $\Gamma = \Gamma_1 + \Gamma_2 + \Gamma_3$, $\theta_a$ denotes the ambient temperature and $h$ the heat transfer coefficient. The superimposed bar denotes a prescribed value and $\partial(\cdot)/\partial n$ the directional derivative along the outward unit normal. In addition, the initial condition at the time $t = 0$ is expressible as

$$\theta|_{t=0} = \theta_0 \tag{3.3}$$

We now introduce the fundamental solution $\theta^*$ to the heat conduction equation (3.1). This solution should satisfy in the infinite domain of the same properties as those of the problem under consideration the following differential equation:

$$\varkappa \nabla^2 \theta^* + \dot{\theta}^* + \delta(x, x')\,\delta(t, \tau) = 0 \tag{3.4}$$

in which $\delta(\cdot)$ denotes the Dirac delta function. We also introduce the heat flux $q^*$ which can be defined as

$$q^* = \partial\theta^*/\partial n. \tag{3.5}$$

Now, we consider a weighted residual statement of the governing equation (3.1) in which the fundamental solution is used as the weight function. It can be given by

$$\int_0^t \int_\Omega (\varkappa \nabla^2 \theta - \dot{\theta})\,\theta^*\, d\Omega\, d\tau = 0. \tag{3.6}$$

From the above expression, we can derive in the usual manner (Brebbia and Walker, 1980) the following boundary integral equation:

$$C\theta + \varkappa \int_0^t \int_\Gamma q^* \theta\, d\Gamma\, d\tau = \varkappa \int_0^t \int_\Gamma \theta^* q\, d\Gamma\, d\tau + \int_\Omega \theta^* \theta_0\, d\Omega \tag{3.7}$$

in which $C$ is the coefficient that depends only on the geometrical properties of the boundary. For a smooth boundary point $C = 0.5$, and for an internal point $C = 1$.

The fundamental solution $\Theta^*$ for a two dimensional isotropic medium can be expressed as (Carslaw and Jaeger, 1978)

$$\theta^*(t, \tau) = \frac{1}{4\pi\varkappa(t-\tau)} \exp\left(-\frac{r^2}{4\varkappa(t-\tau)}\right) \tag{3.8}$$

and

$$q^*(t, \tau) = -\frac{r}{8\pi\varkappa^2(t-\tau)^2} \exp\left(-\frac{r^2}{4\varkappa(t-\tau)}\right) \tag{3.9}$$

It should be noted that the time-dependent fundamental solution has the following property:

$$\theta^*(t, \tau) \equiv 0 \quad \text{for} \quad t \leqq \tau \tag{3.10}$$

which represents the principle of causality. If the boundary integral equation (3.7) is solved for the unknowns $\theta$ and $q$ on the boundary, we can obtain the temperature change at an arbitrary point throughout in the domain by using the integral equation (3.7) with the coefficient $C = 1$.

Now, we can calculate thermal stresses in the elastic domain under consideration, having known the temperature distribution over the whole domain of interest. As has been well known, the governing differential equations of the thermoelastic problems can be expressed as

$$D_{ijkl}\, U_{k,lj}(x) + \Theta_{ij}\, \theta_{,j}(x) = 0 \tag{3.11}$$

where $D_{ijkl}$ and $\Theta_{ij}$ are the well known material constants of the linear elastic body (Malvern, 1969). For the so called Hookean elastic solids $D_{ijkl}$ are expressible in terms of the shear modulus $G$ and Poisson's ratio $\nu$, and the components of $\Theta_{ij}$ are expressed in terms of $G$ and the coefficient of linear thermal expansion $\alpha$.

The boundary conditions for Eq. (3.11) can be prescribed in such a way that

$$\begin{aligned} u_i &= \bar{u}_i && \text{on} \quad \Gamma_u \\ p_i &\equiv n_j\, \sigma_{ij} = \bar{p}_i && \text{on} \quad \Gamma_p \equiv \Gamma - \Gamma_u \end{aligned} \tag{3.12}$$

where $\sigma_{ij}$ denotes the stress tensor and $u_i$ the displacement vector.

We now introduce Kelvin's fundamental solution $U^*_{kl}$ to the governing equation (3.11), which should satisfy the differential equation:

$$D_{lamb}\, U^*_{kl,ab}(x, x') + \tilde{\delta}_{km}\, \delta(x, x') = 0 \tag{3.13}$$

in the infinite domain of the same linear elastic properties as in the problem to be analyzed. Here we denote by $\tilde{\delta}_{km}$ the Kronecker delta.

We also consider a weighted residual statement of the governing equation (3.11) as in the heat conduction problems. That is,

$$\int_\Omega (D_{lamb}\, U_{m,ab} + \Theta_{la}\, \theta_{,a})\, U^*_{kl}\, d\Omega = 0. \tag{3.14}$$

Making use of the Gaussian divergence theorem and taking into account the properties of the Kelvin's fundamental solution, we can obtain the integral equation which holds at an arbitrary internal point $x$ in the domain $\Omega$. Then, taking the limiting process in which the point $x$ approaches the boundary point $y$ and evaluating the singular integrals, we can arrive at the following boundary integral equation:

$$C_{kl}(y)\ u_l(y) + \int_\Gamma P^*_{kl}(y,y')\ u_l(y')\ d\Gamma(y') = \int_\Gamma U^*_{kl}(y,y')\ p_l(y')\ d\Gamma(y') + \omega_k(y) \tag{3.15}$$

where

$$\omega_k(y) = \int_\Omega \Theta_{la}\,\theta_{,a}(x)\ U^*_{kl}(y,x)\ d\Omega(x) - \int_\Gamma \Theta_{lm}\,\theta(y')\ U^*_{kl}(y,y')\ n_m(y')\ d\Gamma(y'). \tag{3.16}$$

The coefficient matrix $C_{kl}(y)$ is a diagonal one, and its components depend only on the geometrical properties of $\Gamma$ at the boundary point $y$. Their values can also be estimated indirectly by considering the conditions of rigid body displacements as shown in Brebbia and Walker (1980).

The fundamental solutions $U^*_{kl}$ and $P^*_{kl}$ are expressed for the 2-D plane strain problems as (Brebbia and Walker, 1980)

$$U^*_{kl} = \frac{1}{8\pi G(1-\nu)}\left[(3-4\nu)\ln\left(\frac{1}{r}\right)\tilde{\delta}_{kl} + \frac{\partial r}{\partial x_k}\frac{\partial r}{\partial x_l}\right] \tag{3.17}$$

and

$$P^*_{kl} \equiv D_{lamb}\, U^*_{ka,b}\, n_m = -\frac{1}{4\pi(1-\nu)\,r}\left[\frac{\partial r}{\partial n}\left\{(1-2\nu)\,\tilde{\delta}_{kl} + 2\frac{\partial r}{\partial x_k}\frac{\partial r}{\partial x_l}\right\} - (1-2\nu)\left(\frac{\partial r}{\partial x_k}n_l - \frac{\partial r}{\partial x_l}n_k\right)\right] \tag{3.18}$$

where $P^*_{kl}$ denote the boundary tractions corresponding to the fundamental solution $U^*_{kl}$.

## 3.3 Computational Scheme

### 3.3.1 Transient Heat Conduction Problems

For the numerical solution of the boundary integral equation (3.7), the boundary $\Gamma$ and the time domain $0 \leqq t \leqq t_F$ are divided into $M$ and $N$ elements, respectively. In order to evaluate the domain integrals, if necessary, the internal domain $\Omega$ is also subdivided into small cells.

In the simplest computational scheme, which can give sufficiently accurate results in most cases of practical use, it is assumed that $\theta$ and $q$ are piecewisely constant on the boundary as shown in Fig. 3.2. Supposed that the time variations of $\theta$ and $q$ are small compared with those of $\theta^*$ and $q^*$, it can be reasonably assumed that $\theta$ and $q$ are also constant at each time step. Then, we can define at the $i$-th step

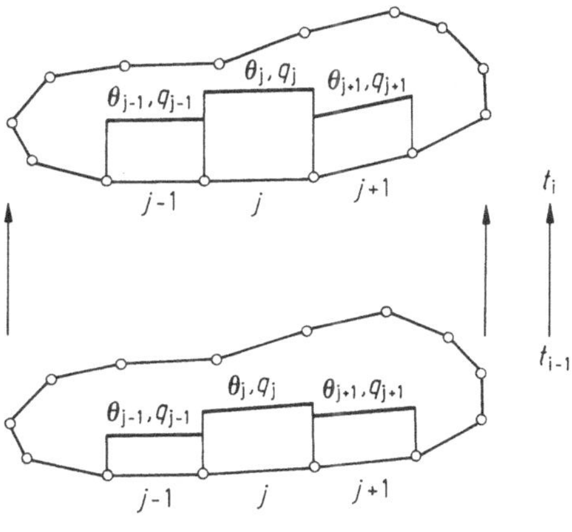


**Fig. 3.2.** Basic idea of time stepping technique

($t = t_i$) the following vectors of nodal boundary values:

$$\{\tilde{\theta}(t_i)\} = \{\theta_1(t_i)\ \theta_2(t_i) \ldots \theta_N(t_i)\}$$
$$\{\tilde{q}(t_i)\} = \{q_1(t_i)\ q_2(t_i) \ldots q_N(t_i)\} \tag{3.19}$$

where $i = 1, 2, \ldots, M$.

Using the above nodal values, we can obtain the discretized version of the boundary integral equation (3.7) as follows:

$$C_l \tilde{\delta}_{jl}\, \theta_j(t_i) + \varkappa \sum_{k=1}^{i} \left[ \sum_{j=1}^{N} \int_{\Gamma_j} \theta_j(t_k) \left\{ \int_{t_{k-1}}^{t_k} q^*(t_i, \tau)\, d\tau \right\} d\Gamma \right]$$
$$= \varkappa \sum_{k=1}^{i} \left[ \sum_{j=1}^{N} \int_{\Gamma_j} q_j(t_k) \left\{ \int_{t_{k-1}}^{t_k} \theta^*(t_i, \tau)\, d\tau \right\} d\Gamma \right] \tag{3.20}$$
$$+ \sum_{j=1}^{N} \int_{\Omega} \theta_0\, \theta_j^*(t_i, 0)\, d\Omega .$$

The time integration in Eq. (3.20) can be analytically carried out, and this yields

$$\int_{t_{k-1}}^{t_k} \theta^*(t_i, \tau)\, d\tau = \frac{1}{4\pi\varkappa} [E_1(a_{k-1}) - E_1(a_k)] \tag{3.21 a}$$

$$\int_{t_{k-1}}^{t_k} q^*(t_i, \tau)\, d\tau = -\frac{1}{2\pi\varkappa r} \left( \frac{\partial r}{\partial n} \right) [\exp(-a_{k-1}) - \exp(-a_k)] \tag{3.21 b}$$

where

$$a_k = \frac{r^2}{4\varkappa(t_i - t_k)} \tag{3.22}$$

and $E_i(\cdot)$ is the exponential-integral function.

If Eq. (3.20) is written for all the nodal points in space and time, we can obtain the following system of linear equations:

$$[H]\{\tilde{\theta}\} = [G]\{\tilde{q}\} + \{\tilde{w}\}. \tag{3.23}$$

The column vector $\{\tilde{w}\}$ in the above equation represents the contributions of the initial condition (3.3). The coefficient matrices $[H]$ and $[G]$ have a triangular form due to the principle of causality, and can be expressed as

$$[H]=\begin{bmatrix} h_{11} & 0 & 0\ldots 0 \\ h_{21} & h_{22} & 0\ldots 0 \\ \vdots & \vdots & \ddots \quad \vdots \\ h_{M1} & h_{M2} & \ldots\ldots h_{MM} \end{bmatrix} \tag{3.24}$$

$$[G]=\begin{bmatrix} g_{11} & 0 & 0\ldots 0 \\ g_{21} & g_{22} & 0\ldots 0 \\ \vdots & \vdots & \ddots \quad \vdots \\ g_{M1} & g_{M2} & \ldots\ldots g_{MM} \end{bmatrix} \tag{3.25}$$

For the solution of Eq. (3.23), we can proceed successively from the first time step. Namely, we first solve the system of equations:

$$[h_{11}]\{\tilde{\theta}(t_1)\}=[g_{11}]\{\tilde{q}(t_1)\}+\{\tilde{w}(t_1)\} \tag{3.26}$$

for the boundary unknowns of $\{\tilde{\theta}(t_1)\}$ and $\{\tilde{q}(t_1)\}$. The column vector $\{\tilde{w}(t_1)\}$ denotes the contribution of the initial condition at the first time step. Once the boundary unknowns at the first time step have been determined, we can proceed to the second time step. For an $i$-th time step, we have

$$[h_{ii}]\{\tilde{\theta}(t_i)\}=[g_{ii}]\{\tilde{q}(t_i)\}+\{\tilde{w}(t_i)\}+\sum_{k=1}^{i-1}([g_{ik}]\{\tilde{q}(t_k)\}-[h_{ik}]\{\tilde{\theta}(t_k)\}). \tag{3.27}$$

Since $\{\tilde{\theta}(t_k)\}$ and $\{\tilde{q}(t_k)\}$ for $k=1,2,\ldots,i-1$ are known from the solutions at previous time steps, we have only to obtain the boundary unknowns at the $i$-th time step.

It should be noted that a six-point Gaussian numerical quadrature can be used for computing the components of the matrices $[H]$ and $[G]$ except the diagonal ones. For evaluating the diagonal components of the sub-matrices $[h_{ii}]$ and $[g_{ii}]$, special care must be taken. The diagonal terms of $[h_{ii}]$ has the singularity of $1/r$, but they vanish due to the orthogonality of $r$ and $n$, which makes $\partial r/\partial n=0$. The diagonal terms of $[g_{ii}]$ have a logarithmic singularity which is directly integrable. As shown in Wrobel and Brebbia (1981), expanding the exponential-integral function in series, we can evaluate the diagonal components $[g_{ii}]_{jj}$ in the following closed form:

$$[g_{ii}]_{jj}=\frac{l_j}{2\pi}\left\{2-\gamma-\ln(\alpha_j)+\sum_{m=1}^{\infty}(-1)^{m-1}\frac{\alpha_j^m}{m(2m+1)\,m!}\right\} \tag{3.28}$$

where

$$\alpha_j=\frac{l_j^2}{4\varkappa(t_i-t_{i-1})} \tag{3.29}$$

in which $\gamma$ is Euler's constant, i.e. $\gamma=0.57721566\ldots$ and $l_j$ is the length of the $j$-th boundary element.

### 3.3.2 Thermoelastic Problems

After having known the temperature distribution in space and time by means of the boundary element analysis mentioned above, we can calculate the domain-integral term $\omega_k$ in Eq. (3.15). This is done by subdividing the internal domain $\Omega$ into small cells and using the standard Gaussian numerical quadrature rule. Then, the usual boundary element method available for elastostatic problems can be applied to the solution of the thermoelastic problems under consideration. Using the same discretization scheme as in the heat conduction problems, we can obtain the following system of linear equations:

$$[A]\{\tilde{u}\} = [D]\{\tilde{p}\} + \{\tilde{f}\} \tag{3.30}$$

where $[A]$ and $[D]$ are coefficient matrices which can be calculated from the fundamental solutions (3.17) and (3.18), respectively. In equation (3.30) the column vectors $\{\tilde{u}\}$ and $\{\tilde{p}\}$ denote the nodal displacement and nodal traction vectors, respectively, and $\{\tilde{f}\}$ the column vector corresponding to the temperature distribution throughout in the domain.

Equation (3.30) is now solved for the unknown components of $\{\tilde{u}\}$ and $\{\tilde{p}\}$ by taking into account the boundary conditions (3.12). Then, we can compute the thermoelastic stresses and strains produced at an arbitrary time. Applying this solution procedure to every time step, we can determine the whole behavior of the thermoelastic problems.

When we are concerned with fracture mechanics application of the above solution procedure, the method of sub-regions is very useful, in particular for the stress intensity factor computation of a crack with arbitrary shape and location.

We now consider a two dimensional domain divided into two sub-regions as shown in Fig. 3.3. The subdivision is made in such a way that the interface $\Gamma_I$ includes the cracked surface. The outer boundaries of the sub-regions (Domain 1 and Domain 2) are denoted by $\Gamma_1$ and $\Gamma_2$, respectively. Denoting by subscript $I$ the

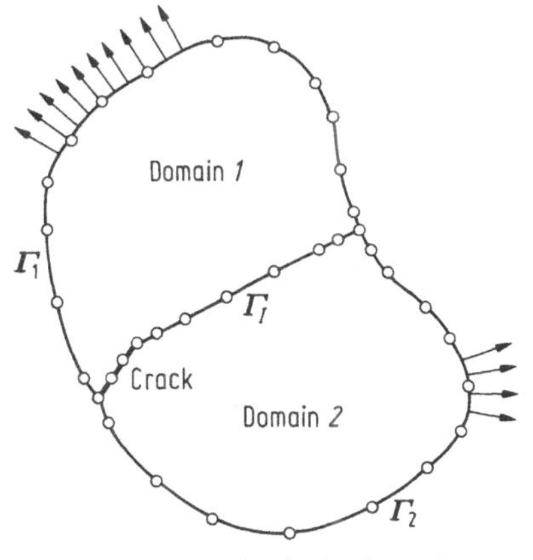


**Fig. 3.3.** Method of sub-regions

variables on the interface, we can express the discretized set of Eq. (3.30) for the two sub-regions. Namely, we have for the sub-region 1

$$[A^1\ A_I^1]\begin{Bmatrix}\tilde{u}^1\\ \tilde{u}_I^1\end{Bmatrix} = [D^1\ D_I^1]\begin{Bmatrix}\tilde{p}^1\\ \tilde{p}_I^1\end{Bmatrix} + \{\tilde{f}_1\} \tag{3.31}$$

and for the sub-region 2

$$[A^2\ A_I^2]\begin{Bmatrix}\tilde{u}^2\\ \tilde{u}_I^2\end{Bmatrix} = [D^2\ D_I^2]\begin{Bmatrix}\tilde{p}^2\\ \tilde{p}_I^2\end{Bmatrix} + \{\tilde{f}_2\} \tag{3.32}$$

where $\{\tilde{u}^i\}$ and $\{\tilde{p}^i\}$ are the nodal displacement and traction vectors on the boundary $\Gamma_i$, respectively.

The nodal displacements $\{\tilde{u}_I^i\}$ and nodal tractions $\{\tilde{p}_I^i\}$ on the interface $\Gamma_I$ must satisfy both the compatibility and equilibrium conditions. These conditions can be expressed as

$$\begin{aligned}\{\tilde{u}_I^1\} &= \{\tilde{u}_I^2\} \equiv \{\tilde{u}_I\}\\ \{\tilde{p}_I^1\} &= -\{\tilde{p}_I^2\} \equiv \{\tilde{p}_I\}\end{aligned} \tag{3.33}$$

Making use of Eq. (3.33) for Eqs. (3.31) and (3.32) and rearranging them, we can obtain

$$\begin{bmatrix}A_1 & A_I^1 & 0\\ 0 & A_I^2 & A_2\end{bmatrix}\begin{Bmatrix}\tilde{u}^1\\ \tilde{u}_I\\ \tilde{u}^2\end{Bmatrix} = \begin{bmatrix}G_1 & G_I^1 & 0\\ 0 & -G_I^2 & G_2\end{bmatrix}\begin{Bmatrix}\tilde{p}^1\\ \tilde{p}_I\\ \tilde{p}^2\end{Bmatrix} + \begin{Bmatrix}\tilde{f}_1\\ \tilde{f}_2\end{Bmatrix}. \tag{3.34}$$

On the cracked surface, however, the compatibility condition (3.33) can not be used. Since in most cases the cracked surface is traction-free, instead of Eq. (3.33) the following conditions are applicable to the cracked surface:

$$\{\tilde{p}_I^1\} = \{\tilde{p}_I^2\} = \{0\}, \qquad \{\tilde{u}_I^1\} \neq \{\tilde{u}_I^2\}. \tag{3.35}$$

Substitution of Eq. (3.35) into Eq. (3.34) leads to the global set of equations for the unknowns on the outer boundary $\Gamma_1 + \Gamma_2$ and also on $\Gamma_I$. Instead of a direct application of the Gaussian elimination to the solution of these equations, the following elimination procedure can be recommended for practical use. Namely, we first eliminate the supplemented unknowns on the interface $\Gamma_I$, and then the real unknowns on the outer boundary $\Gamma_1 + \Gamma_2$. If this stepwise elimination procedure is used in the method of sub-regions, the system matrix to be inverted has the same order as that of the original boundary element equations without sub-regions. When only smaller computers are available, the so-called block elimination procedure, which is one of the standard numerical techniques in the finite element methods, can also be applied to the solution of the boundary element equations associated with the method of sub-regions. It is, however, beyond the scope of this article to describe the details of this procedure.

## 3.4 Numerical Results and Discussion

### 3.4.1 Transient Heat Conduction

In order to demonstrate the potentiality of the boundary element solution procedure developed in this study, we compute a couple of typical transient heat conduction problems. The thermal diffusivity is assumed as $\varkappa = 1.39 \times 10^{-5}$ m$^2$/s and the coefficient of the thermal expansion as $\alpha = 12.0 \times 10^{-6}$ m/°C.

The first example is the infinitely long strip which is initially kept at a uniform temperature, say 100 °C, and then the left hand side is suddenly cooled into 0 °C. In Fig. 3.4 comparison is made between the results obtained by the present method and the analytical ones by Carslaw and Jaeger (1978). Figure 3.5 shows the results obtained for the case of sudden heating. In this case the initial uniform temperature is 0 °C, and both the sides of the strip are suddenly heated into 100 °C.

It is interesting to note that fairly good agreement between the numerical results obtained by the proposed method of solution and the analytical ones can be recognized for the whole time range under consideration.

### 3.4.2 Stress Intensity Factor Computation

The stress intensity factors $K_{\mathrm{I}}$ and $K_{\mathrm{II}}$ for cracking modes I and II are calculated for some selected problems. The material constants are assumed as $G = 7.92 \times 10^4$ MPa and $\nu = 0.3$. The computation was carried out by using both constant and linear boundary elements. There were no appreciable differences between both the results. Hence, we employ in what follows only the constant boundary elements.

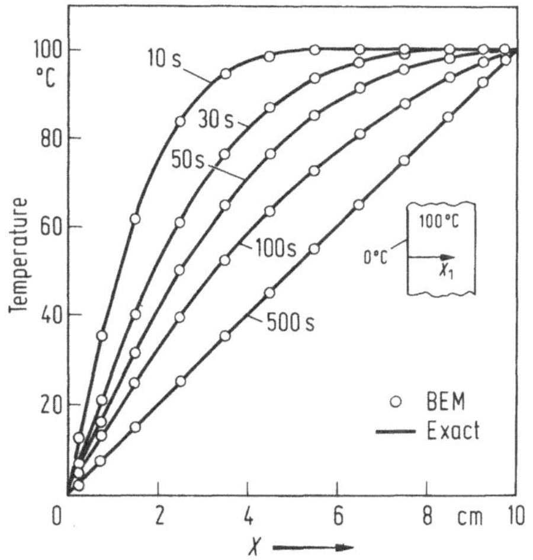


**Fig. 3.4.** Time variations of temperature for case of one-side cooling

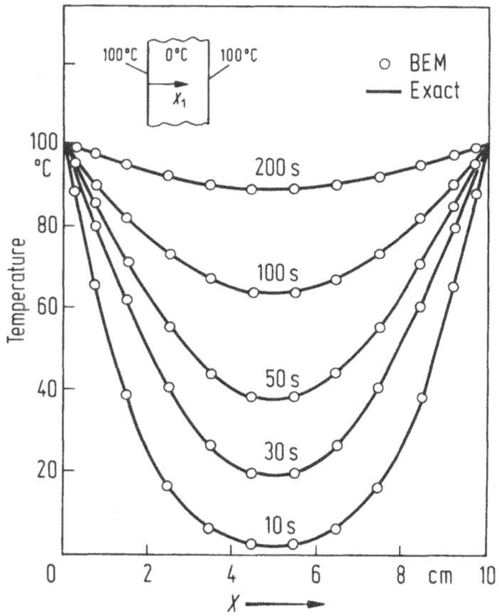


**Fig. 3.5.** Time variations of temperature for case of two-side heating

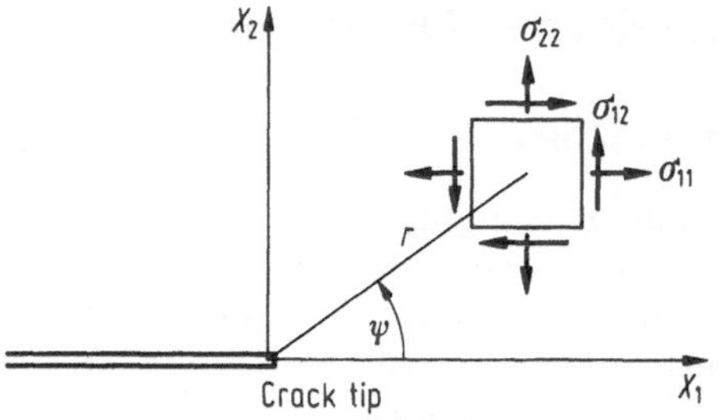


**Fig. 3.6.** Coordinate system at crack tip

We now define the coordinate system at the crack tip as shown in Fig. 3.6. Although we may apply different procedures to the evaluation of the stress intensity factors, we shall use in this study the following two methods. The first one is called the stress method in which the values of stress intensity factors are estimated by linear extrapolation using the stress values in the vicinity of the crack tip. When our attention is paid to the stresses $\sigma_{22}$ and $\sigma_{12}$ on the axis $x_1$ ($\psi = 0$), we can obtain $K_{\mathrm{I}}$ and $K_{\mathrm{II}}$ as the limiting values defined by

$$\left.\begin{aligned} K_{\mathrm{I}} &= \lim_{r \to 0} [\sqrt{2\pi r}\,(\sigma_{22})_{\psi=0}] \\ K_{\mathrm{II}} &= \lim_{r \to 0} [\sqrt{2\pi r}\,(\sigma_{12})_{\psi=0}] \end{aligned}\right\} . \tag{3.36}$$

The other one is named the displacement method in which the crack surface displacements in the vicinity of crack tip ($\psi = \pi$) are used to obtain $K_{\mathrm{I}}$ and $K_{\mathrm{II}}$ by linear extrapolation, i.e.

$$\begin{aligned} K_{\mathrm{I}} &= \frac{2G}{1+\lambda} \lim_{r \to 0} \sqrt{\frac{2\pi}{r}}\,(u_2)_{\psi=\pi} \\ K_{\mathrm{II}} &= \frac{2G}{1+\lambda} \lim_{r \to 0} \sqrt{\frac{2\pi}{r}}\,(u_1)_{\psi=\pi} \end{aligned} \tag{3.37}$$

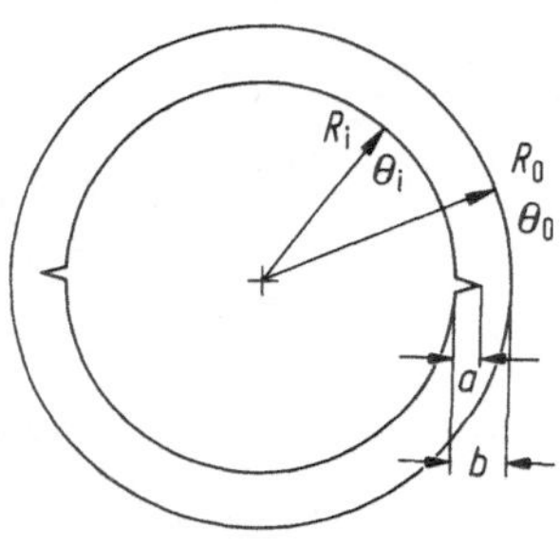


**Fig. 3.7.** Circular cylinder with two opposite radial cracks

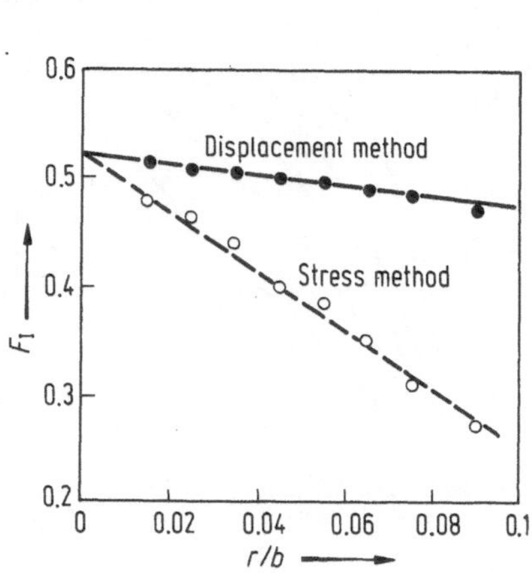


**Fig. 3.8.** Example of linear extrapolation

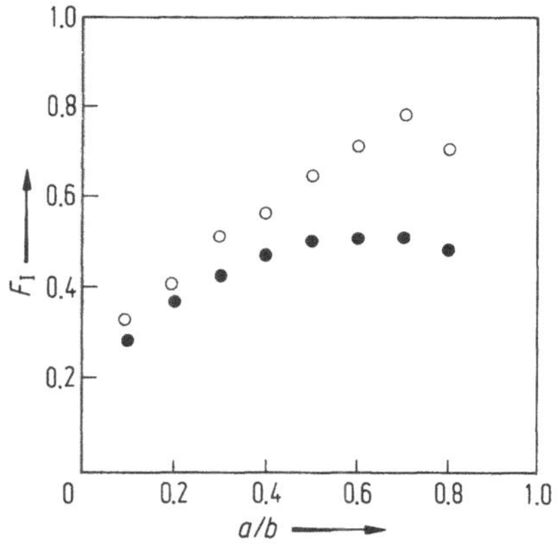


**Fig. 3.9.** Calculated $F_I$ for hollow circular cylinder with two opposite radial cracks along internal surface. ○ $\theta = -100\,°C$ over crack surface; ● zero heat flux over crack surface

where $\lambda$ is defined as $\lambda = 3 - 4\nu$ for the plane strain and as $\lambda = (3-\nu)/(1+\nu)$ for the plane stress.

Attempt is first made for the computation of $K_I$ values. The first example is a thick-walled cylinder with two radial cracks as shown in Fig. 3.7. The cylinder with outer radius $R_0$, inner radius $R_i$ and two opposite internal cracks is subjected to the thermal gradient under the plane strain condition. Because of symmetry, a quarter of the cylinder cross-section is analyzed.

Figure 3.8 shows a typical example of the linear extrapolation for the stress and also displacement methods. It is assumed for this example that $R_i/R_0 = 0.8$ and $a/b = 0.3$, and the heat conduction is in a steady state without any heat generating source in the domain, the outer and inner surfaces of the cylinder being kept at $0\,°C$ and $-100\,°C$, respectively. It can be seen that the estimated values of $K_I$ by the two methods agree well. It is noted here that the stress intensity factor $K_I$ is normalized as $F_I = K_I/K_0$, in which $K_0 = E\,\alpha\,\Delta\theta\,\sqrt{b}/(1-\nu)$, where $E$ is Young's modulus and $\Delta\theta$ is the absolute value of the temperature difference between the inner and outer surfaces of the cylinder.

In Fig. 3.9 are shown the results obtained for $K_I$ values of the cracked cylinder mentioned above. It is assumed in this example that the crack surface has zero heat flux or the temperature of the crack surface is equal to that of the inner surface. Computation is carried out for the case $R_i/R_0 = 0.8$ and $b = 2$ cm.

In Table 3.1 are summarized the results obtained under two different boundary conditions on the crack surfaces. For the same crack length, the difference between the $F_I$ values of an internal and external crack is seen to be small. For an increasing value of $a/b$, the case where the crack surface temperature is equal to that of the inner surface gives larger $F_I$ values than the case of zero heat flux over the crack surface.

We next deal with the rectangular plate with width $b$, length $2L$ and a single edge crack at its left hand side as shown in Fig. 3.10.

Numerical results for various crack lengths are summarized in Fig. 3.11 as well as Table 3.2, and compared with the corresponding finite element solutions. The

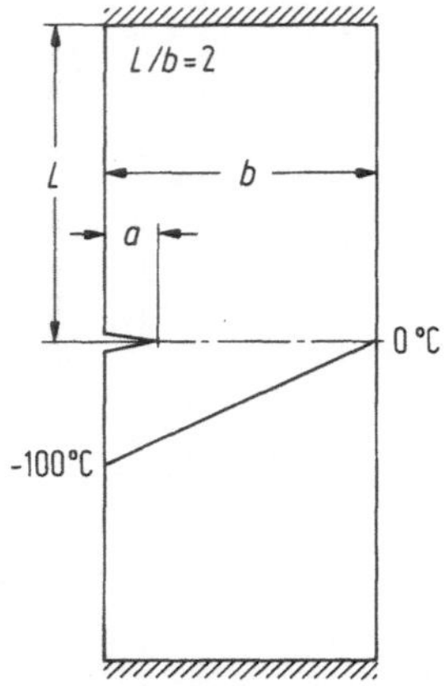


Fig. 3.10. Edge-cracked plate subjected to linear temperature gradient

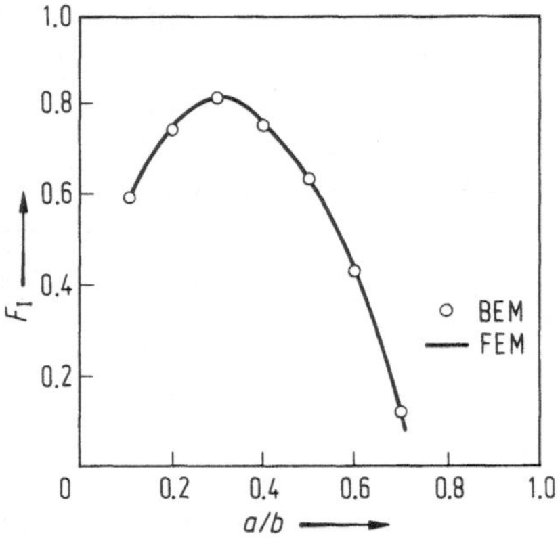


Fig. 3.11. Normalized stress intensity factor $F_I = K / K_0$ of edge-cracked plate

Table 3.1. $K_I/K_0$ for circular cylinder with two opposite radial cracks subjected to thermal gradient

| $a/b$ | $K_I/K_0$ Internal crack | | $K_I/K_0$ External crack | |
|---|---|---|---|---|
| | −100 °C | flux = 0 | −100 °C | flux = 0 |
| 0.1 | 0.330 | 0.280 | 0.332 | 0.280 |
| 0.2 | 0.405 | 0.367 | 0.410 | 0.375 |
| 0.3 | 0.521 | 0.424 | 0.525 | 0.472 |
| 0.4 | 0.571 | 0.471 | 0.600 | 0.516 |
| 0.5 | 0.652 | 0.508 | 0.704 | 0.556 |
| 0.6 | 0.724 | 0.510 | 0.791 | 0.570 |
| 0.7 | 0.788 | 0.512 | 0.854 | 0.576 |
| 0.8 | 0.712 | 0.484 | 0.760 | 0.539 |

Table 3.2. $K_I/K_0$ for edge-cracked plate subjected to thermal gradient

| $a/b$ | $K_I/K_0$ | |
|---|---|---|
| | BEM | FEM |
| 0.1 | 0.584 | 0.580 |
| 0.2 | 0.724 | 0.724 |
| 0.3 | 0.812 | 0.810 |
| 0.4 | 0.726 | 0.724 |
| 0.5 | 0.630 | 0.630 |
| 0.6 | 0.431 | 0.431 |
| 0.7 | 0.100 | 0.098 |

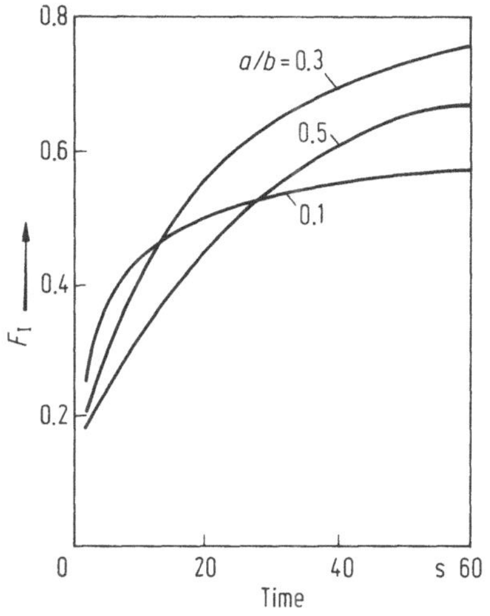


**Fig. 3.12.** Time variations of $F_I$ for thermally shocked plate

**Table 3.3.** $K_I/K_0$ for edge-cracked plate subjected to thermal shock

| t (sec) | $K_I/K_0$ | | |
|---|---|---|---|
| | $a/b$=0.1 | 0.3 | 0.5 |
| 2 | 0.266 | 0.200 | 0.171 |
| 4 | 0.350 | 0.273 | 0.220 |
| 10 | 0.446 | 0.415 | 0.325 |
| 30 | 0.531 | 0.644 | 0.542 |
| 50 | 0.568 | 0.746 | 0.630 |
| ∞ | 0.584 | 0.812 | 0.630 |

heat conduction is also assumed to be in a steady state. Fairly good agreement is recognizable between the *BE* and *FE* solutions. It is interesting to note that the maximum value of $F_I = 0.814$ appears when $a/b$ is approximately equal to 0.3. In this example the heat flux is assumed to be zero over the crack surface.

In Fig. 3.12 are shown the time variations of $F_I$ values for an encastre thermally shocked plate having the same shape as in Fig. 3.10. The results are also summarized in Table 3.3. For this case a steady state is approached after about 50 seconds.

We now compute the cracked cylinder of the same configurations as in Fig. 3.7 subjected to different thermal shocks. The results obtained for cooling shock and for two different thermal shocks are shown in Figs. 3.13 and 3.14, respectively. The results are also summarized in Table 3.4.

Next, let us deal with some examples which exhibit the two cracking modes I and II. Since there are few available results for the stress intensity factors even in

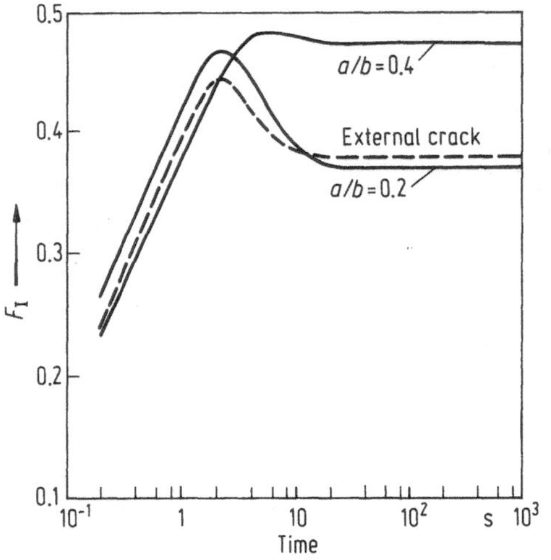


**Fig. 3.13.** Effects of $a/b$ on $F_I$. $R_i/R_0 = 0.8$, on inner surface $\theta = -100\,°C$, on outer surface $\theta = 0\,°C$ and over crack surface heat flux is zero

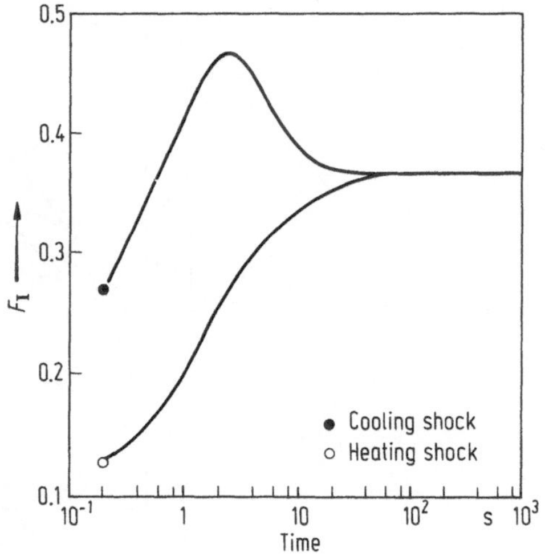


**Fig. 3.14.** Effects of boundary conditions on $F_I$. $R_i/R_0 = 0.8$, $a/b = 0.2$ and heat flux is zero over crack surface; ×, on inner surface $\theta = -100\,°C$ and outer surface $\theta = 0\,°C$; O, on inner surface $\theta = 0\,°C$ and outer surface $\theta = 100\,°C$

**Table 3.4.** $K_I/K_0$ for circular cylinder with two opposite radial cracks subjected to thermal shock. $R_i/R_0 = 0.8$, $a/b = 0.2$ ($b = 2$ cm); zero heat flux over crack surface

| $t$ (sec) | $K_I/K_0$ | | |
|---|---|---|---|
| | Cooling shock | | Heating shock |
| | $\theta_i = -100\,°C$, $\theta_0 = 0\,°C$ | $\theta_i = 0\,°C$, $\theta_0 = -100\,°C$ | $\theta_i = 0\,°C$, $\theta_0 = 100\,°C$ |
| 0.2 | 0.260 | 0.240 | 0.131 |
| 0.4 | 0.335 | 0.300 | 0.150 |
| 1.0 | 0.423 | 0.379 | 0.204 |
| 2.0 | 0.467 | 0.445 | 0.262 |
| 4.0 | 0.445 | 0.412 | 0.300 |
| 10.0 | 0.390 | 0.384 | 0.337 |
| 20.0 | 0.372 | 0.368 | 0.352 |
| 100.0 | 0.367 | 0.367 | 0.367 |

steady heat conduction states, we first calculate the stress intensity factors under mechanical loadings and then proceed to the cases of thermal loadings.

Now, we show the results obtained for the rectangular plate with an oblique edge crack as shown in Fig. 3.15.

In Fig. 3.16 comparison is made between the results obtained by the present BEM and Freese's analytical ones (Sih, 1973) for the case of the plate subjected to

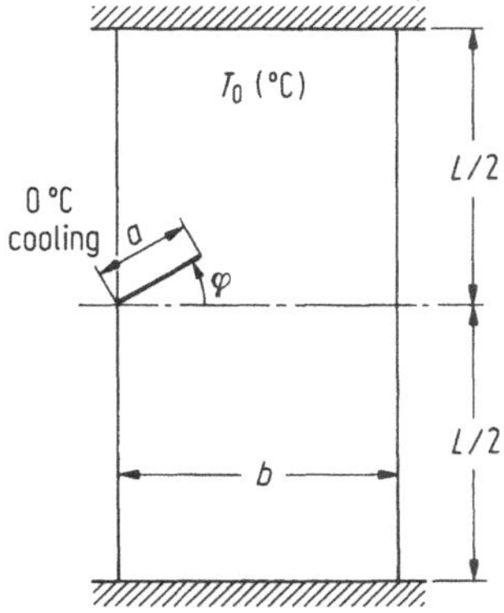


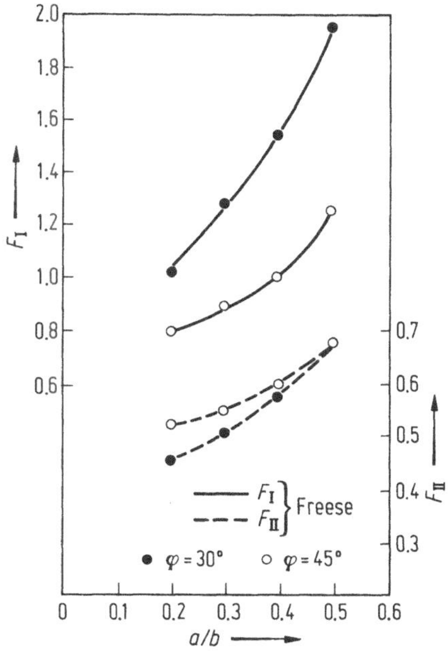


**Fig. 3.15.** Rectangular plate with oblique edge crack

**Fig. 3.16.** Values of $F_I$ and $F_{II}$ for rectangular plate with oblique edge crack subjected to uniaxial tension

**Table 3.5.** Stress intensity factors of rectangular plate with oblique edge crack

| | | $F_I$ $(L/b = 2)$ | | $F_{II}$ $(L/b = 2)$ | |
|---|---|---|---|---|---|
| | | BEM | Freese | BEM | Freese |
| $\varphi = 30°$ | 0.2 | 1.11 | 1.11 | 0.36 | 0.36 |
| | 0.3 | 1.28 | 1.28 | 0.42 | 0.41 |
| | 0.4 | 1.54 | 1.55 | 0.48 | 0.48 |
| | 0.5 | 1.98 | 1.98 | 0.58 | 0.58 |
| $\varphi = 45°$ | 0.2 | 0.80 | 0.80 | 0.43 | 0.41 |
| | 0.3 | 0.90 | 0.90 | 0.45 | 0.45 |
| | 0.4 | 0.99 | 1.02 | 0.50 | 0.50 |
| | 0.5 | 1.25 | 1.27 | 0.58 | 0.58 |

uniaxial tensile stress $\sigma_0$ at its ends. The plate aspect ratio is assumed as $L/b = 2$. The values of $K_I$ and $K_{II}$ are normalized as

$$F_I = K_I/K_0, \quad F_{II} = K_{II}/K_0, \quad K_0 = \sqrt{\pi a}\,\sigma_0. \tag{3.38}$$

There are no appreciable differences between both the results. They are also summarized in Table 3.5.

We now discuss the case where the plate is initially kept at uniform temperature $T_0$ (°C) and suddenly at time $t = 0$ its one side is cooled into 0 °C as shown in Fig. 3.15. It is assmued that $L/b = 4$, $b = 10$ cm and the plate is in a plane strain state. The stress intensity factors are normalized in this example such that

$$F_I = K_I/K_0, \quad F_{II} = K_{II}/K_0, \quad K_0 = E\,\alpha\,T_0\sqrt{\pi a}/(1-\nu). \tag{3.39}$$

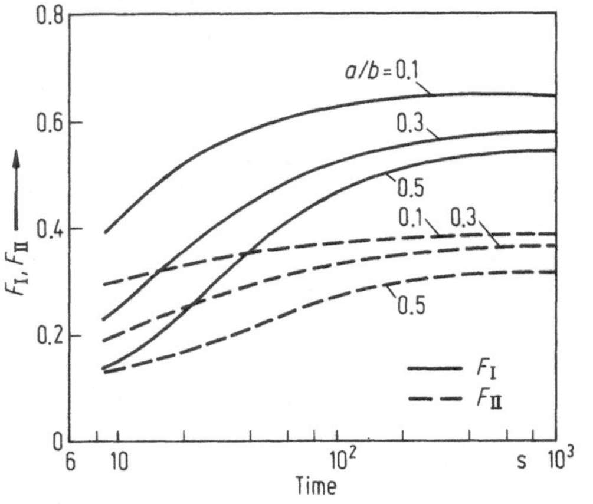


Fig. 3.17. Time variations of $F_I$ and $F_{II}$ for rectangular plate with oblique edge crack subjected to thermal loading

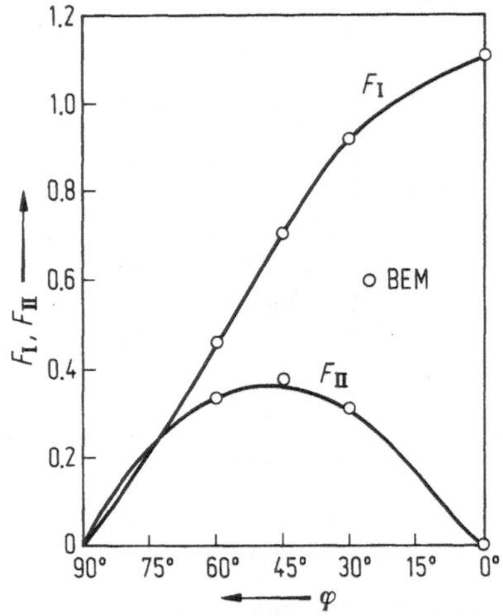


Fig. 3.18. Values of $F_I$ and $F_{II}$ for rectangular plate with oblique edge crack in steady thermal state

**Table 3.6.** Time variations of stress intensity factors in rectangular plate with oblique edge crack

| $t$ (sec) | $a/b$ | | | | | |
|---|---|---|---|---|---|---|
| | 0.1 | | 0.3 | | 0.5 | |
| | $F_I$ | $F_{II}$ | $F_I$ | $F_{II}$ | $F_I$ | $F_{II}$ |
| 10 | 0.423 | 0.300 | 0.251 | 0.211 | 0.155 | 0.147 |
| 50 | 0.602 | 0.358 | 0.475 | 0.303 | 0.396 | 0.231 |
| 100 | 0.630 | 0.375 | 0.526 | 0.334 | 0.468 | 0.275 |
| 500 | 0.654 | 0.380 | 0.590 | 0.364 | 0.546 | 0.322 |

Figure 3.17 shows the time variations of $F_I$ and $F_{II}$ for various aspect ratios when $\varphi = 45°$. The results are also summarized in Table 3.6. In these examples a steady state is approached in about 500 seconds. In Fig. 3.18 comparison is made between the present results and Nishitani's ones (Nishitani, 1975) for the steady state heat conduction.

Finally, we compute the center-cracked rectangular plate as shown in Fig. 3.19. It is assumed that the plate is initially kept at uniform temperature 0 °C and suddenly at time $t = 0$ both the sides are heated into $T_0$ (°C). The plate is in a plane strain state. The stress intensity factors are made nondimensional in the same manner as in Eq. (3.39).

The results obtained are shown in Fig. 3.20 for the case $\varphi = 0°$, and also summarized in Table 3.7 for various values of $\varphi$ and $a/b$.

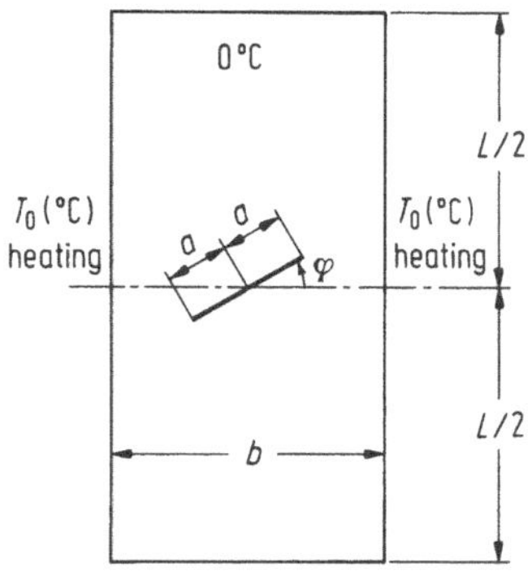


**Fig. 3.19.** Center-cracked rectangular plate

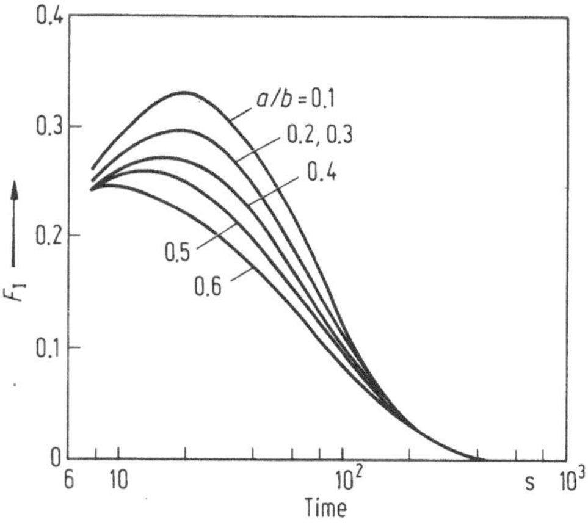


**Fig. 3.20.** Time variations of $F_I$ in case of $\varphi = 0°$ for center-cracked rectangular plate

**Table 3.7.** Center-cracked rectangular plate subjected to heating shock from opposite two sides

| $t$ (sec) | $\varphi = 0°$ | | | $\varphi = 45°$ | | | | | |
|---|---|---|---|---|---|---|---|---|---|
| | $a/b$=0.1 | 0.3 | 0.5 | 0.1 | | 0.3 | | 0.5 | |
| | | | | $F_I$ | $F_{II}$ | $F_I$ | $F_{II}$ | $F_I$ | $F_{II}$ |
| 10 | 0.301 | 0.288 | 0.263 | 0.270 | 0.236 | 0.249 | 0.312 | 0.246 | 0.325 |
| 20 | 0.338 | 0.306 | 0.254 | 0.296 | 0.255 | 0.276 | 0.320 | 0.248 | 0.321 |
| 30 | 0.317 | 0.288 | 0.230 | 0.277 | 0.230 | 0.250 | 0.288 | 0.221 | 0.302 |
| 50 | 0.255 | 0.224 | 0.180 | 0.211 | 0.162 | 0.191 | 0.219 | 0.157 | 0.235 |
| 100 | 0.133 | 0.119 | 0.101 | 0.105 | 0.080 | 0.099 | 0.111 | 0.093 | 0.122 |
| 200 | 0.038 | 0.036 | 0.028 | 0.038 | 0.025 | 0.036 | 0.036 | 0.028 | 0.038 |

## 3.5 Some Considerations on Numerical Accuracy

We first look at the influence of the boundary element discretization on numerical accuracy. An example of the boundary element discretization by means of constant element and also an internal cell subdivision are shown in Fig. 3.21 for the plate with an oblique edge crack treated before. Keeping the number of boundary elements on the outer boundary to be constant, we increase the BE number near the crack tip. One of such results is given in Fig. 3.22. It is interesting to note that a sufficient accuracy can be obtained if the number of BE is chosen as $N \geqq 80$ for this example.

We also investigate some influences of the internal cell subdivision and the interval of time stepping on numerical accuracy. We shall take a rectangular plate

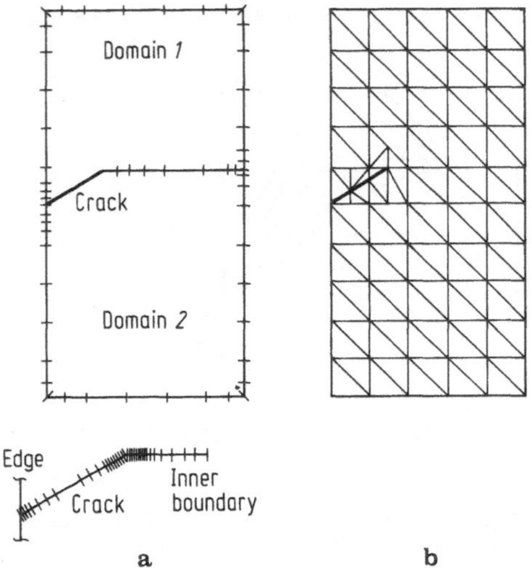


**Fig. 3.21.** Example of boundary element discretization (**a**) and internal cell subdivision (**b**)

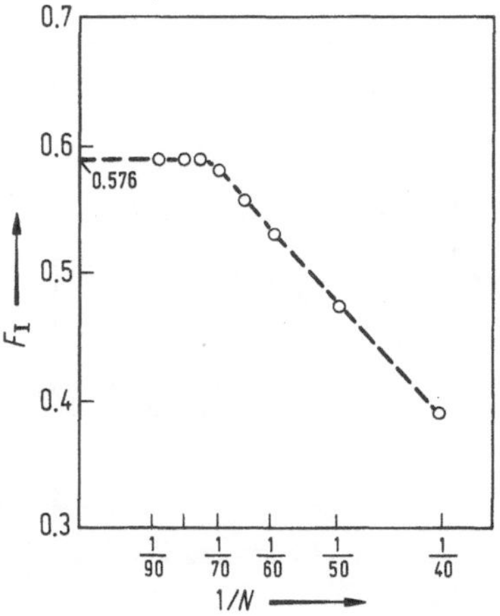


**Fig. 3.22.** Number of boundary elements ($N$) versus numerical accuracy

which is initially kept at uniform temperature 100 °C and suddenly at time $t = 0$ one side of the plate is cooled into 0 °C. This plate is assumed to have no cracks.

Figure 3.23 illustrates an influence of the internal cell subdivision on the result obtained at point $A$, which is made nondimensional by dividing by the corresponding analytical result. We assumed a uniform internal cell subdivision and increased the number of cells. In this example, the steady state was reached in about 500 seconds, and the time stepping interval was chosen as $\Delta t = 50$ seconds.

In Fig. 3.24 is shown an influence of the time interval on the result at point $A$ after 100 seconds. Keeping the numbers of boundary elements and internal cells to be constant, we changed only the time stepping interval. It is worth noting that the numerical error is about 5%, even if the time stepping interval was chosen as $\Delta t =$ 100 seconds for obtaining the result after 100 seconds. That is, one time stepping could provide an accurate result in this case.

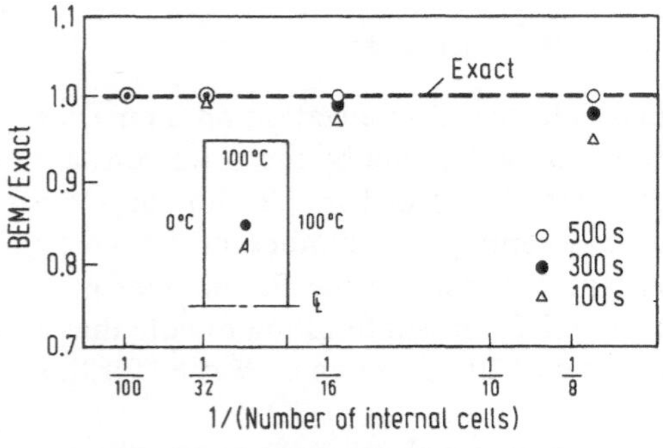


**Fig. 3.23.** Influence of internal cell subdivision on accuracy

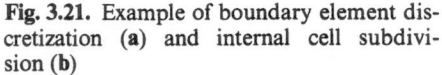

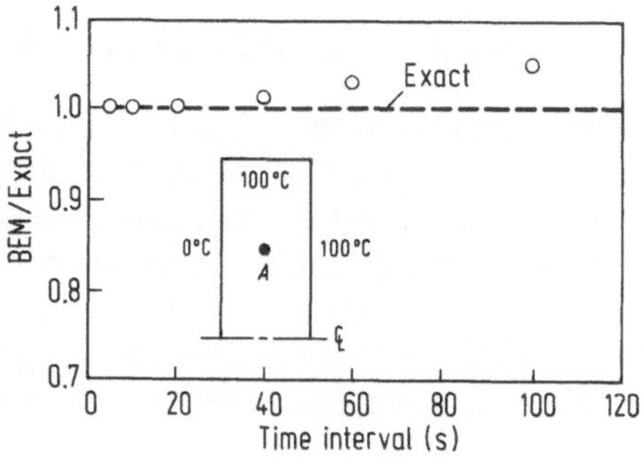


**Fig. 3.24.** Influence of time stepping

## 3.6 Concluding Remarks

The boundary element method has been applied to 2-D thermoelastic problems in steady and nonsteady heat conduction states. Several sample problems were computed by the computer program developed in this study. A series of the stress intensity factors $K_I$ and $K_{II}$ of cracked structures subjected to various thermal loadings have been summarized for practical use.

Although the present investigation is restricted to the 2-D cases, we can make a straightforward extension of the proposed method to the 3-D problems. In the 3-D problems, however, we have to innovate a more efficient numerical procedure for reducing the order of system matrix to be inverted as well as data storage at computation. In authors' opinion, the method of sub-regions which employs a block elimination procedure seems to be very promising.

## References

1. Brebbia, C. A., Walker, S., *Boundary element techniques in engineering,* Butterworths, London 1980
2. Carslaw, H. S., Jaeger, J. C., *Conduction of heat in solids,* 2nd ed., Clarendon Press, Oxford 1978
3. Danson, D. J., *Boundary element methods,* ed. by C. A. Brebbia, pp. 105–122, Springer-Verlag, Berlin 1981
4. Hellen, T. K., Cesari, F., Martin, A., Int. J. Pressure Vessels and Piping **10,** 181–204, 1982
5. Malvern, L. E., *Introduction to the mechanics of a continuous medium,* Prentice-Hall, New York 1969
6. Nishitani, H., Trans. Japan Soc. Mech. Engrs., **41,** 1103, 1975
7. Sih, G. C. (ed.), *Mechanics of Fracture,* Vol. 1, p. 51, Noordhoff/Holland 1973
8. Wrobel, L. C., Brebbia, C. A., *Progress in boundary element methods,* Vol. 1, ed. by C. A. Brebbia, pp. 192–212, Pentech Press, London 1981

## Chapter 4

# Applications of Boundary Element Methods to Fluid Mechanics

*by J. A. Liggett and P. L.-F. Liu*

## 4.1 Introduction

In this chapter we make an effort to review the applications of boundary methods to fluid mechanics. At the outset we wish to give our definition of boundary methods. First, we exclude such techniques from the general class of finite element methods. Although some of the language and a few of the numerical techniques of the two methods have merged, they are historically quite separate.

However, we do classify as boundary methods a number of techniques that have little or no difference in concept, application, or history but appear under different names. These include: the Boundary Integral Equation Method, the Boundary Element Method, the Boundary Integral Method, the Surface-source (or Surface-velocity, Surface-doublet, Surface-singularity) Method, and Panel Methods *. Nor do we wish to differentiate between the "direct method" (where the physical variables appear as solutions of the integral equation) and the "indirect method" (where the integral equation solution leads to the strengths of singularities from which the physical variables can be calculated). We view all these techniques as variations of the method of Trefftz [1].

We are hopeful that this review will show something of the wide range of applications that boundary methods have in fluid mechanics problems, and that it will encourage engineers not using such techniques to begin applying them. We have attempted to cite references of historical importance and those that are recent. Many more references can be found in the papers that are cited herein. For those beginning to apply boundary methods, the emerging books represent the easiest way to become acquainted with the procedures [2, 3, 4, 5, 6, 7, 8].

## 4.2 History

Far from being a new theory, the boundary element method has a long history. Although it became popular in the 1960s, those who developed it for use on computers stood squarely on the shoulders of the applied mathematicians of the

* We concede that Green's function methods and panel methods, for example, are different in their mathematical basis and that some techniques using one method are not readily apparent using the other. Both, however, involve the use of fundamental singularities and integral equations, and in some cases are indistinguishable.

19th and early 20th centuries. The basic integral equations for potential theory were formulated by Green in 1828 [9].

The first application of boundary methods to fluid mechanics (and probably the first numerical calculation using boundary method in any application) was in the amazing paper by Trefftz in 1917 [1]. He solved the problem of a round jet issuing from a hole in an infinite tank with the objective of calculating the contraction coefficient. Such a calculation was not repeated until 1967 (see Hunt, [10]) using modern computing equipment. Thus, Trefftz not only made the first boundary integral calculation, he was first to solve an axisymmetric problem with the $\theta$-integration formulated in terms of elliptic integrals. He used a method of successive approximations to satisfy the integral equation. His calculated contraction coefficient was 0.611, which was assumed correct for many years, although Hunt [10] and others have since corrected it to 0.58.

A doubly symmetric potential flow about an elliptic cylinder was solved by Prager in 1928 [11]. His calculation divides the boundary into elements and uses simultaneous equations to solve the integral equation. [However, the replacement of the integral with a finite sum and the use of simultaneous equations had also been done by Nyström in 1928 [12] who used the integrals of Fredholm [13] to solve a torsion problem.] Due to the double symmetry Prager was able to obtain reasonable results with only three algebraic equations.

Calculations using distributed singularities along the axis of two dimensional or axisymmetric bodies became routine in the early 20th century. That method was, however, restricted to flow around long slender objects and could unexpectedly fail for certain shapes. To remedy this problem Lotz [14] (later Flügge-Lotz, who was to become a distinguished professor at Stanford University) developed the surface-source method. She applied source distributions directly on the surface of a body. Integration was done numerically or graphically and the integral equations were solved by iteration. The method was later adopted by Vandrey in 1951 [15] who used a surface distribution of sources and vortices.

Axially symmetric flow has been a frequent application of boundary methods. Weinstein in 1948 [16] used sources which were placed on rings, disks, or cylinders to solve this problem. Van Tuyl in 1950 [17] and Sadowsky and Sternberg in 1950 [18] continued that development. Landweber [19] used what is essentially a surface-source method with velocity as an unknown in the integral equation. Vandrey [15], [20] attempted to simplify the labor involved for practical calculations.

Thus, it is clear that boundary methods had become well known and were in use during the 1950's. The labor involved in their use was an obstacle to general popularity. The use of the digital computer, just becoming a practical tool at that time, was a natural step which should be classified as evolutionary rather than revolutionary. Such a use was probably first done by Smith and Pierce in 1958 [21] although many groups, especially in the aircraft industry and at the National Advisory Committee on Aeronautics (NACA, the predessor to NASA) had also begun to use the computer. Having resolved the difficulty of the artihmetical labor, the uses of boundary methods began to expand rapidly. Three dimensional problems in potential flow were solved by Hess and Smith in 1962 [22]. Other notable contributions were made by Davenport in 1963 [23], Jaswon in 1963 [24], Symm in 1963 [25], and Chaplin in 1964 [26].

Most of the early developments concerned aerodynamic applications. From the mid-1960s to the present, other applications became prominent. These include free surface flow, flow in porous media, diffusion, wave problems, applications which include viscous flows, porous-elasticity, etc. Several such applications involve nonlinear problems. A brief review of the modern developments in these areas is given in the following sections.

Historically, the relationship between the finite element method and the boundary element method is unclear. If our definitions are accepted, the boundary element method preceeded the finite element method as the basic concepts of the latter were introduced in the 1940s (see, for example, Gallagher in 1975 [27]), while the former began with Trefftz in 1917 [1]. In the 1960s, however, the finite element method became much more popular. When the boundary element method began a surge of popularity, it was promoted by many of the finite element analysts. By the early 1970s boundary methods were influenced by the finite element method, hence the term "boundary element method"; however, it is clear that the essential ideas preceeded such influence. Indeed, Prager in 1928 [11] used "elements" as did Lotz in 1932 [14] and those that followed using "panel methods". Higher order "elements" were used by Chaplin in 1964 [26]. In the opinion of the authors the finite element method has contributed considerably to the jargon but little to the development.

## 4.3 Aerodynamics and Hydrodynamics

The aerodynamic applications dominated the early developments of boundary methods. Reviews are given by Hess in 1980 [28] and Hunt in 1980 [29], the details of which will not be repeated herein. Potential flow problems have been solved for quite complex, three-dimensional shapes which can be composed of thousands of elements. Such solutions include flow around lifting bodies in which the calculation requires an iterative procedure in order to locate the position of the trailing vortex sheets. More recently viscous flows and boundary layer flows have been computed by boundary methods or by hybrid methods which combine boundary techniques with finite differences or finite elements.

The surface-source method, in which source distributions are located on surface elements called "panels" (hence "panel theory"), is the primary technique used in aerodynamic applications. Of course such elements may also contain surface-vortices of surface-doublets. Alternatively, we may formulate the problem as the solution of Poisson's equation

$$\nabla^2 \Phi = P \tag{4.1}$$

which can be converted into the well-known integrals

$$\alpha \Phi = \int_{\Gamma} \left( \Phi \frac{\partial G}{\partial n} - G \frac{\partial \Phi}{\partial n} \right) ds + \int_{R} P G \, dV \tag{4.2}$$

in which $\alpha = 2\pi$ (two dimensional) or $\alpha = -4\pi$ (three dimensional) for points interior to the flow region, $\alpha = \pi$ (two dimensional) or $\alpha = -2\pi$ (three dimensional) for points on a smooth boundary, or $\alpha$ is the included angle or solid angle for points on an unsmooth boundary; $G$ is the free space Green's function, $\ln r$ (two-dimensional) or $1/r$ (three-dimensional); and $\partial/\partial n = \nabla \cdot \mathbf{n}$, where $\mathbf{n}$ is the unit normal vector directed outward from the region $R$ on the boundary $\Gamma$. The right hand side, $P$, represents sources that may be present in any particular problem or nonlinear terms that are to be approximated iteratively (e.g., in compressible flow). For common incompressible potential flow problems $P = 0$ and (4.2) contains only the boundary integral. Using the above formulation, or the panel method, the solution of incompressible potential flow problems has become routine. The extension to compressible flows using a perturbation technique has been formulated by Carey [30] and Carey and Kim [31].

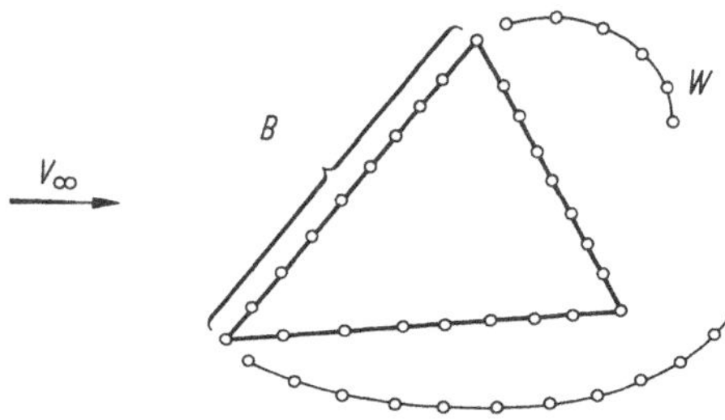


**Fig. 4.1.** Vortices on a triangular body and in the wake (after Inamuro et al., 1982)

Two-dimensional separated flows have been treated by a vortex model. Instead of distributing sources in the surface of a body, vortices are used (Figure 4.1). The potential then becomes (Inamuro, et al. [32])

$$\Phi = V_\infty x + \sum_{i=1}^{N} \frac{\Gamma_i^B}{2\pi} \tan^{-1} \frac{y - y_i}{x - x_i} + \sum_{i=N+1}^{M} \frac{\Gamma_i^w}{2\pi} \tan \frac{y - y_i}{x - x_i} \tag{4.3}$$

in which $V_\infty$ is the free stream velocity, $\Gamma_i^B$ is the strength of the $i$th vortex on the body, and $\Gamma_i^w$ is the strength of the $i$th vortex in the wake. In order to have a single valued potential, the total vorticity must be zero, i.e.,

$$\sum_{i=1}^{N} \Gamma_i^B + \sum_{i=N+1}^{M} \Gamma_i^w = 0 \tag{4.4}$$

The $\Gamma^B$ are adjusted so as to fulfill the boundary condition $\partial\Phi/\partial n = 0$ on the body. The wake vortices are those shed from the corner of the body and acquire the strength of the corner vortices as they are shed. They are advected by a finite difference stepping scheme

$$\mathbf{r}^w(t + \Delta t) = \mathbf{r}^w(t) + \mathbf{u}(r, t)\,\Delta t \tag{4.5}$$

in which $\mathbf{r}^w$ is the position of the vortex and $\mathbf{u}$ is the velocity vector which results from the total collection of vortices and the parallel flow. Inamuro, et al. [32] found it necessary to limit the $\theta$-component of the core velocity of each wake vortex, $i$, as follows

$$u_{i\theta} = \begin{cases} \dfrac{1}{2\pi\varrho}\Gamma_i^w & (\varrho \geqq \sigma) \\[2ex] \dfrac{\varrho}{2\pi\sigma^2}\Gamma_i^w & (\varrho < \sigma) \end{cases} \tag{4.6}$$

in which $\varrho$ is the distance from the vortex center and $\sigma = 2.24\sqrt{\nu t_i^*}$ where $\nu$ is the kinematic viscosity and $t_i^*$ is the time that has elapsed since the $i$th vortex was shed.

Although the vortex separation method treats only inviscid flows, viscous flows have also been calculated by boundary methods. White and Kline [32] used boundary elements to compute flows through an axisymmetric diffuser in which the boundary layer may be either attached or separated. Only the potential part of the flow, however, was computed by using boundary elements. The wall geometry was altered by adding the boundary layer displacement thickeness, $\delta_*$. Separation points were found by standard boundary layer techniques. The flow downstream of a separation point was computed with the streamline through the separation point, $\psi_s$, acting as a free surface (Figure 4.2). Along $\psi_s$ is constant, implying that the velocity is constant. Thus, the potential is taken as

$$\Phi = \Phi_{sp} + V_{sp}(s - s_{sp}) \quad \text{along } \psi_s \tag{4.7}$$

in which $\Phi_{sp}$ is the potential at the separation point, $s - s_{sp}$ is distance along $\psi_s$ from the separation point, $V_{sp}$ is the velocity at the separation point (equal to the exit velocity of the diffuser), and $s_{sp}$ is the position of the separation point. After an initial guess for $\psi_s$, the problem is iterated whereby the normal derivatives along $\psi_s$, computed from the boundary element method, are used to adjust the position of $\psi_s$ so that $\partial\Phi/\partial n \to 0$ along the free streamline. Details of the iteration method are found in White and Kline [33].

In a remarkable series of papers dating from 1970, Wu and his colleagues (see Wu [34], for a review) have applied boundary methods to viscous flow problems. They solve the incompressible Navier-Stokes equations (Wu and Thompson [35])

$$\frac{\partial \mathbf{v}}{\partial t} + (\mathbf{v} \cdot \nabla)\,\mathbf{v} = -\frac{1}{\varrho}\nabla p + \nu \nabla^2 \mathbf{v} \tag{4.8}$$

$$\nabla \cdot \mathbf{v} = 0 \tag{4.9}$$

in which the symbols have their usual meaning of velocity, time, density, pressure, and kinematic viscosity. They divide the flow into its kinematic and dynamic components. The kinematic component is described by the continuity equation (4.9) and the dynamic component is described by the vorticity vector

$$\omega = \nabla \times \mathbf{v} \tag{4.10}$$

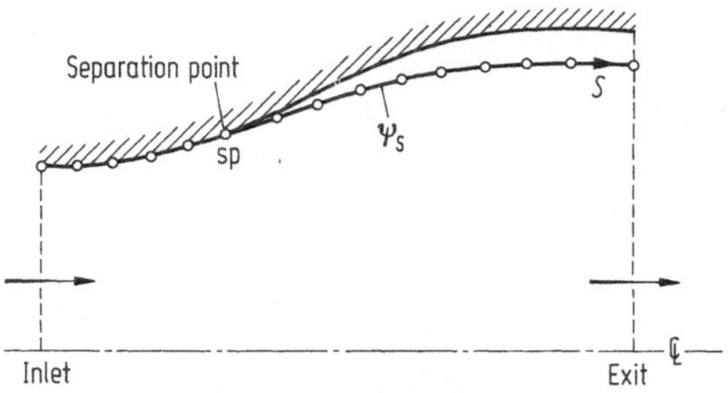


**Fig. 4.2.** Separation in an axisymmetric diffuser (after White and Kline, 1975)

By taking the curl of (4.8), the pressure is eliminated and thus the dynamic component of the problem is governed by

$$\frac{\partial \omega}{\partial t} = \omega \cdot \nabla \mathbf{v} - \mathbf{v} \cdot \nabla \omega + \nu \nabla^2 \omega \tag{4.11}$$

In the velocity and vorticity values are known at some time $k$, (4.11) can be used to advance the vorticity values to time $k+1$. These new values could then be used in (4.8) and (4.9) to compute velocities at time $k+1$. Instead, equations (4.8) and (4.9) are written in integral form in which

$$\alpha \mathbf{v} = -\int_R (\omega \times \nabla \tilde{G})\, dV + \int_\Gamma [\nabla \tilde{G}\, \mathbf{v} \cdot \mathbf{n} + (\mathbf{v} \times \mathbf{n}) \times \nabla \tilde{G}]\, ds \tag{4.12}$$

A corresponding equation for vorticity is

$$\omega = \int_R (\omega \tilde{G})_{t=0}\, dV + \int_0^t dt_0 \int_\Gamma (\omega \cdot \nabla \mathbf{v} - \mathbf{v} \cdot \nabla \omega)\, \tilde{G}\, dV$$
$$+ \nu \int_0^t dt_0 \int_\Gamma (\omega \nabla \tilde{G} - \tilde{G} \nabla \omega) \cdot \mathbf{n}\, ds \tag{4.13}$$

in which $\tilde{G}$ is the Green's function for the diffusion equation,

$$\frac{\partial f}{\partial t} = a \nabla^2 f \tag{4.14}$$

$$\tilde{G} = \frac{1}{[4\pi a (t - t_0)]^{d/2}} \exp \left[ -\frac{r^2}{4a(t-t_0)} \right] \tag{4.15}$$

where $d = 2$ for two-dimensional problems and $d = 3$ for three-dimensional problems. The details of the calculations are given in the many papers by Wu (see Wu [34]). Examples include flows at a wide range of Reynolds numbers which have the complications of turbulence, separation, and wakes.

## 4.4 Porous Media Flow

The macroscopic calculation of flow through a granular matrix is usually governed by Darcy's law

$$\mathbf{v} = -\underline{\mathbf{K}} \cdot \nabla \Phi \tag{4.16}$$

in which $\underline{\mathbf{K}}$ is the permeability tensor and $\Phi$ is the velocity potential. Thus

$$\nabla \cdot \mathbf{v} = -(\nabla \cdot \underline{\mathbf{K}}) \cdot \nabla \Phi - \underline{\mathbf{K}} : \nabla \nabla \Phi = 0 \tag{4.17}$$

In the usual groundwater application $\underline{\mathbf{K}}$ is known as a function of the space coordinates only in a regional sense; that is, there are "zones" in which $\underline{\mathbf{K}}$ is taken as a constant which may differ from the value in adjacent "zones". Such a regional concept fits well the usual vertical structure of the soil in which beds or layers often have sharply defined boundaries (see Figure 4.3). It is somewhat less well suited to the horizontal structure, but because the data are almost always scarce, it fits as well as any other model. With $\nabla \cdot \underline{\mathbf{K}} = \mathbf{0}$, with rotation of the coordinate axes to coincide with the principal axes of $\underline{\mathbf{K}}$, with proper stretching of the coordinate

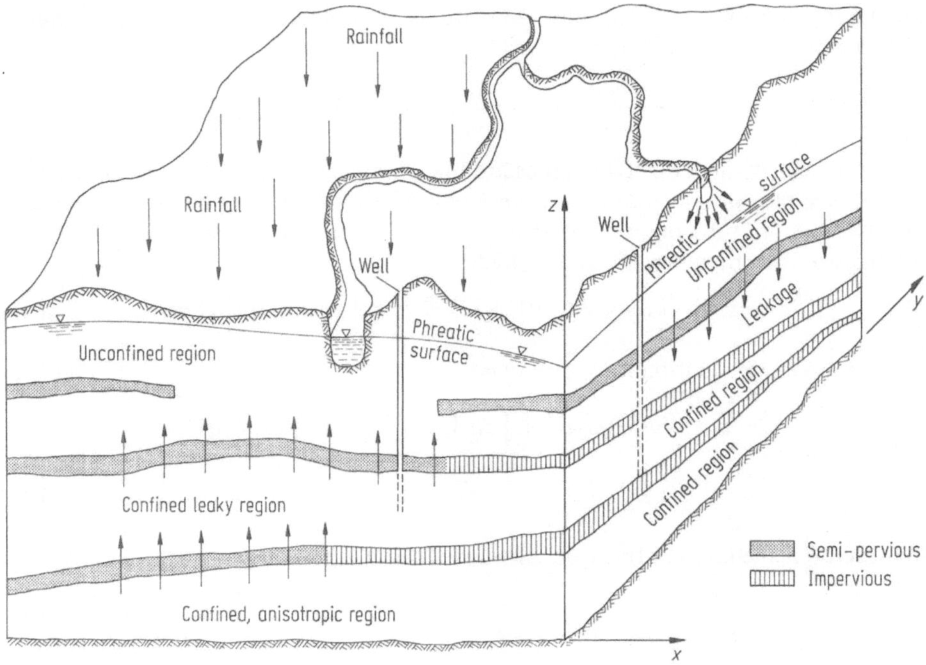


**Fig. 4.3.** A complex groundwater basin (after Liggett and Liu, 1983)

axes, and with the equations of continuity $\nabla \cdot \mathbf{v} = \mathbf{0}$, (4.17) can be written simply (Liggett and Liu, [6])

$$\nabla^2 \Phi = 0 \tag{4.18}$$

in which the operator $\nabla$ now applies to the rotated and stretched coordinate system.

Boundary conditions for a porous media flow consist of the usual Neumann and Dirichlet conditions except where a free surface exists. In the latter case, the elevation, $\eta$, of the free surface becomes a part of the problems. Then the boundary conditions become

$$\Phi = \eta \tag{4.19a}$$

$$\frac{\partial \eta}{\partial t} = -(1 + |\nabla_2 \eta|^2)^{1/2} (\nabla \Phi \cdot \mathbf{n}) \tag{4.19b}$$

in which $\nabla_2$ signifies the horizontal vector operator.

Two other important cases are necessary in order to complete the specification of flow in porous media. First is the case in which the compressibility of the granular matrix is important. The governing equation then becomes the diffusion equation

$$\nabla^2 \Phi = S \frac{\partial \Phi}{\partial t} \tag{4.20}$$

in which $S$ is the "storativity" of the media.

A commonly used approximation, the Dupuit-Forchheimer approximation, assumes that the flow is "nearly horizontal" so that vertical pressure gradients can be neglected compared to horizontal pressure gradients. Then the "transmissivity" is defined as $T = K\Phi$ where $\Phi$ is now the thickness of the fluid bearing media measured from an impermeable base to the free surface. Assuming a constant permeability, $K$, but not constant transmissivity, the governing equation becomes

$$\nabla_2^2 \Phi^2 = n_e \frac{\partial \Phi}{\partial t} \tag{4.21}$$

in which $n_e$ is the "effective porosity" of the medium.

Finally, many problems are solved by assuming two-dimensional flow between confining layers (or a confining lower layer and a free surface) but allowing the confining layers to "leak" fluid to or from the medium. This leakage is assumed proportional to the potential in the medium so that

$$\nabla_2^2 \Phi - \alpha \Phi = 0 \tag{4.22}$$

where $\alpha$ is a constant that defines leakage as proportional to the potential.

Thus, the porous media problem consists of solving either (4.18), (4.20), (4.21), or (4.22) under some combination of Dirichlet, Neumann, or free surface (4.19a, b) boundary conditions. The medium may be divided into zones in two or three dimensions and various combinations of the foregoing problems are possible within the same aquifer (Figure 4.3). Laplace's equation (4.1) is solved by the boundary integral (4.2) whereas the diffusion equation is solved by

$$\Phi = \int_0^t dt_0 \int_\Gamma \left[ \tilde{G} \frac{\partial \Phi}{\partial n} - \Phi \frac{\partial \tilde{G}}{\partial n} \right] ds + \int_R \Phi \tilde{G} \bigg|_{t=0} dV \tag{4.23}$$

in which $\tilde{G}$ is given by (4.15) with $a = S^{-1}$.

An alternative method of dealing with (4.20) is to remove time by use of the Laplace transform,

$$\tilde{\Phi}(x, y, q) = \int_0^\infty \Phi \, e^{-qt} \, dt \tag{4.24}$$

Then (4.20) becomes

$$\nabla_2^2 \tilde{\Phi} - q S \tilde{\Phi} = - S \Phi(x, y, 0) = - S \Phi_0 \tag{4.25}$$

in which $\Phi(x, y, 0) = \Phi_0$ is the initial condition. The corresponding boundary integral is

$$\alpha \tilde{\Phi} = \int_R S K_0(\sqrt{S q}\, r) \, \Phi_0 \, dV + \int_\Gamma \left[ \sqrt{S q} \;\; \tilde{\Phi} \, K_1(\sqrt{S q}\, r) \frac{\partial r}{\partial n} + K_0(\sqrt{S q}\, r) \frac{\partial \tilde{\Phi}}{\partial n} \right] ds \tag{4.26}$$

The foregoing equations and boundary conditions represent a "mix and match" situation. In groundwater calculations several different such situations in which different equations and/or boundary conditions can apply to portions of the overall problem. These include situations which are basically two-dimensional, axisymmetric or three-dimensional. Many such situations have been assembled into a comprehensive program which zones in both the horizontal and vertical directions (Liggett and Liu [6]). The zones are connected by equating pressures and flow along the interzonal boundaries.

Many special situations exist in porous media calculation. For example, the calculations in the vicinity of wells is important in most situations. For regional calculations wells appear as singularities and may be represented as point or line sinks. At the other extreme the detailed flow in the vicinity of a well can be treated by the solution of an axisymmetric problem with or without a free surface depending on the nature of the aquifer (Lennon et al. [37], [38]). The intermediate case of a well field, in which several wells interact, is a full three-dimensional problem (Lennon et al. [38]). In all of these cases boundary methods appear to offer outstanding advantages over domain methods. In the first instance the boundary method makes dealing with both the singularity and the far field especially easy. In the last case the ease of handling the free surface and the efficiency in three-dimensional problems become important.

One important problem that has not been solved by boundary methods is the case of flow in unsaturated media. That problem is so highly nonlinear that it appears boundary methods have no advantage over a domain method. In the case of saturated-unsaturated flow, however, boundary and domain methods can be combined so as to take maximum advantage of each (Liggett and Liu [6])

## 4.5 Free Boundary Problems

The complexities introduced by a free boundary are often more serious than nonlinearities in the governing differential equations. The nonlinearities in the boundary conditions contribute a further complication to hydrodynamic problems. Such problems are usually governed by Laplace's equation (4.18) with free surface conditions

$$\frac{\partial \eta}{\partial t} = \frac{\partial \Phi}{\partial z} - \frac{\partial \Phi}{\partial x}\frac{\partial \eta}{\partial x} - \frac{\partial \Phi}{\partial y}\frac{\partial \eta}{\partial y} \tag{4.27}$$

$$\frac{\partial \Phi}{\partial t} = B - g\,\eta - \frac{1}{2}\left[\left(\frac{\partial \Phi}{\partial x}\right)^2 + \left(\frac{\partial \Phi}{\partial y}\right)^2 + \left(\frac{\partial \Phi}{\partial z}\right)^2\right] \tag{4.28}$$

in which $\eta$ is the free surface elevation and $B$ is a constant (the "Bernoulli constant"). The first equation simply states that a fluid particle on the free surface follows the free surface movement. The second equation is a consequence of having a constant pressure on the free surface as expressed by Bernoulli's equation. Both conditions are nonlinear. Further, there is no damping in the system so that any input of energy from a numerical strategy or any entrapment of energy (e.g., in a small, $2\,\Delta x$, wavelength or in a larger wavelength) leads to a divergent calculation.

In steady state calculations a stream function formulation is often superior to that of the velocity potential. Using the stream function in (4.27) and (4.28) with the time dervative set to zero, these equations become, respectively,

$$\psi = 0 \tag{4.29}$$

$$\frac{1}{2}\left(\frac{\partial \psi}{\partial n}\right)^2 + g\,\eta = B \tag{4.30}$$

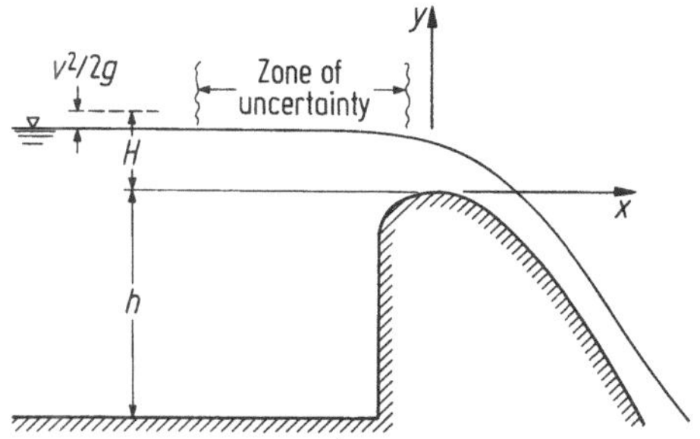


**Fig. 4.4.** Flow over a spillway (after Cheng et al., 1981)

Thus the kinematic condition (4.27) has become linear. The dynamic condition (4.28) is simplified and only contains the normal derivative, which is a natural consequence of the calculation. Boundary derivatives in directions other than normal are awkward to compute.

A typical problem is the two dimensional flow over a spillway (Fig. 4.4). The solution region is terminated at the right and left boundaries where uniform horizontal flow is assumed, $\partial\psi/\partial n = 0$, and $\psi$ is linear along the boundary. Thus the solution consists of finding the following $n_s + n_f + 1$ unknowns (assuming a known flow rate, $q$):

On the free surface: $\left(\dfrac{\partial y}{\partial n}\right)_i$, $\eta_i$, $\quad i = 1, 2, \ldots, n_f$

On the right boundary: $\psi_i$, $\quad i = n_f + 1, n_f + 2, \ldots, n_r - 1$

On the solid boundary: $\left(\dfrac{\partial y}{\partial n}\right)_i$, $\quad i = n_r, n_r + 1, \ldots, n_s$

The Bernoulli constant: $B$

in which $n_f$ is the number of free surface points, $n_r$ is the number of points as the solid boundary. The relevant equations are (a) the matrix form of the boundary integral ($n_s$ equations),

$$[R]\{\psi\} = [A]\left\{\frac{\partial\psi}{\partial n}\right\} \tag{4.31}$$

(b) Bernoulli's equation (4.30) at each free surface point ($n_f$ equations), and (c) the specification of uniform flow through the left boundary, $(\partial\psi/\partial n)_s = (\partial y/\partial n)_i$ (in which the normal is with respect to the lower boundary and the free surface, respectively).

Because Bernoulli's equation is nonlinear and because $\eta$ is unknown, implying nonlinearity in the boundary equations, the solution must be interative. After an initial guess the solution can be improved by a standard Newton-Raphson technique. Cheng et al. [39], however, found poor convergence in a zone behind the spillway crest. Better convergence was achieved by creating a group of "pseudo-nodes" on the free surface and satisfying Bernoulli's equation at the real nodes plus the pseudo-nodes in a least squares sense. A similar result could also be achieved by guaranteeing smoothness of the free surface through the use of cubic spline interpolation functions on the free surface (Liggett and Salmon [40]).

## 4.6 Unsteady Free Boundaries – Water Waves

The water-wave problems considered in this section are also described by Laplace's equation with free surface boundary conditions given by (4.27) and (4.28). We will consider the linearized wave problem in the next section, which not only linearizes (4.27) and (4.28) but also applies these equations on the equilibrium surface, and thus is not truly a free surface problem.

The nonlinear boundary conditions (4.27) and (4.28) are often written in terms of normal, $n$, and tangential, $s$, coordinates as

$$\frac{\partial \eta}{\partial t} = \frac{1}{\cos \beta} \frac{\partial \Phi}{\partial n} \tag{4.32}$$

$$\left(\frac{\partial \Phi}{\partial t}\right)_x = B - g\,\eta - \frac{1}{2}\left[\left(\frac{\partial \Phi}{\partial s}\right)^2 - \left(\frac{\partial \Phi}{\partial n}\right)^2 - 2\left(\frac{\partial \Phi}{\partial s}\right)\left(\frac{\partial \Phi}{\partial n}\right)\tan \beta\right] \tag{4.33}$$

in which $\beta$ is the angle the free surface makes with horizontal. The subscript on the derivative $(\partial \Phi/\partial t)_x$ indicates that this derivative is taken with the $x$-coordinate held constant but the derivative follows the free surface vertically; that is, it is the rate of change of $\Phi$ on the free surface at a fixed horizontal point.

Equations (4.32) and (4.33) are suited for a time stepping scheme. Beginning with an initial free surface at time $k\,\Delta t$, (4.32) can be used to find the free surface position at time $(k+1)\,\Delta t$ and (4.33) gives the potential at time $(k+1)\,\Delta t$. A possible centered finite difference scheme is

$$\eta^{k+1} = \eta^k + \frac{2(\Delta t)}{\cos \beta^k + \cos \beta^{k+1}}\left[\theta_1\left(\frac{\partial \Phi}{\partial n}\right)^k + (1-\theta_1)\left(\frac{\partial \Phi}{\partial n}\right)^{k+1}\right] \tag{4.34}$$

$$\begin{aligned}\Phi^{k+1} = \Phi^k &+ \left[B - \frac{g}{2}(\eta^k + \eta^{k+1})\right]\Delta t - \frac{1}{2}\left\{\theta_2\left[\left(\frac{\partial \Phi^k}{\partial s}\right)^2 - \left(\frac{\partial \Phi^k}{\partial n}\right)^2\right.\right. \\ &\left. - 2\frac{\partial \Phi^k}{\partial n}\frac{\partial \Phi^k}{\partial s}\tan \beta^k\right] + (1-\theta_2)\left[\left(\frac{\partial \Phi^{k+1}}{\partial s}\right)^2 - \left(\frac{\partial \Phi^{k+1}}{\partial n}\right)^2\right. \\ &\left.\left. - 2\frac{\partial \Phi^{k+1}}{\partial n}\frac{\partial \Phi^{k+1}}{\partial s}\tan \beta^{k+1}\right]\right\}\end{aligned} \tag{4.35}$$

in which $\theta_1$ and $\theta_2$ are weighting factors. In the actual calculation (4.34) and (4.35) must be linearized with respect to the unknown time level, $k+1$. There appears to be a number of ways to accomplish the linearization (e.g. Liu and Liggett [41]). For example, all factors superscripted with $k+1$ can initially be approximated with the value at time level $k$ and then they can be updated using the initial solution in an iterative scheme.

A further question arises in free boundary problems as to where the free surface nodes should be placed. In most of the papers dealing with wave problems or problems in porous media the boundary element equations (4.31) are written at time $k+1$ whereas the boundary conditions (4.32), (4.33) are centered between $k$ and $k+1$. That is not, however, a centered scheme. Presumably the proper surface to discretize is somewhere between the two time levels. In an iteration scheme an intermediate level is easily used but a new integration would have to be done at each iteration. The tradeoffs in order to achieve a given accuracy between size of time step and the linearization-iteration method have not been investigated. Yet an efficient method is important because the non-linear problem can be expensive, even using efficient boundary methods. Since the discretized boundary must follow the free surface, an integration is required at each time step (if not at each iteration) which changes the coefficient matrix in the simultaneous equations requiring, in turn, a new matrix inversion or decomposition. Further, in wave problems the number of free surface nodal points is often large, on the order of a dozen nodes per wave length (using linear interpolation functions) if the wave is smooth and many more for sharp crested or nearly breaking waves.

There are a number of examples of nonlinear wave calculations using boundary techniques. The two-dimensional problem of sloshing in a tank was studied by Nakayama and Washizu [42], Washizu [43], and Faltinsen [44]. Simulation of a sea bed rise was calculated by Liu and Liggett [41]. Using complex variables in a transformed plane Longuet-Higgins and Cokelet [45] and Cokelet [46] were able to compute the breaking of waves in water of inifinte depth. Vinje and Brevig [47], still using complex notation but in the physical plane, extended the results of Longuet-Higgins and Cokelet so that results could be obtained in water of finite depth and so that forces on objects could be calculated (Vinje and Brevig [48]). Three-dimensional nonlinear wave problems have rarely been solved, probably because of the expense due to computing time and storage requirements and because of stability consideration. However, Isaacson [49] reports on the calculation of forces on fixed and floating bodies using a three-dimensional formulation. The three-dimensional solution region must be broken up into a large number of panels for the purposes of integration and the satisfaction of the free surface boundary conditions. The problem is ripe in the possibility of instabilities due to the numerical scheme. Moreover, physical instabilities (caustics) often appear in three-dimensional problems. They represent infinite (or large) wave amplitudes (which may be indistinguishable from a numerical instability) which would effectively destroy the calculation.

An unsolved problem exists in connection with nonlinear wave problems in an unbounded sea. Since the computational boundaries must be at finite distances, unwanted wave reflections contaminate the solution and effectively reduce the time that the computation is valid. Since the wave problem as formulated is totally undamped, moving the computational boundaries further apart only increases the time of valid solution at considerable computational expense. For linear problems absorbing boundary conditions can sometimes be devised (Engquist and Majda [50]; Smith [51]) but none has been formulated for nonlinear problems. Damping can be provided by adding a term to (4.33) which becomes (Betts and Mahamad [52]; LeMéhauté [53])

$$\left(\frac{\partial \Phi}{\partial t}\right)_x = B - g\eta - \frac{1}{2}\left[\left(\frac{\partial \Phi}{\partial s}\right)^2 - \left(\frac{\partial \Phi}{\partial n}\right)^2 - 2\left(\frac{\partial \Phi}{\partial s}\right)\left(\frac{\partial \Phi}{\partial n}\right)\tan\beta\right] + f\Phi \qquad (4.36)$$

in which $f$ is the damping coefficient. Betts and Mohamad solved the problem of a progressive wave in a channel with a numerical wave maker in one end. For a few nodes at the end opposite to the wave maker they used equation (4.36) with $f \neq 0$, thus providing a wave absorber which reduced reflection. That, however, does not solve the usual problem where the effects of a wave on a structure are studied. In that case reflection from the structure would again reflect from the wave maker and contaminate the solution. Better transmitting or absorbing boundary conditions, which allow the outgoing wave to pass while not effecting the incoming wave, are urgently needed.

## 4.7 Linear Waves

In the small amplitude wave problem the free surface boundary conditions (4.32) and (4.33) are linearized to give

$$\frac{\partial \eta}{\partial t} = \frac{\partial \Phi}{\partial z} \tag{4.37}$$

$$\frac{\partial \Phi}{\partial t} = -g\,\eta \tag{4.38}$$

in which $z$ is the vertical coordinate. Moreover, these conditions are applied at the equilibrium surface so that the point of application of the boundary conditons is no longer a part of the solution.

Several different cases of linear wave problems are summarized herein. Additional detail is found in the references below and in Liu and Liggett [41].

Assuming monochromatic waves of frequency $\omega$, the substitutions

$$\eta = \zeta(x, y)\, e^{-i\omega t} \quad \Phi = \varphi(x, y, z)\, e^{-i\omega t} \tag{4.39}$$

lead to

$$\nabla^2 \varphi = 0 \tag{4.40}$$

with the free surface boundary condition

$$\frac{\partial \varphi}{\partial z} = \frac{\omega^2}{g}\,\varphi \tag{4.41}$$

In the above equations $i = \sqrt{-1}$ and the final solution results from taking the real (or imaginary) part.

Two cases of two-dimensional flow are important. Vertical plane problems assume a uniform behavior in the horizontal $y$-direction. In the case where rigid surfaces are uniform in the $y$-direction

$$\varphi(x, y, z) = \varphi_*(x, z)\, e^{ik_* y} \tag{4.42}$$

leading to a modified Helmholtz equation

$$\frac{\partial^2 \varphi_*}{\partial x^2} + \frac{\partial^2 \varphi_*}{\partial z^2} - k_*^2\, \varphi_* = 0 \tag{4.43}$$

The appropriate free space Green's function is

$$G_* = -K_0(k_* r) \tag{4.44}$$

in which $K_0$ is a modified Bessel function of order zero. The boundary integral is given by (4.26) without the area integration.

Horizontal plane problems assume a constant water depth and all structures must penetrate the water full depth to the bottom. In such a case

$$\varphi(x, y, z) = -\left[\frac{i\,g\,A}{\omega}\frac{\cosh k(z+h)}{\cosh k\,h}\,e^{ikx} + \varphi_*(x, y)\right] \tag{4.45}$$

in which $g$ is the acceleration of gravity, $A$ is the amplitude of the incident wave (assumed to travel in the positive $x$-direction), $k$ is the wave number of incident wave, and $h$ is the water depth. The equation for $\varphi_*$ is now a Helmholtz equation,

$$\frac{\partial^2 \varphi_*}{\partial x^2} + \frac{\partial^2 \varphi_*}{\partial y^2} + k^2\,\varphi_* = 0 \tag{4.46}$$

The boundary integral is

$$\alpha\,\varphi_* = \int_\Gamma \left[k\,H_1(k\,r)\,\varphi_*\,\frac{\partial r}{\partial n} - H_0(k\,r)\,\frac{\partial \varphi_*}{\partial n}\right] ds \tag{4.47}$$

where $H_1$ and $H_0$ are Hankle functions of the first kind of order one and zero.

For transient linear wave problems a time stepping technique along the lines of that cited for nonlinear waves is easy to devise. Fortunately, the stability and accuracy of such a linear scheme can be analyzed (Salmon et al. [54]). Three-dimensional linear wave solutions are a solution of Laplace's equations under the transient, free surface boundary conditions (Lennon et al. [55]).

Even in the linear case suitable radiation boundary conditions are difficult to find and are needed to limit the solution region. For shallow water such conditions are given by Lennon et al. [55] but the shallow condition is inaccurate for deep water waves. At a penalty in computing time the method of Smith [51] can be used.

## 4.8 Stoke's Flow

If all inertia terms are neglected in the Navier-Stokes equation and a stream function formulation is used, the viscous flow statisfies the biharmonic equation

$$\nabla^4 \psi = 0 \tag{4.48}$$

By use of the Rayleigh-Green identity the equation analogous to (4.2) (without the volume integral) is (Jaswon and Symm [2], p. 271)

$$\beta\,\psi(p) = -\int_\Gamma \left[\frac{\psi}{r^2}\frac{\partial r}{\partial n} + \frac{1}{r}\frac{\partial \psi}{\partial n} + r\frac{\partial}{\partial n}(\nabla^2\psi) - (\nabla^2\psi)\frac{\partial r}{\partial n}\right] dA \tag{4.49}$$

in which $\beta = 4\pi$ for $p$ in $R$ and $\beta = 2\pi$ if $p$ is on a smooth part of $\Gamma$.

Creeping flow inside a cylinder was computed by Mills [56] who used a more elaborate Green's functions which was limited to the cylinder. A general formulation is given by Okabe [57].

## 4.9 Porous-Elasticity

Boundary methods as applied to fluid mechanics are similar to these applied to solid mechanics. In this last section we wish to briefly mention one of the areas where fluid and solid mechanics come together, the problem of the fluid flow in a porous, elastic medium. Such problems are governed by the equations of Biot [58], [59], which assume a linear elastic response of the solid matrix and flow through the matrix which is governed by Darcy's equation (see equation 4.16). The applications appear in a wide variety of engineering problems including soil consolidation, seabed response to wave action, land subsidence from water or oil removal, earthquake analysis, geothermal energy recovery, hydrofracturing for enhanced production of water or oil wells, etc. Obviously these applications have great economic consequences.

The porous-elastic problem is large, usually larger than either the comparable elastic problem or flow through porous media. Thus excessively expensive analysis is required by finite element methods whereas boundary methods offer possibility to decrease computer time and storage to reasonable levels. A boundary solution was formulated by Cleary [60], [61]. In applying the direct boundary integral method, the fundamental solutions (essentially Green's functions) for the Biot equations must be developed. These were presented by Cleary [62] in integral form and by Cheng and Liggett [63], [64] in closed form. The latter papers applied integral methods to the soil consolidation and fracture problems.

Although some work has been done in this important area where solid and fluid mechanics converge, the practical problems of greatest economic consequence remain unsovled. This is a most fruitful area for future research.

## 4.10 Concluding Remarks

The boundary element method is now well established as a numerical technique for the solution of fluid mechanics problems. Once thought to be applicable only to linear, homogeneous equations with constant coefficients, the boundary element method now has wide applications for beyond that currently discarded limitation. And such applications are expanding rapidly.

The popularity of boundary methods comes, primarily, from its computational efficiency. Other advantages, however, may ultimately be more important. Liggett [65] presented an ordered list of advantages of the method. For various applications items in this list maybe more or less important, thus changing the order. Nevertheless, it is worth repeating:

*1. User Convenience.* There are two subheadings under this item which are: (a) The user can choose the boundary grid largely from his intuitive knowledge of the problem rather than from mathematical considerations which he may not understand. That is, if the instructions state that the program interpolates linearly, the user can choose a nodal spacing such that the actual boundary conditions and the expected solution are approximated satisfactorily, according to the user's criteria, by piecewise linear functions. (b) The amount of data which must be supplied is minimized by boundary methods relieving the user of elaborate data input. Similarly, but of less importance, it is easy for the user to choose the output that he needs.

*2. Special Situations.* It appears to be easier to design boundary methods to handle automatically singularities, moving boundaries and infinite boundaries.

*3. Immediate Results.* The variables of interest can be obtained from the method as opposed to some domain methods where an additional numerical step must be performed (e.g. in domain methods the potential flow velocity usually comes from a numerical differentiation of the potential leading to a degradation of accuracy).

*4. Efficiency.*

*5. Use on Mini- and Micro-Computers.*

Not explicitly in the above list may be the biggest advantage of all – the ability to solve huge problems. Even though computing is becoming less expensive and computer storage more abundant, the shear size of many practical problems form the greatest obstacle to their solutions. Several of the problems mentioned on the previous pages fall into that category. Size may be a result of physical complexity or of nonlinearities which require fine discretization, small time steps, and repeated iterations. For that class of problem boundary methods may be the only economically feasible technique.

## References

1 Trefftz, E., Über die Kontraktion kreisförmiger Flüssigkeitsstrahlen. Z. Math. Phys., **64,** 34, 1917
2 Jaswon, M. A., Symm, G. T., *Integral Equation Methods in Potential Theory and Elastostatics.* Academic Press, 1977
3 Brebbia, C. A., *The Boundary Element Method for Engineers.* John Wiley and Sons, 1978
4 Brebbia, C. A., Walker, S., *Boundary Element Techniques in Engineering.* Newnes-Butterworths 1980
5 Banerjee, P. K., Butterfield, R., *Boundary Element Methods in Engineering Science.* McGraw-Hill 1981
6 Liggett, J. A., Liu, P. L.-F., *The Boundary Integral Equation Method for Porous Media Flow.* George Allen Unwin, 1983
7 Crouch, S. L., Starfield, A. M., *Boundary Methods in Solid Mechanics.* George Allen and Unwin, 1983
8 Mukherjee, S., Boundary element methods in creep and fracture. Applied Science Publishers, Ltd, Essex, 1982, Also Elsevier Science Publishing, Co, N.Y.
9 Green, G., *An Essay on the Application of Mathematical Analysis to the Theories of Electricity and Magnetism,* Nottingham 1828
10 Hunt, B. W., Numerical solutions of an integral equation for flow from a circular orifice. Jour. Fluid Mech., **31,** 361–377, 1968
11 Prager, W., Die Druckverteilung an Körpern in ebener Potentialströmung. Physik. Zeitschr., **29,** 865–869, 1928
12 Nyström, E. J., Über die praktische Auflösung von linearen Integralgleichungen mit Anwendungen auf Randwertaufgaben der Potentialtheorie. Soc. Sci. Fennica, Comment. Physico-Math., **4,** 15, 1–52, 1928
13 Fredholm, I., Sur une classe d'equations fonctionelles. Acta Math., **27,** 365–390, 1903

14 Lotz, I., Calculation of potential past airship bodies in yaw. NACA TM 675, 1932 [also Ingenieur-Archiv, Vol. II, 1931]
15 Vandrey, F., A direct iteration method for the calculation of velocity distribution of bodies of revolution and symmetrical profiles. Admiralty Research Lab. Rept. R 1/G/HY/12/2, 1951
16 Weinstein, A., On axially symmetric flows. Quarterly of Applied Math., **5**, No. 4, 1948
17 Van Tuyl, A., On the axially symmetric flow around a new family of half bodies. Quarterly of Applied Math., **7**, No. 4, 1950
18 Sadowsky, M. A., Sternberg, E., Elliptic integral representation of axially symmetric flows. Quarterly of Applied Math. **8**, No. 2, 1950
19 Landweber, L., The axially symmetric potential flow about elongated bodies of revolution. David Taylor Model Basin Rep., 761, 1951
20 Vandrey, F., On the calculation of the transverse potential flow past a body of revolution with the aid of the method of Mrs. Flügge-Lotz. Astia AD-40089, 1951
21 Smith, A. M. O., Pierce, J., Exact solution of the Neumann problem. Calculation of plane and axially symmetric flows about or within arbitrary boundaries. Douglas Aircraft Company Report No. 26988, 1958 [Summary in Proc. of the *3rd* Int. Congress of Applied Math., Brown University, 1958]
22 Hess, J. L., Smith, A. M. O., Calculation of nonlifting potential flow about arbitrary three-dimensional bodies. ES 40622, Douglas Aircraft Corp., Long Beach, Calif. 1962 (Also in Jour. of Ship Research **8**, No. 2, Sept. 1964)
23 Davenport, F. J., Singularity solutions to general potential flow airfoil problem. D6-7207, Boeing Airplane Co., Seattle, Wash. 1963
24 Jaswon, M. A., Integral equation methods in potential theory: I. Proc. Royal Soc. A, **275**, 23–32, 1963
25 Symm, G. T., Integral equation methods in potential theory; II. Proc. Royal Soc. A, **275**, 33–46, 1963
26 Chaplin, H. R., A method for numerical calculation of slip stream contraction of a shrouded impulse disk in the static case with application to other axisymmetric potential flow problems. David Taylor Model Basin Report No. 1857, 1964
27 Gallagher, R. H., *Finite Element Analysis Fundamentals,* Prentice Hall 1975
28 Hess, J. L., Review of integral-equation techniques for solving potential-flow problems with complicated boundaries. Innovative Numerical Analysis for the Applied Engineering Sciences. University Press of Virginia, Charlottesville, 131–143, 1980
29 Hunt, B., The mathematical basis and numerical principles of the boundary integral equation method for incompressible potential flow over 3-D aerodynamic configurations. *Numerical Methods in Applied Fluid Dynamics* (B. Hunt, ed.), Academic Press, 49–135, 1980
30 Carey, G. F., Extension of boundary elements to lifting compressible aerodynamics. *Finite Element Flow Analysis* (T. Kawai, ed.), Univ. of Tokyo Press, 939–943, 1982
31 Carey, G. F., Kim, S. W., Extension of boundary element method to lifting subcritical flows. 19th Annual Meeting, Society of Engineering Science, University of Missouri-Rolla, 1982
32 Inamuro, T., Adachi, T., Sakata, H., A numerical analysis of unsteady separated flow by discrete vortex model using boundary element method. *Finite Element Flow Analysis* (T. Kawai, ed.), University of Tokyo Press, 931–938, 1982
33 White, J. W., Kline, S. J., A calculation method for incompressible axisymmetric flows, including unseparated, fully separated, and free surface flows. Report MD-35, U.S. Air Force Office of Scientific Research Mechanics Divison, Contract AF-F44620-74-C-0016; Thermosciences Division, Department of Mechanical Engineering, Stanford University, 1975
34 Wu, J. C., Problems of general visous flow. In: *Developments in Boundary Element Methods – 2* (P. K. Banerjee and R. P. Shaw, eds.). Applied Science Publishers, 69–109, 1982
35 Wu, J. C., Thompson, J. F., Numerical solutions of time-dependent incompressible Navier-Stokes equations using an integro-differential formulation. Computers and Fluids **1**, 197–215, 1973

36 Lennon, G. P., Liu, P. L-F., Liggett, J. A., Boundary integral equation solution to axisymmetric potential flows, 1, Basic formulation. Water Resources Research, **15** (5), 1102–1106, 1979

37 Lennon, G. P., Liu, P. L-F., Liggett, J. A., Boundary integral equation solution to axisymmetric flows, 2, Recharge and well problems in porous media. Water Resources Research, **15** (5), 1107–1115, 1979

38 Lennon, G. P., Liu, P. L-F., Liggett, J. A., Boundary integral solutions to three-dimensional unconfined Darcy's flow. Water Resources Research, **16** (4), 651–658, 1980

39 Cheng, A. H-D., Liu, P. L-F., Liggett, J. A., Boundary calculations of sluice and spillway flows. J. of the Hydraulics Division, ASCE, **107**, (HY 10), 1163–1178, 1981

40 Liggett, J. A., Salmon, J. R., Cubic spline boundary elements. Int. J. for Numerical Methods in Engineering, **17**, 543–556, 1981

41 Liu, P. L-F., Liggett, J. A., Applications of boundary element methods to problems of water waves. In: *Developments in Boundary Element Methods – 2* (P. K. Banerjee and R. P. Shaw, eds.), Elsevier's Applied Science Publishers, Ltd., 37–67, 1982

42 Nakayama, T., Washizu, K., The boundary element method applied to the analysis of two-dimensional nonlinear sloshing problems. Int. J. Num. Meth. Engrg., **17**, No. 11, 1631–1646, 1981

43 Washizu, K., Some applications of finite element techniques to nonlinear free surface fluid flow problems. *Finite Element Flow Analysis* (T. Kawai, ed.). Univ. of Tokyo Press, 3–15, 1982

44 Faltinsen, O. M., A numerical nonlinear method of sloshing in tanks with two-dimensional flow. J. Ship Research, **23** (3), 193–202, 1978

45 Longuet-Higgins, M. S., Cokelet, E. D., The deformation of steep surface waves on water. I. *A numerical method of computation.* Proc. of the Royal Society, **A 350**, 1–25, 1976

46 Cokelet, E. D., Breaking waves – the plunging jet and interior flow-field. Proc. Sym. on Mechanics of Wave-induced Forces on Cylinders, Bristol, 1978

47 Vinje, T., Brevig, P., Numerical solution of breaking waves. Adv. Water Resources, **4**, 77–82, 1981

48 Vinje, T., Brevig, P., Numerical calculations of forces from breaking waves. Preprints, Int. Sym. on Hydrodynamics in Ocean Engineering, Trondheim, Norway, 547–566, 1981

49 Issacson, M. de St. Q., Nonlinear-wave effects on fixed and floating bodies. J. Fluid Mechanics, **120**, 267–281, 1982

50 Engquist, B., Majda, A., Absorbing boundary conditions for the numerical simulation of waves. Math. and Computers, **31**, 629–651, 1977

51 Smith, W. D., A non-reflecting boundary for wave propagation problems. J. Computational Physics, **15**, 492–503, 1974

52 Betts, P. L., Mohamad, T. T., Water waves: A time-varying unlinearized boundary element approach. *Finite Element Flow Analysis* (T. Kawai, ed.), University of Tokyo Press, 923–929, 1982

53 LeMéhauté, B., Progressive wave absorber. J. Hyd. Res., **10** (2), 153–169, 1972

54 Salmon, J. R., Liu, P. L-F., Liggett, J. A., Integral equation method for linear water waves. J. Hydraulics Division, ASCE, **106**, (HY 12), 1995–2010, 1980

55 Lennon, G. P., Liu, P. L-F., Liggett, J. A., Boundary integral solutions of water wave problems. J. Hydraulics Division, ASCE, **108** (HY 8), 921–931, 1982

56 Mills, R. D., Computing internal viscous flow problems for the circle by integral methods. J. Fluid Mech., **79** (3), 609–624, 1977

57 Okabe, M., A boundary element approach in the incompressible viscous flow. *Finite Element Flow Analysis* (T. Kawai, ed.). Univ. of Tokyo Press, 915–922, 1982

58 Biot, M. A., General theory of three-dimensional consolidation. J. of Applied Physics, **12**, 155–164, 1941

59 Biot, M. A., Theory of elasticity and consolidation for a porous anisotropic solid. J. of Applied Physics, **26**, 182–185, 1955

60 Cleary, M. P., Fundamental solutions for a fluid-saturated porous solid. Int. J. of Solids and Structures, **13**, 785–806, 1977

61 Cleary, M. P., Moving singularities in elasto-diffusive solids with applications to fracture propagation. Int. J. of Solids and Structures, **14**, 81–97, 1978

62 Cleary, M. P., Fundamental solutions for fluid-saturated porous media and applications to local rupture phenomena. Thesis submitted in partial fulfillment of requirements for for the Ph.D. degree, Brown University, 1976

63 Cheng, A. H-D., Liggett, J. A., Boundary integral equation method for linear porous-elasticity with applications to fracture propagation. Int. J. for Numerical Methods in Engineering, (In press), 1983

64 Cheng, A. H-D., Liggett, J. A., Boundary integral equation method for linear porous-consolidation. Int. J. for Numerical Methods in Engineering, (In press), 1983

65 Liggett, J. A., Hydrodynamic calculations using boundary elements. *Finite Element Flow Analysis* (T. Kawai, ed.), Univ. of Tokyo Press, 889–896, 1982

# Chapter 5

# Water Waves Analysis

*by M. C. Au and C. A. Brebbia*

## 5.1 Introduction

Over the last decades the study of water waves diffraction and radiation problems, especially in offshore structures has interested many researchers. The problem is sometimes solved by using the finite element method as shown by Newton [1]; Berkhoff [2]; Bai [3]; Yue, Chen and Mei [4], and more recently Zienkiewicz and Bettess [5]. Alternatively boundary integral equations have been used as illustrated by the work of Garrison [6]; Hogben and Standing [7]; Eatock-Taylor [8] and Isaacson [9].

The boundary element method [10] is a combination of classical boundary integral techniques with finite element concepts and thus, combines some of the advantages of the two methods. In this chapter boundary elements are used to solve wave diffraction-radiation problems in a simple and general way. The fundamental solution applied here are $1/r$ or $\ln(1/r)$ for three and two dimensional problems rather than the more complex solutions applied by other authors [6], [7]. The advantage of this simple approach is that it becomes easier to write a general program as those solutions are accurate and give stable results. The disadvantage however is that more elements are usually required to solve a problem and because of this taking advantage of the system symmetry becomes more important. Several types of symmetry are presented here to indicate how to reduce computational effort. The examples included in this chapter serve to illustrate the applicability and potentialisies of the technique.

## 5.2 Basic Theory

*Governing Equations*

For the wave structure interaction problem described in Fig. 5.1, the fluid is usually assumed to be incompressible and inviscid. Under these conditions the fluid will remain irrotational and a velocity potential $U_t$ can be defined such that,

$$\mathbf{V} = \nabla \cdot U_t \tag{5.1}$$

where $\mathbf{V}$ is the velocity vector. In this case the continuity equation can be written as,

$$\nabla^2 U_t = 0 \tag{5.2}$$

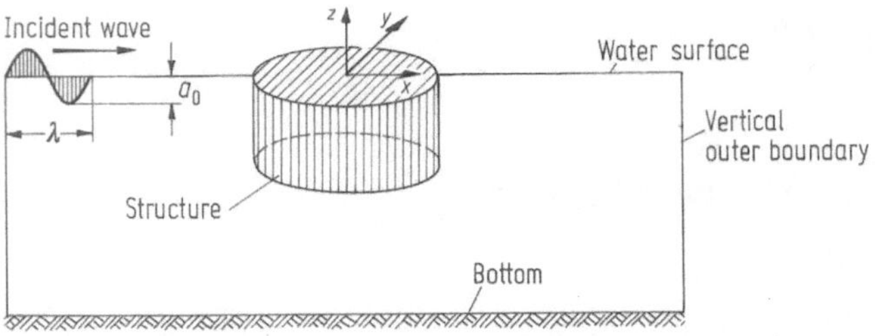


**Fig. 5.1.** Problem definition

The water surface condition for small amplitude waves can be expressed in function of the potential as,

$$\frac{\partial^2 U_t}{\partial t^2} + g\frac{\partial U_t}{\partial n} = 0 \tag{5.3}$$

where $g$ is the acceleration of the gravity and $n$ is the normal to the water surface.

A moving-body surface will produce a normal velocity given by,

$$\frac{\partial U_t}{\partial n} = \mathbf{n} \cdot \dot{\mathbf{X}} \tag{5.4}$$

where $\mathbf{n}$ is the unit normal vector to the surface and $\dot{\mathbf{X}}$ is the velocity vector of the structure.

*Incident Wave*

The time dependent surface elevation, $\xi_0$, due to a monochromatic wave can be written as

$$\xi_0 = \text{Re}\,\{a_0 \cdot e^{i(k_1x+k_2y-\omega t)}\} \tag{5.5}$$

where $a_0$ is the surface incident wave amplitude; $\omega$ is the wave frequency and Re indicates the real part of the complex value between brackets. $k_1$ and $k_2$ are the wave number components in $x$ and $y$ directions.

The incident wave potential in the fluid can be expressed as

$$U_0 = \text{Re}\,\{u_0\, \mathrm{e}^{-i\omega t}\} \tag{5.6}$$

where

$$u_0 = -\frac{i\,g\,a_0}{\omega}\,\frac{\cosh k\,(z+h)}{\cosh k\,h}\,e^{i(k_1x+k_2y)}$$

with $k_1 = k\cos\alpha$ and $k_2 = k\sin\alpha$; where $\alpha$ is the angle of incident wave with respect to $x$-axis; $h$ is the depth of the water; $k$ is the wave number which can be defined by the dispersion expression as

$$k\tanh k\,h = \frac{\omega^2}{g} \tag{5.7}$$

*Boundary Value Problem*

The presence of the structure (Fig. 5.1) will disturb the incident wave and the total potential of the fluid can be written in terms of the incident and diffracted wave potential, i.e.

$$U_t = U_0 + U \tag{5.8}$$

with

$$U = \mathrm{Re}\,(u\, e^{-i\omega t})$$

where $u$ is the diffracted wave potential which is governed by the continuity equation, i.e.

$$\nabla^2 u = 0 \tag{5.9}$$

with the following boundary conditions:

(i) On the $\Gamma_1$ surface of the structure,

$$\frac{\partial u}{\partial n} + \frac{\partial u_0}{\partial n} = \mathbf{n} \cdot \mathbf{V} \quad \text{on } \Gamma_1 \tag{5.10}$$

(ii) On the horizontal sea bottom $\Gamma_2$,

$$\frac{\partial u}{\partial n} = 0 \quad \text{on } \Gamma_2 \tag{5.11}$$

(iii) Sufficiently far from the body the diffracted outgoing wave can be approximated by the radiation condition:

$$\frac{\partial u}{\partial n} - i\, k\, u = 0 \quad \text{on } \Gamma_3 \tag{5.12}$$

The boundary $\Gamma_3$ should be vertical and at a distance away from the source of disturbance.

(iv) The water surface condition (5.3) can be written as

$$\frac{\partial u}{\partial n} - \frac{\omega^2}{g} u = 0 \quad \text{on } \Gamma_4 \tag{5.13}$$

$\Gamma_4$ is the undisturbed water surface.

The above four different types of boundary conditions can be summarized by the general expression

$$\frac{\partial u}{\partial n} + p\, u - q = 0 \quad \text{on } \Gamma \tag{5.14}$$

where $\Gamma = \Gamma_1 + \Gamma_2 + \Gamma_3 + \Gamma_4$. $p$ and $q$ are parameters defined by the above conditions (i) to (iv).

## 5.3 Boundary Element Formulation

The boundary value problem defined by equations (5.9) and (5.14) can be formulated in terms of a weighted residual expression [10] and solved approximately. Given an approximation for the $u$ potential and a weighting or distribution function $u^*$, one can write,

$$\int_\Omega (\nabla^2 u)\, u^*\, d\Omega = \int_\Gamma \left( \frac{\partial u}{\partial n} + p\, u - q \right) u^*\, d\Gamma \tag{5.15}$$

The $u^*$ function in boundary elements is the fundamental solution of the equation:

$$\nabla^2 u^* + \Delta_i = 0 \tag{5.16}$$

where $\Delta_i$ is the Dirac delta function. This gives

$$u^* = \frac{1}{2\pi} \ln \frac{1}{r} \quad \text{in 2-D}$$

or

$$u^* = \frac{1}{4\pi r} \quad \text{in 3-D} \tag{5.17}$$

where

$$r = \sqrt{(x-x_i)^2 + (z-z_i)^2} \quad \text{in 2-D}$$
$$r = \sqrt{(x-x_i)^2 + (y-y_i)^2 + (z-z_i)^2} \quad \text{in 3-D} \tag{5.18}$$

After integrating twice by parts equation (5.15) and applying formula (5.16) the integral expression for a point $i$ can be written as:

$$c_i u_i + \int_\Gamma \left( \frac{\partial u^*}{\partial n} + p\, u^* \right) u \, d\Gamma = \int_\Gamma u^* q \, d\Gamma \tag{5.19}$$

where

$c_i = \frac{1}{2}$ for the point '$i$' on the smooth boundary $\Gamma$;
$c_i = 1$ if the point '$i$' is in $\Omega$ and
$c_i = 0$ if '$i$' is external to $\Omega$ domain and boundary $\Gamma$.

$\Gamma$ can now be discretized into a series of $N$ boundary elements as shown in figure 5.2. Hence,

$$\Gamma = \sum_{e=1}^{N} \Gamma_e \tag{5.20}$$

The potential within each element can be defined as

$$u = [\Phi]\,\{u_e^n\} \tag{5.21}$$

where $[\Phi]$ is the interpolation function vector and $\{u_e^n\}$ is a vector defining the potential at the nodes (Fig. 5.2) of elements $\Gamma_e$. Similarly, the flux can be written as

$$q = [\Phi]\,\{q_e^n\} \tag{5.22}$$

where $\{q_e^n\}$ is the vector of nodal fluxes.

Therefore equation (5.19) can be written as

$$c_i u_i + \sum_e^N \int_{\Gamma_e} \left( \frac{\partial u^*}{\partial n} + p\, u^* \right) [\Phi] \, d\Gamma \, \{u_e^n\} = \sum_e^N \int_{\Gamma_e} u^* [\Phi] \, d\Gamma \, \{q_e^n\} \tag{5.23}$$

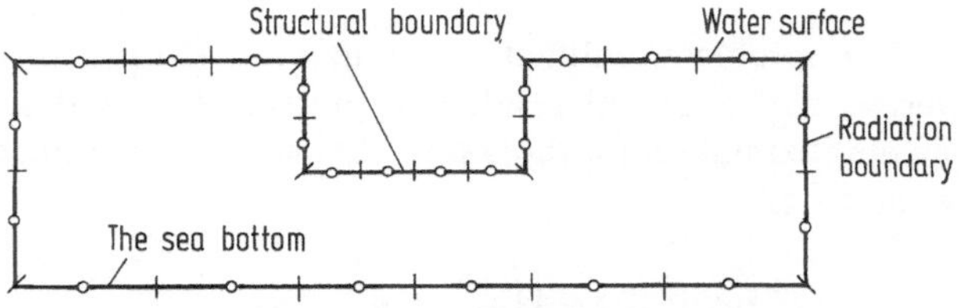


**Fig. 5.2.** A boundary element mesh

which can be expressed in the form,

$$c_i u_i + \sum_e^N [\hat{H}_{ie}] \{u_e^n\} = \sum_e^N [G_{ie}] \{q_e^n\} \tag{5.24}$$

where

$$[\hat{H}_{ie}] = \int_{\Gamma_e} \left( \frac{\partial u^*}{\partial n} + p\, u^* \right) [\Phi]\, d\Gamma$$

$$[G_{ie}] = \int_{\Gamma_e} u^* [\Phi]\, d\Gamma$$

*Matrix Formulation*

By considering that the point '$i$' of equation (5.24) refers to every nodal point on $\Gamma$, a complete matrix system of equations can be formed such as

$$[H]\{u\} = [G]\{q\} \tag{5.25}$$

where $[H] = [\hat{H}] + [C]$. Note that $[C]$ is a diagonal matrix formed by the $c_i$ coefficients of equation (5.24).

One can now apply the four different types of boundary conditions to equation (5.25) and obtain, after suitable rearrangement, the following system.

$$\left[ \mathbf{H}_1, \mathbf{H}_2, \mathbf{H}_3 - i\, k\, \mathbf{G}_3, \mathbf{H}_4 - \frac{\omega^2}{g} \mathbf{G}_4 \right] \begin{Bmatrix} \mathbf{u}_1 \\ \mathbf{u}_2 \\ \mathbf{u}_3 \\ \mathbf{u}_4 \end{Bmatrix} = [G_1]\{q_1\} \tag{5.26}$$

or

$$[A]\{u\} = \{B\}$$

where the matrices $[H_j]_{ie}$ and $[G_j]_{ie}$ are due to the integrals of the type

$$\int_{\Gamma_j \cap \Gamma_e} \frac{\partial u^*}{\partial n} [\Phi]\, d\Gamma \quad \text{and} \quad \int_{\Gamma_j \cap \Gamma_e} u^* [\Phi]\, d\Gamma$$

respectively.

The $j$ in $\Gamma_j$ ($j = 1, 2, 3, 4$) represent the four types of boundary conditions given by (5.10) to (5.13) and the integrations are to be carried out over the element $\Gamma_e$, usually using a system of local coordinates as in finite elements.

The $\{q_1\}$ vector in equation (5.26) can be given by the incident wave and the body surface velocities, i.e. at any point,

$$q_1 = q_0 + \mathbf{n} \cdot \{\dot{\mathbf{X}}\} \tag{5.27}$$

where $q_0 = \dfrac{\partial u_0}{\partial n}$ is given by the incident field. In matrix form,

$$\{q_1\} = \{q_0\} + [Q]\{\dot{X}\} \tag{5.28}$$

where

$$[Q] = [G_1][n_x, n_z, n_{xz}] \quad \text{in 2-D}$$

or

$$[Q] = [G_1][n_x, n_y, n_z, n_{yz}, n_{zx}, n_{xy}] \quad \text{in 3-D}$$

$(n_x, n_y, n_z)$ is the unit normal vector on $\Gamma_1$

$$(n_{yz}, n_{zx}, n_{xy}) \quad \text{is} \quad \{(x - x_0), (y - y_0), (z - z_0)\} \times \{n_x, n_y, n_z\}$$

and $(x_0, y_0, z_0)$ is the centre of moment. Hence,

$$\boldsymbol{B} = [G_1]\{q_1\} = -[G_1]\{q_0\} + [Q]\{\dot{X}\} \tag{5.29}$$

For a fixed structure the $\{\dot{X}\}$ vector is null and the equation for the diffraction potential is given by

$$[A]\{u\} = \{B\}, \quad \{B\} = -[G_1]\{q_0\} \tag{5.30}$$

For radiation potential problems due to the oscillation of a structure, equation (5.29) can be written as,

$$[A]\{u\} = [Q]\{\dot{X}\} \tag{5.31}$$

from which the hydrodynamic loading contribution due to added mass and damping can be computed.

*Wave Forces*

Once the potentials around the structure are known, one can compute the pressures by using the linearized version of Bernoulli's equation, i.e.

$$\{P\} = -\varrho \frac{\partial}{\partial t}\{U_t\} \tag{5.32}$$

where $\varrho$ is the density of the fluid and $U_t$ are the total potentials i.e. $U_t = U_0 + U$. For harmonic motion,

$$\boldsymbol{p} = i\,\varrho\,\omega\,(\boldsymbol{u}_0 + \boldsymbol{u})$$

and

$$\boldsymbol{P} = \mathrm{Re}\,\{p\,e^{-i\omega t}\} \tag{5.33}$$

After the pressures are known the forces acting on the structure can be computed by integration over the surface, i.e.

$$\underset{6\times 1}{\{f\}} = i\,\varrho\,\omega \int_{\Gamma_1} \underset{6\times 1}{\{n\}}\, U_t\, d\Gamma$$

and

$$\{F\} = \mathrm{Re}\,[\{f\}\,e^{-i\omega t}] \tag{5.34}$$

Finally the water surface elevation can be obtained at any point. It is given by the kinematic condition, which for harmonic motion gives,

$$\xi = \mathrm{Re}\left[\frac{i\,\omega}{g}(u_0 + u)\,e^{-i\omega t}\right] \tag{5.35}$$

## 5.4 Special Structural Types

In this section, several special cases will be studied, starting by the vertically integrated structure, i.e. structures of constant horizontal sections and for which the problem can be integrated with respect to depth; this effectively reduces the problem to two dimensions. Other types of two dimensional structures are those which can be studied on a vertical plane because they are sufficiently long in the direction normal to that plane. Fully three dimensional structures are also considered in detail, stressing the importance of their rotational symmetry.

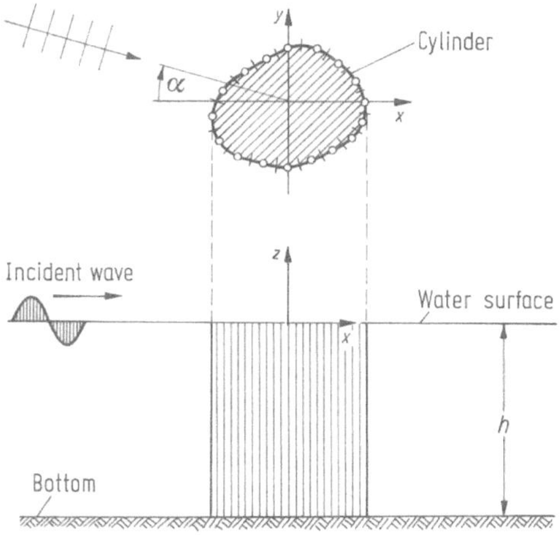


Fig. 5.3. The vertical cylinder

*(i) Vertically Integrated Structures.* This type of structure has a constant horizontal section throughout the water depth as shown in Figure 5.3. The incident wave potential – equation (5.6) – can be written as,

$$u_0 = \bar{u}_0(x, y)\frac{\cosh k(z+h)}{\cosh k\,h} \tag{5.36}$$

where $\bar{u}_0(x, y) = -\dfrac{i\,g\,a_0}{\omega}\,e^{i(k_1x+k_2y)}$. Provided that the $X - Y$ plane formulation for the part of the velocity potential independent of depth $Z$. The diffracted potential can be similarly written as

$$u = \bar{u}(x, y)\frac{\cosh k(z+h)}{\cosh k\,h} \tag{5.37}$$

Notice that (5.36) and (5.37) satisfy the boundary conditions on the water surface and on the sea bottom. The problem can then be expressed in terms of the diffracted potential $\bar{u}(x, y)$ and the modified boundary value problem for the $x-y$ plane becomes

$$(\nabla^2 + k^2)\,\bar{u} = 0 \quad \text{in } \Omega \tag{5.38}$$

with boundary conditions,

$$\frac{\partial \bar{u}}{\partial n} + \frac{\partial \bar{u}_0}{\partial n} = 0 \quad \text{on } \Gamma$$

and

$$\frac{\partial \bar{u}}{\partial n} - i\,k\,\bar{u} = 0 \quad \text{on } \Gamma' \tag{5.39}$$

or in general

$$\frac{\partial \bar{u}}{\partial n} + p\,\bar{u} = q \quad \text{on } \Gamma + \Gamma'$$

By using the weighted residual technique one can write,

$$\int_\Omega u^*\,[(\nabla^2 + k^2)\,\bar{u}]\,d\Omega = \int_{\Gamma+\Gamma'} u^*\left(\frac{\partial \bar{u}}{\partial n} + p\,\bar{u} - q\right)d\Gamma \tag{5.40}$$

After integrating equation (5.40) by parts twice, one finds,

$$\int_{\Omega} \bar{u}\,[(\nabla^2 + k^2)\,u^*]\,d\Omega = \int_{\Gamma+\Gamma'} \bar{u}\left(\frac{\partial u^*}{\partial n} + p\,u^*\right) d\Gamma + \int_{\Gamma+\Gamma'} u^*\,q\,d\Gamma \tag{5.41}$$

The fundamental solution corresponding to the above problem is given by the solution of

$$(\nabla^2 + k^2)\,u^* + \Delta_i = 0 \tag{5.42}$$

and can be written as

$$u^* = \frac{i}{4} H_0'(k\,r) \tag{5.43}$$

where $r = \sqrt{(x - x_i)^2 + (y - y_i)^2}$

Notice that solution (5.43) automatically satisfies the boundary condition on $\Gamma'$, i.e.

$$\lim_{r\to\infty} \int_{\Gamma'} u\left(\frac{\partial u^*}{\partial n} + p\,u^*\right) d\Gamma = 0 \tag{5.44}$$

Now equation (5.41) can be written for every point '$i$' on $\Gamma$ as,

$$c_i\,u_i + \int_{\Gamma} \frac{\partial u^*}{\partial n}\,u\,d\Gamma = \int_{\Gamma} u^*\,q\,d\Gamma \tag{5.45}$$

where $\Gamma$ is the "internal" boundary usually defined by a clock-wise direction loop over the structural section. The matrix form of equation (5.45) can now be produced assuming that $\Gamma$ is discretized into $N$ elements. This gives

$$[H]\,\{\bar{u}\} = [G]\,\{q\} \tag{5.46}$$

where for a constant element,

$$H_{ij} = \int_{\Gamma_j} \frac{\partial}{\partial n}\left[\frac{i}{4} H_0'(k\,r)\right] d\Gamma + \delta_{ij}\,C_i$$

$$G_{ij} = \int_{\Gamma_j} \frac{i}{4} H_0'(k\,r)\,d\Gamma$$

$$q_j = -\left\{\frac{\partial \bar{u}_0}{\partial n}\right\}_j$$

Notice that the elements of the vector $\{q\}$ are equal to the incident potential derivatives computed at the structural nodes.

Equation (5.46) can then be solved and the potential can be computed as

$$u_t = (\bar{u}_0 + \bar{u})\,\frac{\cosh k\,(z + h)}{\cosh k\,h} \tag{5.47}$$

The total three dimensional forces on the vertical structure can be obtained by integrating (5.47) in the vertical direction which gives

$$\begin{Bmatrix} f_x \\ f_y \\ f_{xy} \end{Bmatrix} = i\,\varrho\,\omega\,\frac{\tanh k\,h}{k} \int_{\Gamma} \begin{Bmatrix} n_x \\ n_y \\ n_{xy} \end{Bmatrix} \bar{u}_t\,d\Gamma \tag{5.48}$$

and

$$\begin{Bmatrix} f_{yz} \\ f_{zx} \end{Bmatrix} = \left[ z_0 + \frac{1}{k} \tanh \frac{k\,h}{2} \right] \begin{Bmatrix} f_y \\ -f_x \end{Bmatrix}$$

where $z_0$ is the centre of the moments $f_{yz}$ and $f_{zx}$.

*(ii) Structure with a Vertical Axis or Plane of Symmetry.* A structure as the one shown in Fig. 5.4 has a vertical axis of symmetry with two symmetrical parts $\Gamma_1$ and $\Gamma_2$. If the boundary conditions on the $z$-axis are not known, a full discretization of boundary elements will be required in $\Gamma_1$ and $\Gamma_2$ This gives a matrix equation of the form,

$$\begin{bmatrix} \boldsymbol{H}_{11} & \boldsymbol{H}_{12} \\ \boldsymbol{H}_{21} & \boldsymbol{H}_{22} \end{bmatrix} \begin{Bmatrix} \boldsymbol{u}_1 \\ \boldsymbol{u}_2 \end{Bmatrix} = \begin{bmatrix} \boldsymbol{G}_{11} & \boldsymbol{G}_{12} \\ \boldsymbol{G}_{21} & \boldsymbol{G}_{22} \end{bmatrix} \begin{Bmatrix} \boldsymbol{q}_1 \\ \boldsymbol{q}_2 \end{Bmatrix} \tag{5.49}$$

where the subindex 1 and 2 in $\{u\}$ and $\{q\}$ refers to values on $\Gamma_1$ and $\Gamma_2$ respectively. It can be seen by inspection that $\boldsymbol{H}_{11} = \boldsymbol{H}_{22}$ and $\boldsymbol{H}_{21} = \boldsymbol{H}_{12}$ and similarly for the terms in $[G]$.

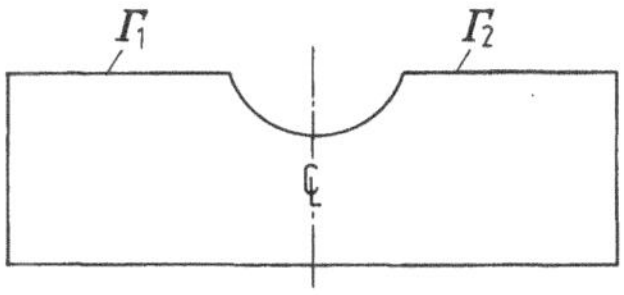


**Fig. 5.4.** A symmetric 2-D domain

In order to reduce the system of equations one can apply a transformation technique to replace each variable by a symmetric and an antisymmetric components. That is

$$u_s = \frac{u_1 + u_2}{2}; \quad u_a = \frac{u_1 - u_2}{2}$$
$$q_s = \frac{q_1 + q_2}{2}; \quad q_a = \frac{q_1 - q_2}{2} \tag{5.50}$$

where '$s$' refers to the symmetrical and '$a$' to the antisymmetrical component.

Substituting (5.50) into (5.49) and rearranging the submatrices one obtains

$$\begin{bmatrix} (\boldsymbol{H}_{11} + \boldsymbol{H}_{12}) & \boldsymbol{0} \\ \boldsymbol{0} & (\boldsymbol{H}_{11} - \boldsymbol{H}_{12}) \end{bmatrix} \begin{Bmatrix} \boldsymbol{u}_s \\ \boldsymbol{u}_a \end{Bmatrix} = \begin{bmatrix} (\boldsymbol{G}_{11} + \boldsymbol{G}_{12} & \boldsymbol{0} \\ \boldsymbol{0} & (\boldsymbol{G}_{11} - \boldsymbol{G}_{12}) \end{bmatrix} \begin{Bmatrix} \boldsymbol{q}_s \\ \boldsymbol{q}_a \end{Bmatrix} \tag{5.51}$$

The full matrix equations presented in (5.49) can then be reduced to a diagonal form as shown in (5.51). Therefore the solution for the symmetric and anti-symmetric components can be obtained independently. Once they are computed one can find the values of $\{u_1\}$ and $\{u_2\}$. However, if the wave forces are the only result required, they can be obtained from $\{u_s\}$ and $\{u_a\}$ as follows,

$$\begin{Bmatrix} f_x \\ f_z \\ f_{zx} \end{Bmatrix} = 2\,i\,\varrho\,\omega \int_{\Gamma_1} \begin{bmatrix} 0 & n_x \\ n_z & 0 \\ 0 & n_{zx} \end{bmatrix} \begin{Bmatrix} u_s \\ u_a \end{Bmatrix} d\Gamma \tag{5.52}$$

Using this technique the computer time required to solve a two dimensional problem can be reduced by approximately one third. For some three dimensional structures, one or more planes of symmetry can be considered and much larger savings in computer time achieved. Some of the examples presented in section 4 require only 15% of the solution time for the fully discretized problem. I.e. where symmetry is not taken into consideration. For further details see reference [11].

*(iii) Structures with Rotational Symmetry.* Consider a domain with rotational symmetry about the $z$-axis as shown in Figure 5.5. The total boundary surface $\Gamma$ can be represented by

$$\Gamma = \sum_{l=1}^{n} \Gamma_l \tag{5.53}$$

where $\Gamma_l$ is a symmetrical component of $\Gamma$; $n$ is the total number of components and $\theta$ is the angle defined by each component (Fig. 5.5), such that

$$\theta = 2\pi/n \tag{5.54}$$

That is, the domain is symmetric for a $\theta$ angle.

A rotational transformation can relate any component to the 1st component such that

$$\Gamma_l = C_n^l \cdot \Gamma_1 \tag{5.55}$$

where $C_n^l$ indicates a rotation transformation through an angle of $l\,\theta$ in the anti-clockwise direction around the $z$-axis. A matrix representation for the rotation $C_n^l$ is given by

$$[C_n] = \begin{bmatrix} \cos l\,\theta & -\sin l\,\theta \\ \sin l\,\theta & \cos l\,\theta \end{bmatrix} \tag{5.56}$$

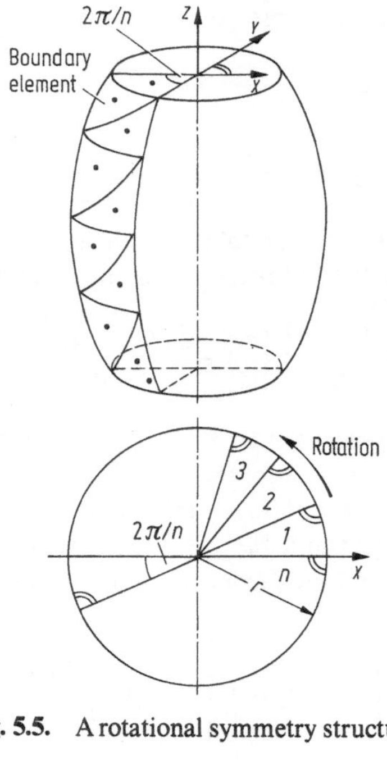


**Fig. 5.5.** A rotational symmetry structure

If the boundary element mesh has the same type of symmetry (i.e. $N$ elements on every $\Gamma_l, l = 1, 2, \ldots, n$) the matrix equivalent of equation (5.23) can now be formed by considering the point $i$ only on the $\Gamma_1$ segment. This gives

$$\underset{N\times nN}{[\boldsymbol{h}_1, \boldsymbol{h}_2, \ldots, \boldsymbol{h}_n]} \underset{nN\times 1}{\begin{Bmatrix} \boldsymbol{u}_1 \\ \boldsymbol{u} \\ \vdots \\ \boldsymbol{u}_n \end{Bmatrix}} = \underset{N\times nN}{[\boldsymbol{g}_1, \boldsymbol{g}_2, \ldots, \boldsymbol{g}_n]} \underset{nN\times 1}{\begin{Bmatrix} \boldsymbol{q}_1 \\ \boldsymbol{q}_2 \\ \vdots \\ \boldsymbol{q}_n \end{Bmatrix}} \tag{5.57}$$

where the submatrices $[h_l]$ and $[g_l]$ are of dimension $N \times N$ whose components are given by

$$[h_l]_{ij} = \int_{\Gamma_j^l} \left( \frac{\partial u^*}{\partial n} + p\, u^* \right) [\Phi]\, d\Gamma \tag{5.58}$$

and

$$[g_l]_{ij} = \int_{\Gamma_j^l} u^* \, [\Phi]\, d\Gamma$$

where $\Gamma_j^l = C_n^l \Gamma_j^1$. $\Gamma_j^l$ is the $j^{\text{th}}$ element in the $l^{\text{th}}$ symmetrical component, $l = 1, 2, \ldots, n$ and $j = 1, 2, \ldots, N$.

Alternatively $[h_l]_{ij}$ and $[g_l]_{ij}$ can also be defined by

$$[\boldsymbol{h}_l]_{ij} = \int_{\Gamma_j^1} \left( \frac{\partial u^*}{\partial n} + p\, u^* \right) [\Phi]\, d\Gamma \tag{5.59}$$

and

$$[g_l]_{ij} = \int_{\Gamma_j^1} u^* \, [\Phi]\, d\Gamma$$

with the following transformation to the '$i$' point

$$\begin{Bmatrix} x_i \\ y_i \end{Bmatrix}_l = [C_n^l]^{-1} \begin{Bmatrix} x_i \\ y_i \end{Bmatrix}_1$$

where the right hand side vector represents the coordinates of the $i$ point on $\Gamma_l$, $l = 1, 2, \ldots, n$.

Considering now the full matrix for all positions of the '$i$' point on the complete boundary $\Gamma$, a matrix similar to (5.27) will be obtained, such that,

$$[H] = \begin{bmatrix} \boldsymbol{h}_1 & \boldsymbol{h}_2 & \boldsymbol{h}_3 \cdots \boldsymbol{h}_n \\ \boldsymbol{h}_n & \boldsymbol{h}_1 & \boldsymbol{h}_2 \cdots \boldsymbol{h}_{n-1} \\ \boldsymbol{h}_{n-1} & \boldsymbol{h}_n & \boldsymbol{h}_1 \cdots \boldsymbol{h}_{n-2} \\ \vdots & \vdots & \vdots \quad\; \vdots \\ \boldsymbol{h}_2 & \boldsymbol{h}_3 & \boldsymbol{h}_4 \cdots \boldsymbol{h}_1 \end{bmatrix} \tag{5.60}$$

Similarly for $[G]$. A transformation can now be applied to this type of matrix such that [12]

$$\{u\} = [R]\, \{\bar{u}\}$$

and

$$\{q\} = [R]\, \{\bar{q}\} \tag{5.61}$$

where $[R]$ is a $n\,N \times n\,N$ matrix with submatrices

$$\underset{N\times N}{[R_{lm}]} = \underset{N\times N}{[I]}\; e^{i(l-1)(m-1)\theta} \quad \text{for} \quad l = 1, \ldots, n, \quad m = 1, \ldots, n \tag{5.62}$$

A new form of equation (5.27) will be written as

$$[\bar{H}]\{\bar{u}\} = [\bar{G}]\{\bar{q}\} \tag{5.63}$$

where

$$[\bar{H}] = \frac{1}{\beta}[R^*][H][R], \quad [\bar{G}] = \frac{1}{\beta}[R^*][G][R] \tag{5.64}$$

where $\beta$ is a normalizing factor to be defined in (5.71) and $[R^*]$ is the complex conjugate matrix of $[R]$. That is

$$[R^*_{lm}] = [I]\, e^{-i(m-1)(l-1)\theta} \tag{5.65}$$

Hence a submatrix of $[\bar{H}]$ will be defined by

$$[\bar{H}_{kj}] = \frac{1}{\beta}\sum_l \sum_m [R^*_{kl}][H_{lm}][R_{mj}] = \frac{1}{\beta}\sum_l \sum_m [H_{lm}]\, e^{i[(m-1)(j-1)-(l-1)(k-1)]\theta} \tag{5.66}$$

The summation defined in (5.66) is over $n^2$ terms, i.e. $l$ and $m$ (both varying from 1 to $n$); each term $[H_{lm}]$ in the equation needs to be considered once. If the order of summation is altered in such a way that the direction of the summation will be in the diagonal way, once can do this by replacing the subindex $m$ by $m'$ such that

$$m' = \begin{cases} m+l-1 & \text{if} \quad m' \leqq n \\ m+l-1-n & \text{if} \quad m' > n \end{cases} \tag{5.67}$$

The submatrix

$$[H_{lm'}] = [H_{1m'}] = [H_{1m}] \tag{5.68}$$

The corresponding factor to be multiplied for $[H_{lm'}]$ will be

$$e^{i[(m'-1)(j-1)-(l-1)(k-1)]\theta}$$

and as $\theta = 2\pi/n$, the submatrices $[\bar{H}_{kj}]$ in (5.66) will be given by

$$\begin{aligned} [\bar{H}_{kj}] &= \frac{1}{\beta}\sum_l \sum_m [H_{1m}]\, e^{i[(m-1+l-1)(j-1)-(l-1)(k-1)]\theta} \\ &= \frac{1}{\beta}\sum_m [H_{1m}]\, e^{i(m-1)(j-1)\theta} \sum_l e^{i(l-1)(j-k)\theta} \end{aligned} \tag{5.69}$$

Taking advantage of the relationships

$$\sum_l e^{i(l-1)(j-k)\theta} = n\,\delta_{kj} \tag{5.70}$$

equation (5.69) can be further simplified to,

$$[\bar{H}_{kj}] = \delta_{kj}\sum_m [H_{1m}]\, e^{i(m-1)(j-1)\theta} \tag{5.71}$$

for $\beta = n$. Therefore, using the submatrices defined in equation (5.60) one can write,

$$\underset{N\times N}{[\bar{H}_{kk}]} = \sum_m \underset{N\times N}{[h_m]}\, e^{i(m-1)(k-1)\theta} \tag{5.72}$$

Hence the $[\bar{H}]$ matrix has been reduced to a new matrix with submatrices only on the diagonal position; i.e.

$$[\bar{H}]=\begin{bmatrix}\bar{H}_{11} & 0 & 0 & \cdots & 0\\ 0 & \bar{H}_{22} & 0 & \cdots & 0\\ 0 & 0 & \bar{H}_{33} & \cdots & 0\\ \vdots & \vdots & \vdots & & \vdots\\ 0 & 0 & 0 & \cdots & \bar{H}_{nn}\end{bmatrix} \tag{5.73}$$

The same consideration can be applied to the matrix $[\bar{G}]$. The final form of equation (5.73) can be written as,

$$[\bar{H}_{kk}]\{\bar{u}_k\}=[\bar{G}_{kk}]\{\bar{q}_k\} \tag{5.74}$$

$k=1, 2, \ldots, n$ and the dimension of the matrices are $N\times N$. The unknown vectors are given by

$$\{\bar{u}_k\}=\sum_j [R_{kj}]^{-1}\{u_j\}$$

$$\{\bar{q}_k\}=\sum_j [R_{kj}]^{-1}\{q_j\} \tag{5.75}$$

The $[R]^{-1}$ matrix can be obtained from;

$$[R]^{-1}=\frac{1}{\beta}[R^*] \tag{5.76}$$

After obtaining the solution of equation (5.75) the wave forces can be calculated from;

$$\underset{6\times 1}{\{f\}}=i\varrho\omega\int_{\Gamma_l}\underset{6\times 1}{\{n\}}\,u_t\,d\Gamma \tag{5.77}$$

where $\Gamma_l$ is the $l^{\text{th}}$ symmetric component of the structural surface boundary. Considering the transforming matrix $[C_n]$ defined by (5.56) the normal vector $\{n\}$ is given by

$$\begin{Bmatrix}n_x\\ n_y\\ n_z\\ n_{yz}\\ n_{zx}\\ n_{xy}\end{Bmatrix}_l=\begin{bmatrix}C^l_n & 0\\ 0 & C^l_n\end{bmatrix}\begin{Bmatrix}n_x\\ n_y\\ n_z\\ n_{yz}\\ n_{zy}\\ n_{xy}\end{Bmatrix}_1 \tag{5.78}$$

One can find the summation for the forces in (5.77) by considering an identity similar to (5.70), i.e.

$$\sum_l\begin{Bmatrix}1\\ \cos(l-1)\,\theta\\ \sin(l-1)\,\theta\end{Bmatrix}e^{i(l-1)(m-1)}=\frac{n}{2}\begin{Bmatrix}2\,\delta_{m2}\\ (\delta_{m2}+\delta_{mn})\\ i\,(\delta_{m2}-\delta_{mn})\end{Bmatrix} \tag{5.79}$$

and the relation defined in (5.61). The summation becomes, after rearranging the terms,

$$\{f\}=i\varrho\omega\int_{\Gamma_1}[N]\begin{Bmatrix}\bar{u}_1\\ \bar{u}_2\\ \bar{u}_n\end{Bmatrix}d\Gamma \tag{5.80}$$

where

$$[N] = \frac{n}{2}\left[\begin{array}{c|c|c} 0 & (n_x - i\, n_y) & (n_x + i\, n_y) \\ 0 & (i\, n_x + n_y) & (-\, i\, n_x + n_y) \\ \hline 2n_z & 0 & 0 \\ \hline 0 & (n_{yz} - i\, n_{zx}) & (n_{yz} + i\, n_{zx}) \\ 0 & (n_{xz} + i\, n_{yz}) & (n_{zx} - i\, n_{yz}) \\ \hline 2n_{xy} & 0 & 0 \end{array}\right]_{6\times 3}$$

and $\{\bar{u}_1\}$, $\{\bar{u}_2\}$ and $\{\bar{u}_n\}$ are the solutions from the equation (74) for $k = 1, 2$ and $n$ respectively. In such a way only three sets of equations are required to be solved.

## 5.5 Equations of Motion for the Structure

The equations of motion for the structure can now be included in the analysis to determine the movements of the system and how the forces change during motion.

Consider the following system

$$[M]\{\ddot{X}\} + [C]\{\dot{X}\} + [K]\{X\} = \{f\}\, e^{-i\omega t} \tag{5.81}$$

with $\{X\} = \{X_0\}\, e^{-i\omega t}$. The mass matrix $[M]$ is given by the structural mass matrix $[M_0]$ and the added mass $[M_a]$; i.e. $[M] = [M_0] + [M_a]$. The damping matrix is $[C]$ and together with $[M_a]$ can be obtained using (5.30) and (5.33) as

$$[M_a]\{\ddot{X}\} + [C]\{\dot{X}\} = -\, i\, \varrho\, \omega \int_{\Gamma_1} \{n\}\, [H]^{-1}\, [Q]\, d\Gamma\, \{\dot{X}\} \tag{5.82}$$

$[K]$ is the total stiffness matrix due to the contribution of the hydrostatic force and those due to the mooring system. The time independent version of (5.81) is

$$\{-\, \omega^2 [M] - i\, \omega\, [C] + [K]\}\, \{X_0\} = \{f\} \tag{5.83}$$

whose solution is given by

$$\{X_0\} = [\boldsymbol{K} - \omega^2\, \boldsymbol{M} - i\, \omega\, \boldsymbol{C}]^{-1}\, \{f\} \tag{5.84}$$

## 5.6 Numerical Examples

Several examples are considered here to illustrate the application of the above theory and show the advantages of using boundary elements, expecially for those cases where full advantage can be taken of symmetry. The examples cover a variety of important problems and are

(i) Submerged half-cylinder,
(ii) Fully submerged cylinder,

(iii) Hydrodynamic coefficients for a half-floating cylinder,
(iv) Moored floating cylinder,
(v) Surface wave on a sloping bottom,
(vi) Square caisson,
(vii) Elliptic cylinder,
(viii) Submerged storage tank and tower,
(ix) Floating hemisphere,
(x) Surface elevation around cylinder,
(xi) Condeep type platform.

*(i) Submerged Half-Cylinder.* This example considers the two dimensional case of a submerged half-cylinder resting on the bottom of the sea (Figure 5.6). 50 constant boundary elements were used to discretize half of the total boundary as shown in Figure 5.7. The results are given in terms of non-dimensional horizontal and vertical forces. They are compared against those given by Naftzer and Chakrabarti [13] and are shown in Figure 5.8.

*(ii) Fully Submerged Cylinder.* This example is similar to (i) but considers that the cylinder is fully submerged as shown in Figure 5.9. The boundary was discretized using 14 quadratic elements as shown in Figure 5.10. Results for horizontal and vertical wave forces are given in Figure 5.11. The numerical results compare well against those by Naftzer and Chakrabarti [13].

*(iii) Hydrodynamic Coefficients for a Half-Floating Cylinder.* The added mass and damping coefficients for the half-cylinder of Fig. 5.12 were computed using,

$$\underset{3\times3}{[C_m]} = -\frac{[M_a]}{\varrho(\pi a^2/2)}; \qquad \underset{3\times3}{[C_d]} = -\frac{[C]}{\varrho(\pi a^2/2)\,\omega} \tag{5.85}$$

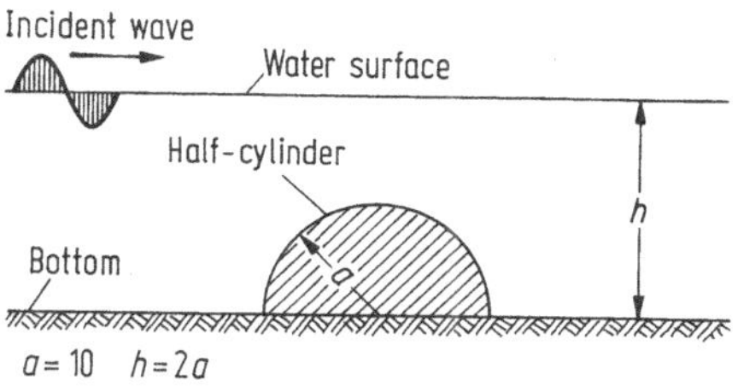


**Fig. 5.6.** The submerged half-cylinder on the sea bottom

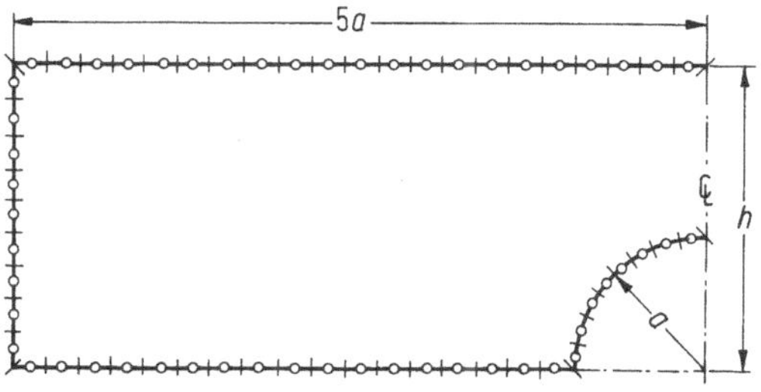


**Fig. 5.7.** Boundary element mesh for submerged half-cylinder

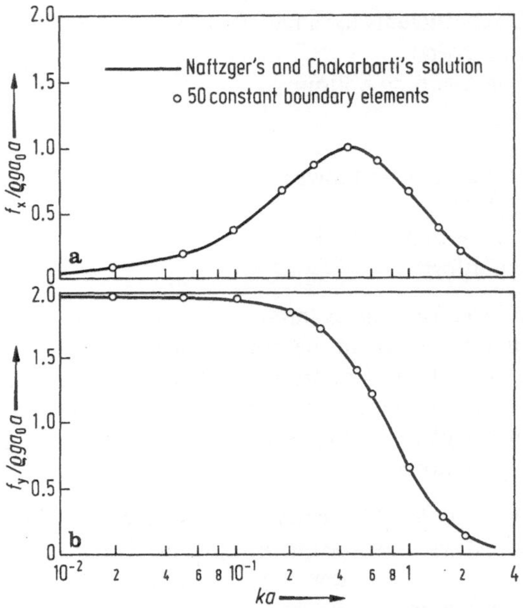


**Fig. 5.8.** The wave forces on submerged half-cylinder by 50 elements. **a** Horizontal force. **b** Vertical force

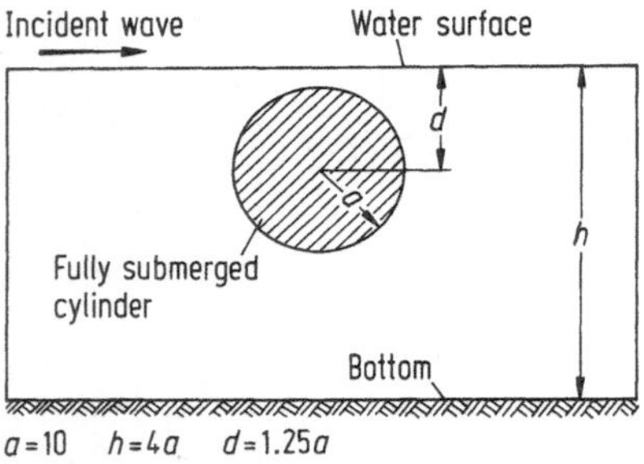


**Fig. 5.9.** The fully submerged cylinder

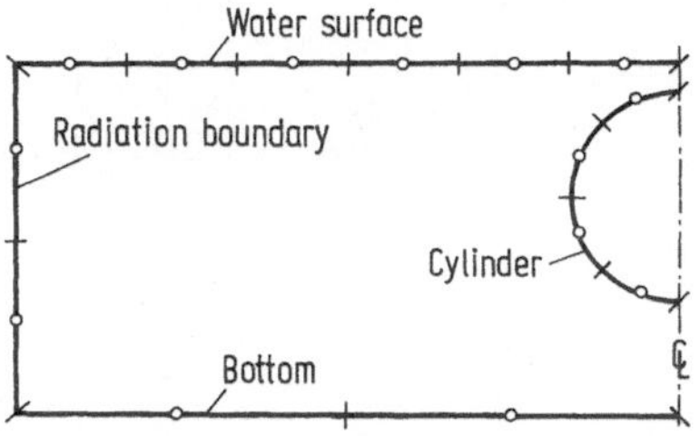


**Fig. 5.10.** The half discretization with 14 quadratic elements

Only 50 constants elements were used on half the total boundary as shown in the figure. The results due to the swaying and heaving motion are given in Fig. 5.13 and compared against those by Newton [1].

*(iv) Moored Floating Cylinder* [14]. The moored floating cylinder of Fig. 5.14 has been analysed assuming that mooring lines remain straight. This type of structure has been used as a breakwater and is normally moored in the way showing in the figure. Amplitude of motion computed using 60 constant elements are given in Figure 5.15.

*(v) Surface Waves on a Sloping Bottom.* The case of surface wave amplitude was studied in the example described in Figure 5.16. The bottom is sloping towards the land and two distinctive parts were considered i.e. the straight and sloping

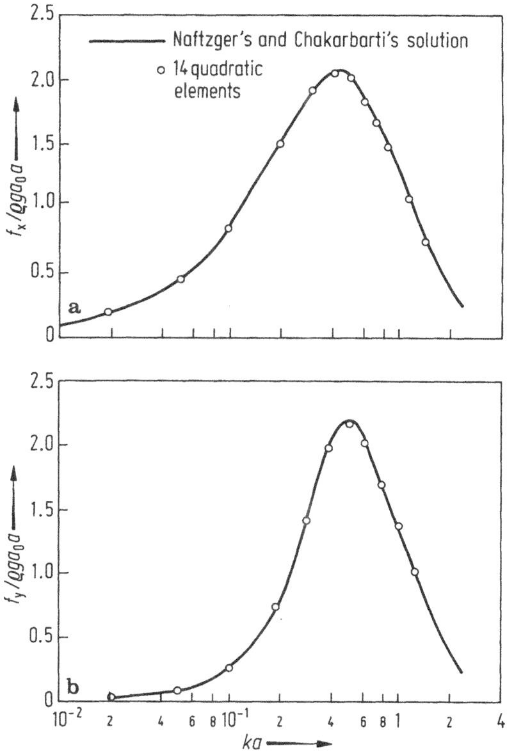


**Fig. 5.11.** Wave forces on the fully submerged cylinder obtained using 14 quadratic elements. **a** Horizontal force. **b** Vertical force

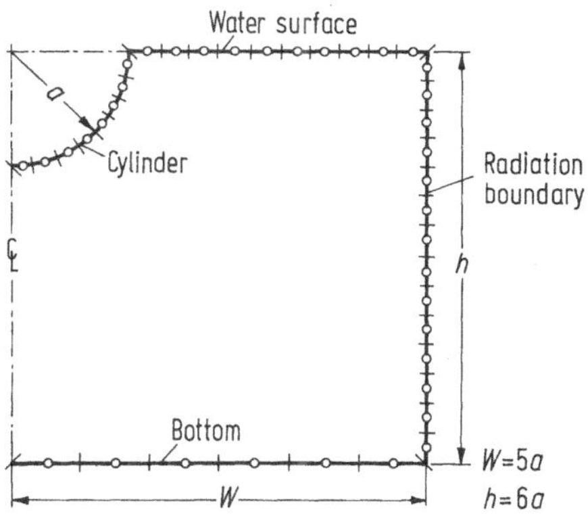


**Fig. 5.12.** A boundary element mesh for the half-floating cylinder

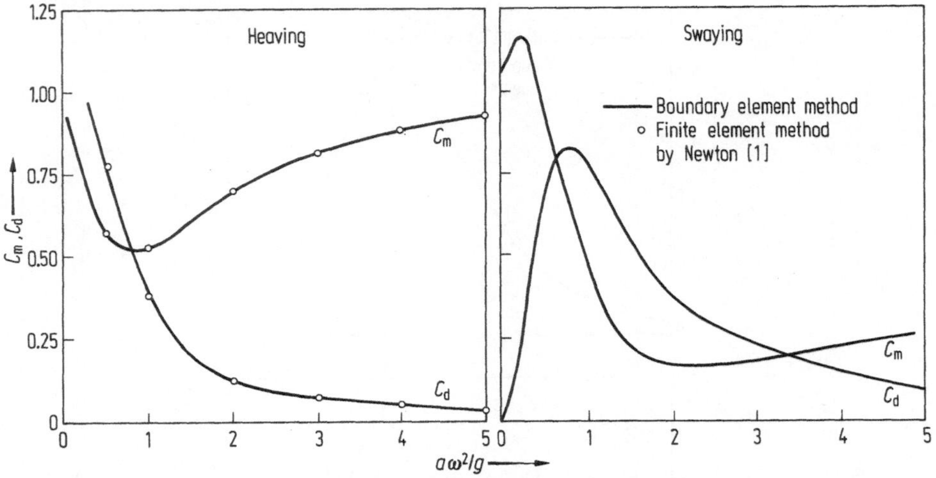


**Fig. 5.13.** The hydrodynamic coefficients for the heaving and the swaying half-cylinder

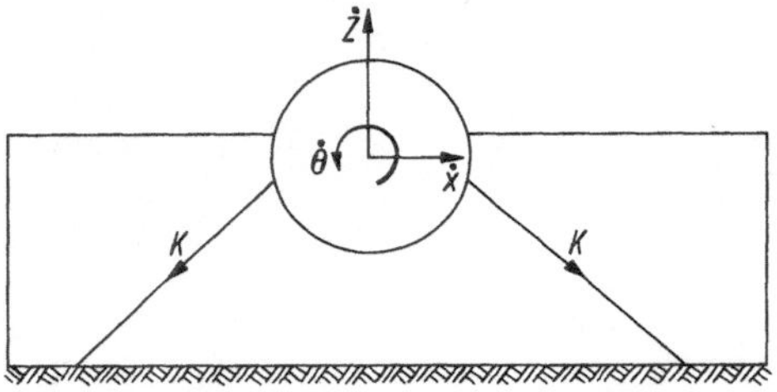


**Fig. 5.14.** Moored floating cylinder. **a** Fixed cylinder; **b** flexible cylinder. $a = 16$; $h = 35$; $d = 20$; $w = 80$; $k/l\,g = 2.632$ and $k$ = wave number

$$\frac{m}{\varrho g} = 0.4782, \quad \frac{I}{\varrho g} = 69.832$$

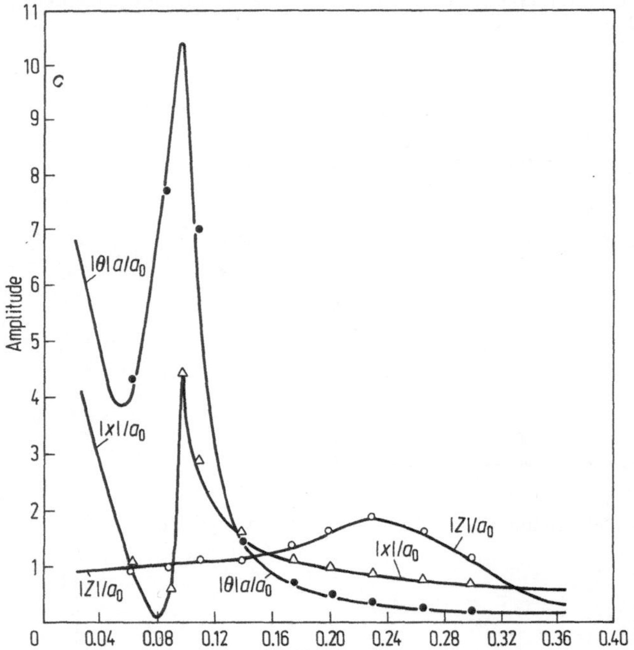


**Fig. 5.15.** Amplitude of motion ($\dot{x}, \dot{Z}, \dot{\theta}$). (—) computed result from 60 elements. (△), (○) and (●) are computed from Ijima [14] for $|x|/a_0$, $|Z|/a_0$ and $|\theta|a/a_0$ respectively

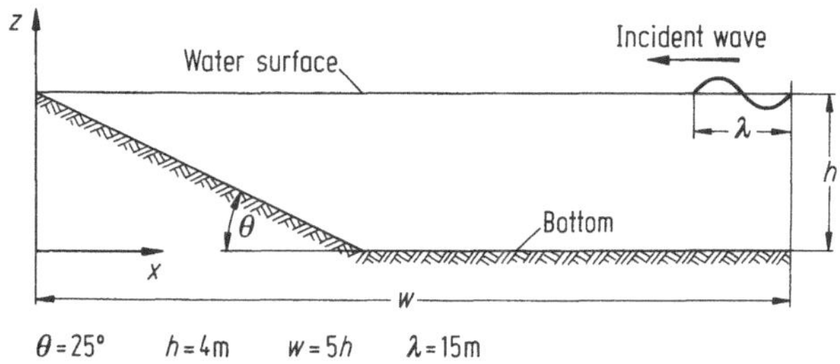


**Fig. 5.16.** The profile of a sloping beach

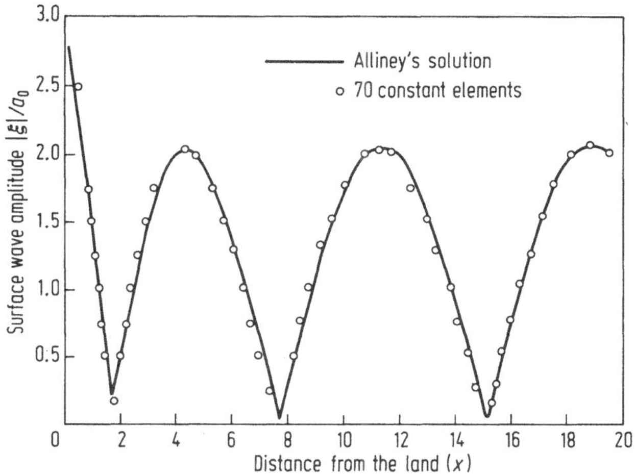


**Fig. 5.17.** The wave amplitude on the free water surface

portions. The horizontal sea bottom was defined as $\Gamma_2$ boundary and the sloping part as $\Gamma_1$. 70 elements were used to discretize the complete boundary and the results for wave amplitude compared against those by Alliney [15] in Figure 5.17.

*(vi) The Square Caisson.* The square caisson of Fig. 5.18 has been studied by Mogridge and Jamieson [16]. They computed the wave force by a circular cylinder approach and compared them to the experimental results. In this chapter, the theory described in section 4 (i) can be applied. 24 constant elements were used and two extreme cases for $\alpha$ angle of incidence were defined ($\alpha = 0°$ and $\alpha = 45°$). The results produced agreed with those of reference [16] as shown in Figures 5.19(a) and 5.19(b).

*(vii) Elliptical Cylinder.* Another case similar to (vi) is the elliptical cylinder of Fig. 5.20, which was studied by Goda and Yoshimura [17] using an analytical method. 16 quadratic elements have been applied here to discretize the boundary and the angles of incidence of the wave was $\alpha = 30°$ and $\alpha = 60°$. Results are given in Fig. 5.21 where they are shown to agree well with those of Goda and Yoshimura.

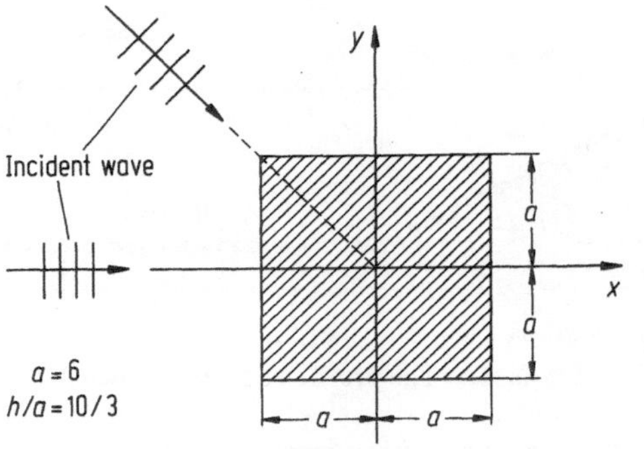


**Fig. 5.18.** The square caisson

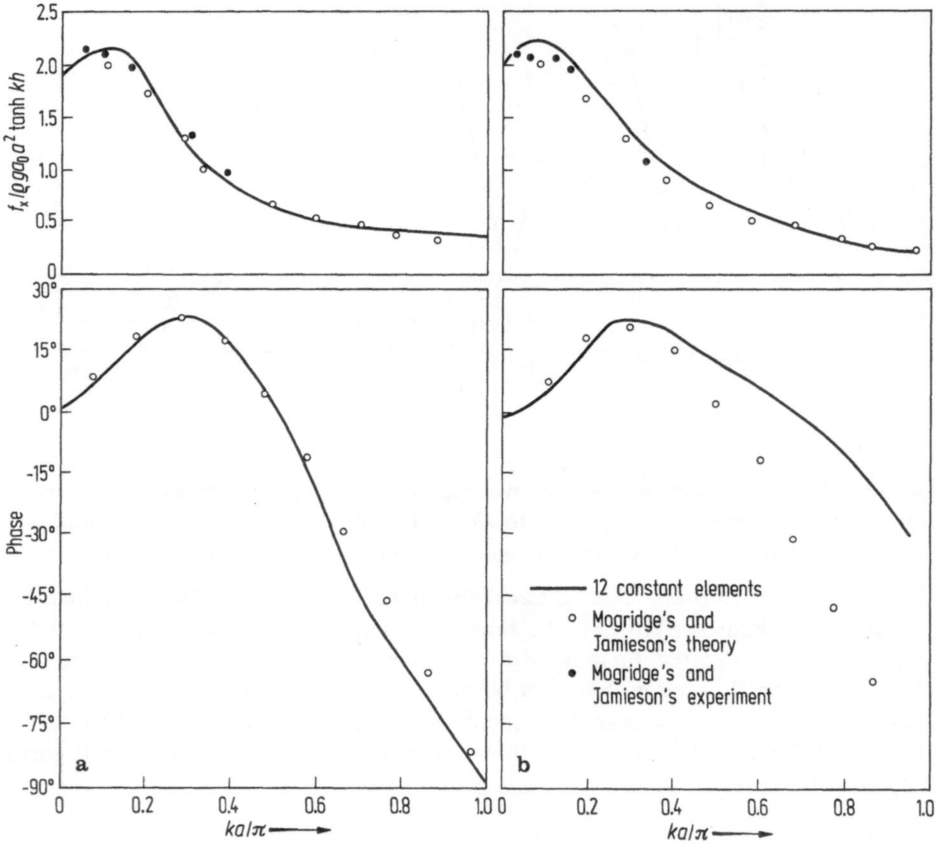


**Fig. 5.19.**
The magnitude and the phase of the horizontal force on a square caisson. **a** $\alpha = 0°$; **b** $\alpha = 45°$

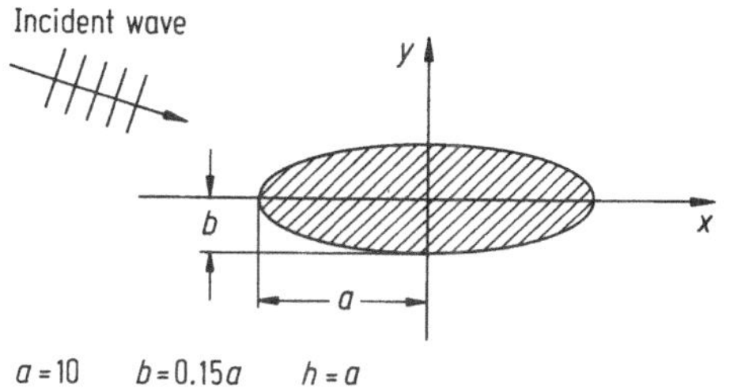


**Fig. 5.20.** Elliptical section $\left(\frac{x}{a}\right)^2 + \left(\frac{y}{b}\right)^2 = 1$

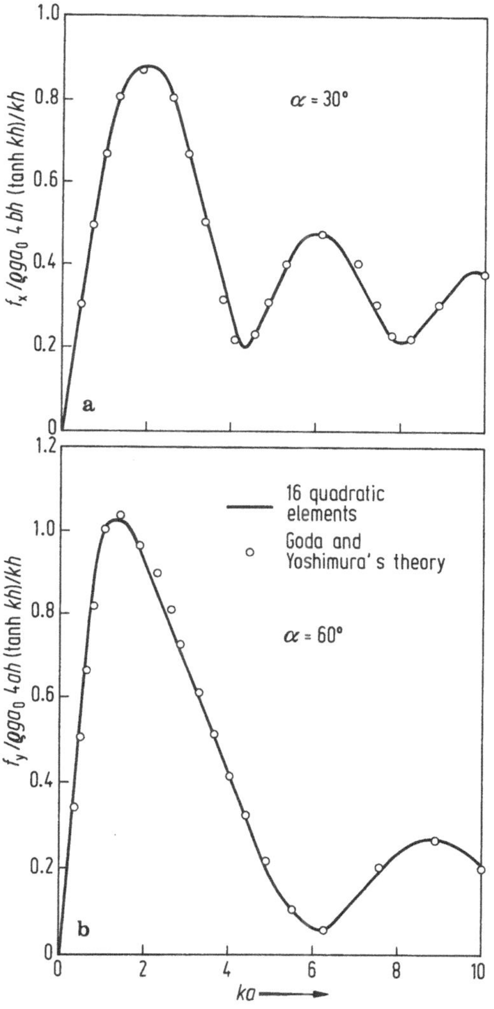


**Fig. 5.21.** The $f_x$ force (**a**) and the $f_y$ force (**b**) on elliptical cylinder by quadratic element

*(viii) Submerged Storage Tank and Tower.* The geometry of this three dimensional structure is given in Fig. 5.22 and in Fig. 5.23 the discretization shows how the planes of symmetry at 60° have been used. The 56 constant triangular elements applied on the one-sixth of the total boundary were used to define the structural surface, the sea bottom, water surface and radiation boundary. Results obtained using boundary elements are compared in Fig. 5.24 with finite element solution due to Zienkiewicz and Bettess [5].

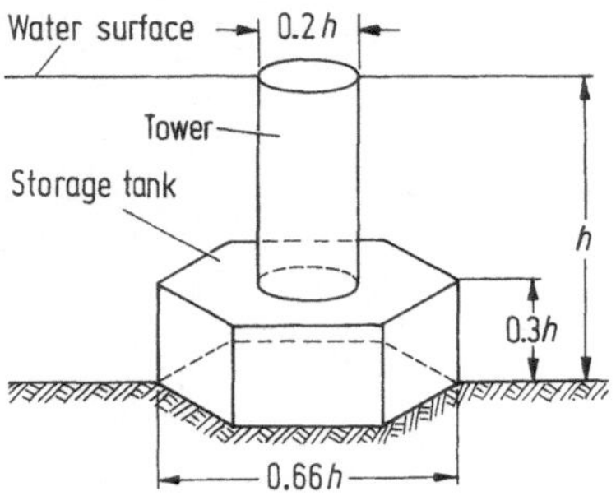


**Fig. 5.22.** The submerged storage tank and tower

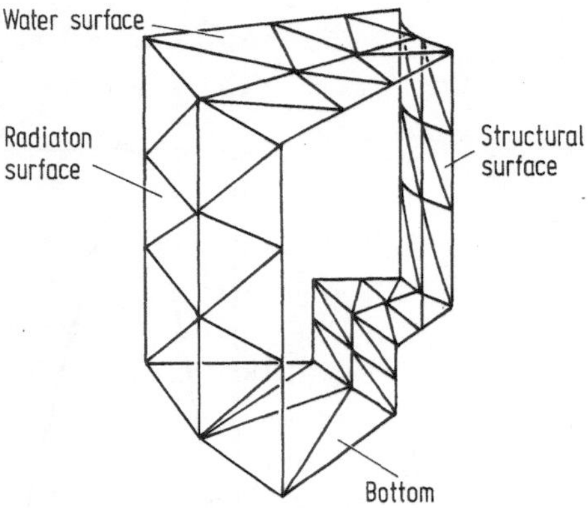


**Fig. 5.23.** The discretization for the submerged storage tank and tower with the relevant boundary conditions

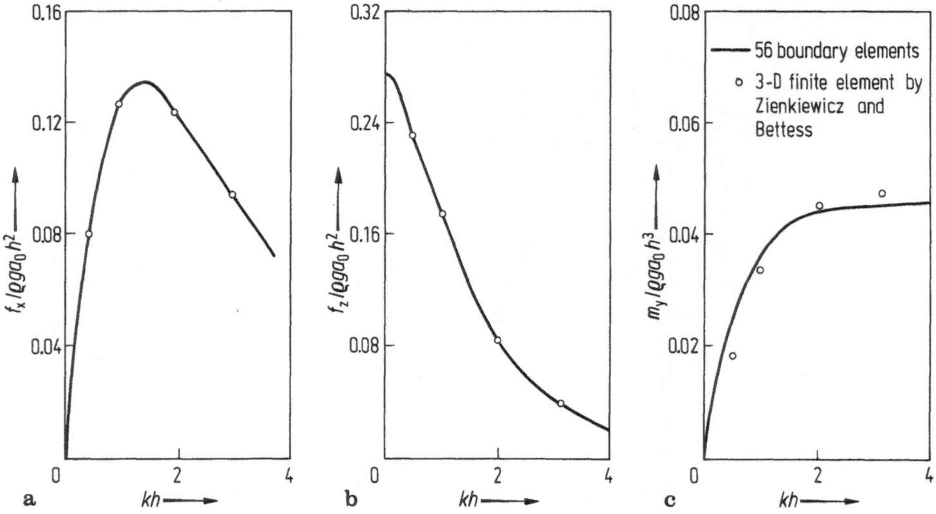


**Fig. 5.24.** The wave forces on the submerged storage tank and tower. **a** Horizontal force. **b** Vertical force. **c** Bending moment about the bottom

*(ix) Floating Hemisphere.* This example has been solved by Garrison [6] using a source distribution technique with a complex fundamental solution. In the present case only the geometry of a vertical section was required and discretization was applied as in Fig. 5.26, making use of rotational symmetry. The three dimensional boundary element mesh for one symmetrical component covers 36° and can be easily obtained by hand or by using an automatic mesh generating code. The results obtained using 30 constant triangular elements are given in Fig. 5.27 and made non-dimensional by the factor $(1/\varrho\, g\, a_0\, a^2)$.

*(x) Surface Elevation Around a Cylinder.* The surface wave elevation around the vertical cylinder in Fig. 5.28 was considered next. In this example, the values of the potential on the water surface are required and it would be difficult to work with a complex fundamental solution as in [6]. By using the theory presented in

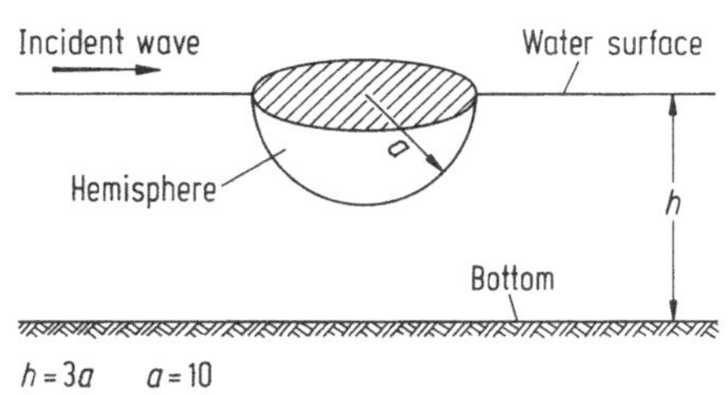


**Fig. 5.25.** The floating hemisphere at the free surface

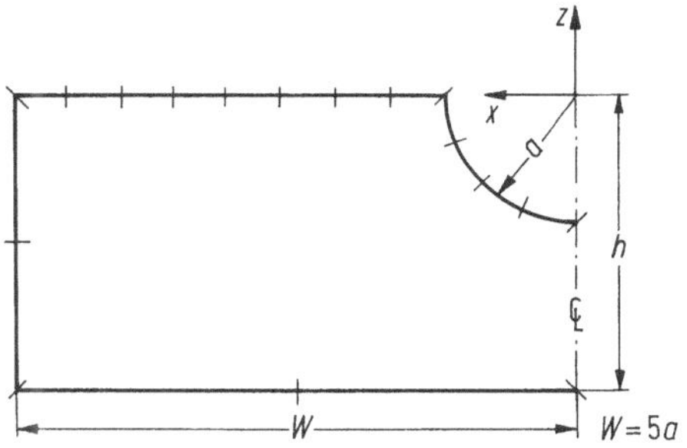


**Fig. 5.26.** A vertical section for the floating hemisphere

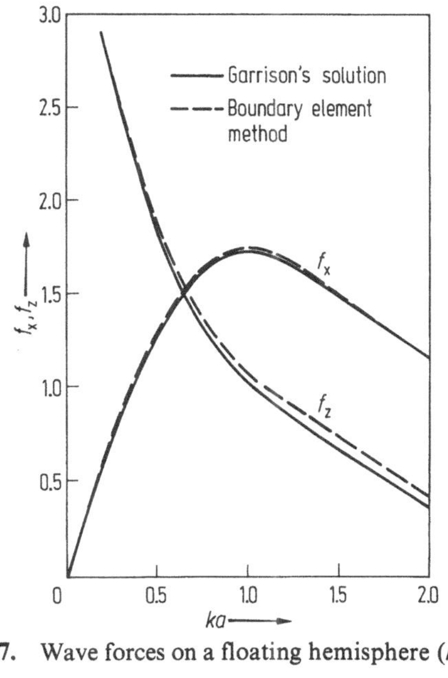


**Fig. 5.27.** Wave forces on a floating hemisphere ($h/a = 3$)

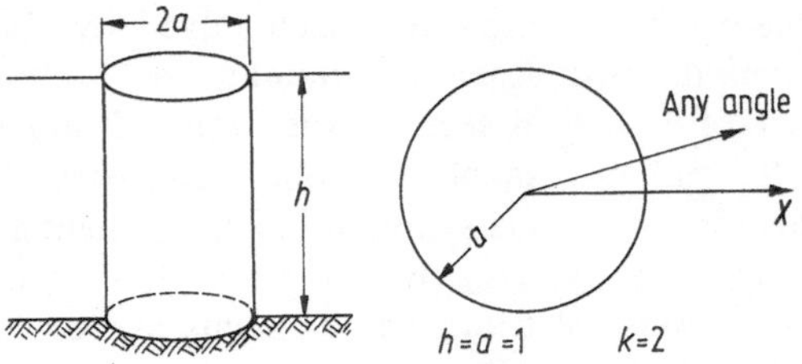


**Fig. 5.28.** The vertical cylinder

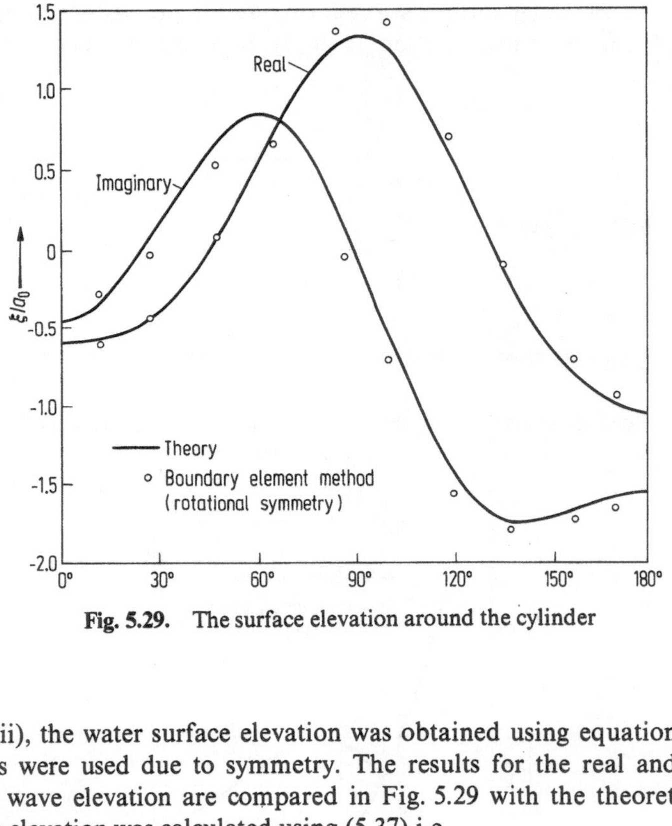


**Fig. 5.29.** The surface elevation around the cylinder

section 4 (iii), the water surface elevation was obtained using equation (5.34) and 40 elements were used due to symmetry. The results for the real and imaginary part of the wave elevation are compared in Fig. 5.29 with the theoretical values. The surface elevation was calculated using (5.37) i.e.

$$\frac{\xi}{a_0} = \text{Real} \times \cos \omega t + \text{Imaginary} \times \sin \omega t$$

*(xi) Condeep Type Platform.* The example shown in Fig. 5.30 represents the discretization into boundary elements of 1/6th of a Condeep type structure. The analysis was carried out using 74 constant elements. The wave force on the top part of the cylinder was obtained analytically using Morison's equation and the analysis concentrated in the bottom portion of the platform. Total forces are presented in Fig. 5.31 for a wave period $T = 15.5$ seconds i.e. for the 100 years wave.

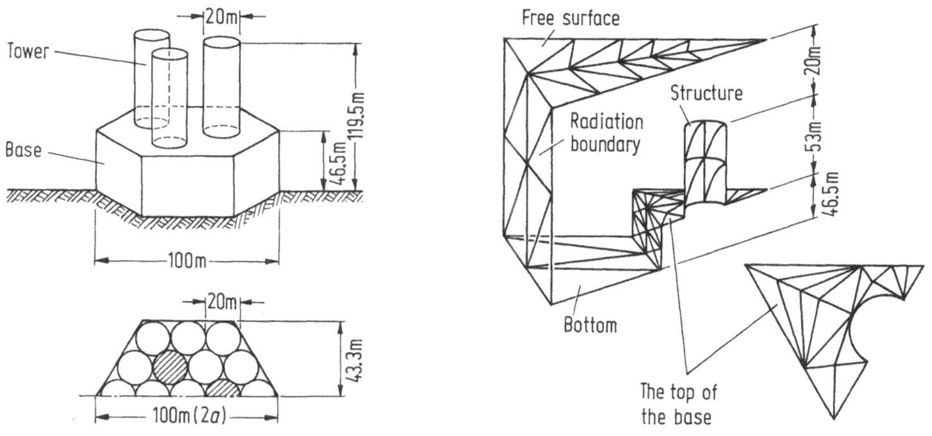


**Fig. 5.30.** The boundary element mesh for the Condeep type structure

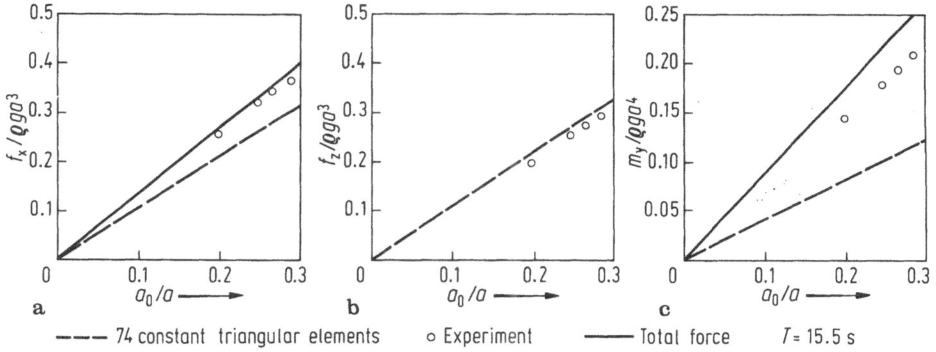


**Fig. 5.31.** The wave forces on a Condeep platform. **a** Horizontal force. **b** Vertical force. **c** Moment about bottom

# 5.7 Conclusions

This chapter shows how the boundary element method can be applied to solve wave diffraction-radiation type problems. By using the fundamental solution corresponding to Laplace's equation results such as pressure, wave forces, hydrodynamic coefficients, water surface elevation and motion response of structures can be easily obtained.

Two dimensional problems can generally be solved without great usage of computer time. However, computer time becomes crucial when solving three dimensional cases. For such cases the codes can be made more efficient and easy to use (i.e. less data is required) by exploiting the symmetry of the structure. The chapter shows how this can be done in practice.

Several examples are considered at the end of the chapter to illustrate the applications of the theory presented and to demonstrate some of the special advantages of using boundary elements for these cases. The examples cover a variety of important two and three dimensional problems, with and without symmetry.

## References


1 Newton, R. E., Finite element analysis of two dimensional added mass and damping. Ch. 11, In: *Finite Elements in Fluids – Vol. 1* ed. by R. H. Gallagher et al. A Wiley-Interscience publication, 219–232, 1975

2 Berkhoff, J. C. W., Computation of combined refraction-diffraction. Ch. 24, 13th Conf. on Coastal Eng., **1,** 471–490, 1972

3 Bai, K. J., Diffraction of oblique waves by an infinite cylinder. J. Fluid Mech., **68,** 513–535, 1975

4 Yu, D. K. P., Chen, H. S., Mei, C. C., A hybrid element method for diffraction of water-waves by three-dimensional bodies. Int. J. Num. Meth. Eng., **12,** 245–266, 1978

5 Zienkiewicz, O. C., Bettess, P., Fluid structura dynamic interaction and wave forces – An introduction to numerical treatment. Int. J. Num. Meth. Eng., **13,** 1–16, 1978

6 Garrison, C. J., Hydrodynamic loading of large offshore structures: three dimensional source distribution methods. Ch. 3, *Numerical Methods in Offshore Engineering,* ed. by O. C. Zienkiewicz et al., A Wiley-Interscience Publication, 1978

7 Hogben, N., Standing, R. C., Wave loads on large bodies. Int. Symposium on the dynamics of marine vehicles and structures in waves, Inst. Mech. Eng., 258–277, London 1974

8 Eatock-Taylor, R., *Generalised Hydrodynamic Forces on Vibrating Offshore Structures by Wave Diffraction Technique.* Offshore Structures Engineering edited by F. L. L. B. Carneiro et al., Pentech Press, London, 249–274, 1977

9 Isaacson, M. de St. Q., Vertical cylinder of arbitrary section in waves. J. Waterway, Port, Coastal and Ocean Div., ASCE, 309–324, 1978

10 Brebbia, C. A., *The Boundary Element Method for Engineers.* Pentech Press, London 1978

11 Au, M. C., Brebbia, C. A., Computation of Wave Forces on three dimensional offshore structures. Proc. of 4th Int. Seminar on Boundary Elements, Southampton and *Boundary Element Methods in Engineering.* Springer-Verlag Berlin, Heidelberg, New York 1982

12 Zheng Xiaozhong, Bao Gang, Sun Shuxun, Application of Group theory to vibrational analysis of shell structures with space rotation symmetry. Proc. Int. Conf. of Finite Element Methods, 1982, Shanghai, China, edited by He Guangquian and Y. K. Cheung

13 Naftzger, R. A., Chakrabarti, S. K., Scatterin of waves by two-dimensional circular obstacles in finite water depths. J. Ship Research, **23,** No. 1 March, 32–42, 1979

14 Ijima, T., Chou, C. R., Yoshida, A., Method of analysis for two-dimensional water wave problems. Pap. 156, Proc. 15th Coastal Eng. Conf., ASCE, 2717–2736

15 Alliney, S., A numerical study of water waves on sloping beaches. Appl. Math. Modelling, **5,** Oct. 321–328, 1981

16 Mogridge, G. R., Jamieson, W. W., Wave forces on square caissons. Pap. 133, Proc. 15th Coastal Eng. wlonf., ASCE, 2271–2289

17 Goda, Y., Yoshimura, T., Wave forces on a vessel tied at offshore dolphins. Pap. 96, Proc. 13th Coastal Eng. Conf. Vol. IV, 1723–1742, 1972

18 Garrison, C. J., Tørum, A., Iversen, C., Leivseth, S., Ebbesmeger, C. C., Wave forces on large volume structures – a comparison between theory and model tests. Proc. 7th Offshore Technology Conf., Paper No. 2137, Houston 1974

# Chapter 6

# Interelement Continuity in the Boundary Element Method

*by C. Patterson and M. A. Sheikh*

## 6.1 Introduction

It is well known that homogeneous elliptic field problems may be alternatively posed as infinite systems of boundary integral equations obtained by using a suitable family of kernel functions and integration by parts [1], [2]. A determinate system of non-homogeneous linear algebraic equations is obtained therefrom by discretizing the system, using elements defined after the manner of finite elements and a finite member of Kernel functions [1]. The result is a practical and powerful numerical method for the solution of elliptic field problems which may easily be generalized to rival the finite element method in its range of applicability.

As compared with the finite element method the boundary element method is attractive in that freedoms occur only on the boundary of the problem lending relative ease in problem definition and a possible reduction in overall freedoms and calculational workload required to achieve a practical numerical solution.

Whilst there are obvious similarities between the two methods they are distinctly different in principle, particularly as regards valid discretization of the initial problem. In the finite element method certain domain integrals describe energies, such as strain energy, which physically are positively definite and have to remain so in the discretization procedure [3]. This requirement mandates minimal interelement continuity which if not realized leads to indefinite integrals and an indeterminate algebraic system of equations. In the boundary element method there is no such constraint on the discretization process; indeed, interelement continuity is not a necessary feature. That this is so is abundantly evident in that constant boundary elements, which necessarily do not respect interelement continuity, may succesfully be used.

An ascending hierarchy of boundary elements, linear, quadratic etc., may be defined just as with finite elements. Until recently it has been the widely accepted practice to impose interelement continuity when using such boundary elements, again, presumably, in analogy with finite elements.

Whilst this action is usually satisfactory throughout the boundary element mesh it can lead to modelling difficulties. Thus, at a geometric singularity, corner, edge etc., the normal to the boundary is not defined and yet is required in defining the flux-like variable of the problem, such as heat flux, traction etc. Again, where there is an abrupt change in the imposed boundary variable, say from temperature to heat flux, there is a difficulty in nodal assignment. Also, a difficulty arises when using subdomains at points where several elements meet [4].

These difficulties arise when using continuous elements because freedom nodes must be taken at the singular locations. If complete interelement continuity is abandoned then freedom nodes may be moved inside the boundary element so that they do not occur at the previously cited locations. The above mentioned problems then disappear.

In this chapter suitable discontinuous elements are presented for two and three dimensional applications and case studies are given to demonstrate their efficacy. Additionally, in the case of three dimensional problems abandonment of interelement continuity allows freely graded, mismatched, meshes to be used advantageously.

## 6.2 Continuous Elements

*Boundary Equations*

The Laplacian problem is typical of elliptic problems and is used here as an example of the development of boundary integral equations and of boundary elements. Suppose a problem is defined on a domain $\Omega$ with boundary $\Gamma$. Then the problem is stated by the requirement that the governing homogeneous field equations be satisfied in $\Omega$ together with appropriate boundary conditions viz:

$$\nabla^2 u(\mathbf{x}) = 0, \quad \mathbf{x} \in \Omega \tag{6.1}$$

$$\begin{aligned} u(\mathbf{x}) &= \bar{u}(\mathbf{x}), & \mathbf{x} \in \Gamma_1 \\ \frac{\partial u}{\partial n}(\mathbf{x}) &= \overline{\frac{\partial u}{\partial n}}(\mathbf{x}) & \mathbf{x} \in \Gamma_2 \end{aligned} \tag{6.2}$$

Here $\Gamma_1 + \Gamma_2 = \Gamma$, the derivative is with respect to the normal, and $\bar{u}$ and $\frac{\partial \bar{u}}{\partial n}$ are the given boundary values. Let $w$ satisfy the governing equation in $\Omega$;

$$\nabla^2 w(\mathbf{x}) = 0, \quad \mathbf{x} \in \Omega \tag{6.3}$$

Then

$$\int_\Omega w \, \nabla^2 u \, d\Omega = 0 \tag{6.4}$$

and after integrating equation (6.4) by parts twice and taking account of equation (6.3) it follows that

$$-\int_\Gamma u \frac{\partial w}{\partial n} d\Gamma + \int_\Gamma \frac{\partial u}{\partial n} w \, d\Gamma = 0 \tag{6.5}$$

Equation (6.5) is a boundary integral equation which is satisfied by the unknown function. A system of boundary integral equation is obtained if a sequence of linearly independent functions $w$ is taken. The infinite system which results on taking a different kernel function $w$ for each point of $\Gamma$ is determined for $u$ on $\Gamma$ and results in an alternate statement of the original field problem. In the Boundary

Element Method such a system of integral equations is found by suitable choice of kernel function. This infinite system is then discretized by discretizing the unknown functions, $u$, $\frac{\partial u}{\partial n}$, using finite elements and discretizing the integral equation system by taking as many kernel functions as there are undetermined shape functions.

In the usual form of the boundary element method the kernel functions are obtained using the fundamental solution $u^*$ which satisfies

$$\nabla^2 u^*(\mathbf{x}) = -\delta(\mathbf{x}-\mathbf{x}') \tag{6.6}$$

where $\delta(\mathbf{x}-\mathbf{x}')$ is the Dirac delta function. Explicitly in three dimensions

$$u^*(\mathbf{x}) = \frac{1}{4\pi r(\mathbf{x},\mathbf{x}')} \tag{6.7}$$

where $r$ is the distance between $\mathbf{x}$ and $\mathbf{x}'$.

Clearly, the fundamental solution is singular at the source point $\mathbf{x}'$. If the source point is taken on the boundary the resultant boundary integral equation is singular and takes the form

$$Cu(\mathbf{x}') + \int_\Gamma u \frac{\partial u^*}{\partial n} d\Gamma - \int_\Gamma \frac{\partial u}{\partial n} u^* \delta\Gamma = 0 \tag{6.8}$$

where the discrete term is the Cauchy prinipal value of the singular integral and $c$ is a known constant dependent on the dimensionality of the problem and the regularity, or otherwise, of the surface at $\mathbf{x}'$. On taking account of the known boundary values (eqn. (6.2)), eqn. (6.8) becomes

$$-\int_{\Gamma_2} u \frac{\partial u^*}{\partial n} d\Gamma + \int_{\Gamma_1} \frac{\partial u}{\partial n} u^* d\Gamma - C u(\mathbf{x}') = \int_{\Gamma_1} \bar{u} \frac{\partial u^*}{\partial n} d\Gamma - \int_{\Gamma_1} \overline{\frac{\partial u}{\partial n}} u^* d\Gamma \tag{6.9}$$

where: the right hand side integrals are known, the discrete term is known if $\mathbf{x} \in \Gamma_1$ and not otherwise, and the left hand side integrals involve the unknown $u$, $\frac{\partial u}{\partial n}$. On taking a finite element representation of $u$, $\frac{\partial u}{\partial n}$ equation (6.9) yields an inhomogeneous linear algebraic equation among the unknown expansion coefficients and on locating the singular point $\mathbf{x}'$ in turn at each freedom node a determinate linear algebraic system ensues.

It is evident that the particular choice of kernel function, the fundamental solution with singularity located on the boundary, is not unique and that other strategies may be employed. If the singular point of the kernel function is taken outside $\Omega$ then the resultant integral equation is *regular* and is given by equation (6.8) with $C = 0$. Again, on taking a distinct location of the singular point for each freedom node, a determinate linear algebraic system results. This strategy has already been successfully employed [5], [6], [7]. On taking the singular point outside the domain a family of higher order kernels can be used, which would in the original approach give divergent integrals. Thus if $D$ is any differential operator which commutes with the operator of the governing equations ($\nabla^2$ here) $D u^*$ is a possible higher order kernel function.

*Discretization*

In modelling a planar problem the boundary is partitioned into a series of segments which are either straight or curved. The straight segments are most easily defined but usually at the cost of poor geometric representation of the problem geometry. Constant or polynomial shape functions for the freedom pairs $u, \dfrac{\partial u}{\partial n}$ (or displacement and traction in a stress problem) are easily defined (Fig. 6.1 a).

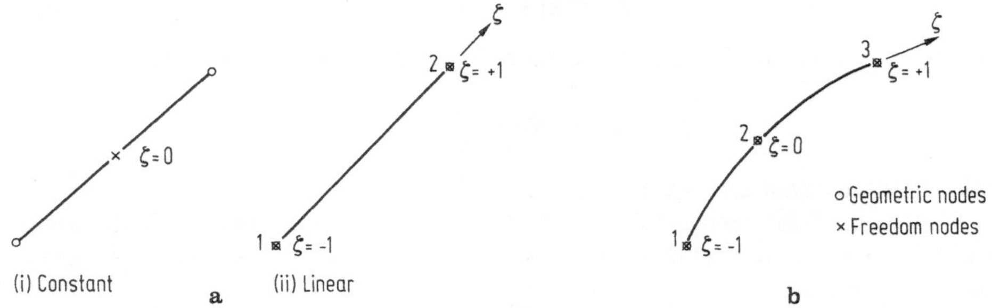


**Fig. 6.1** **a** Straight line element. **b** Quadratic geometric element

Usually a better geometric representation results if the geometric elements are themselves defined using shape functions thereby allowing the use of curved elements. A common choice is a quadratic geometric element (Fig. 6.1 b) defined by three geometric nodes $\mathbf{x}_i^{\varkappa}$; $\varkappa = 1, 2, 3$ and associated shape functions $M^{\varkappa}(\zeta)$. Then the geometry of the element is given by [9]

$$x_i(\zeta) = \sum_{\varkappa=1}^{3} M^{\varkappa}(\zeta)\, x_i^{\varkappa}, \quad i \in (1, 2), \quad \zeta \in [-1, +1] \tag{6.10}$$

where

$$\begin{aligned} M^1(\zeta) &= \tfrac{1}{2}\zeta(\zeta - 1) \\ M^2(\zeta) &= 1 - \zeta^2 \\ M^3(\zeta) &= \tfrac{1}{2}\zeta(\zeta + 1) \end{aligned} \tag{6.11}$$

An ascending hierarchy of shape functions may be used [10] for the freedom functions. Here a linear variation of the field quantities $u, \dfrac{\partial u}{\partial n}$ is assumed. Then on taking freedom nodes at the element ends ($\zeta = \pm 1$) the assumed forms of the freedom functions are

$$\begin{aligned} u(\zeta) &= F_1 u_1 + F_2 u_2 \\ \frac{\partial u}{\partial n}(\zeta) &= F_1 \frac{\partial u_1}{\partial n} + F_2 \frac{\partial u_2}{\partial n} \end{aligned} \tag{6.12}$$

where $u_1, \dfrac{\partial u_1}{\partial n}$ etc. are the nodal values and

$$F_1 = \tfrac{1}{2}(1 - \zeta); \quad F_2 = \tfrac{1}{2}(1 + \zeta) \tag{6.13}$$

Suppose that the boundary is discretized into $N$ elements, then the discretized form of the singular boundary integral equation (Eqn. 6.8) is

$$c^i u^i + \sum_{j=1}^{N} \int_{\Gamma_j} u^{(j)} \frac{\partial u^*}{\partial n} d\Gamma = \sum_{j=1} \int_{\Gamma_j} \frac{\partial u^{(j)}}{\partial n} u^* \, d\Gamma \tag{6.14}$$

where $u^*$, $\frac{\partial u^*}{\partial n}$ is the fundamental solution with singular point at $\mathbf{x}^i$; $u^{(j)}$, $\frac{\partial u^{(j)}}{\partial n}$ are the interpolation field functions of the $j^{\text{th}}$ element $\Gamma_j$.

As previously mentioned it is not necessary to locate the singular point of the kernel function $u^*$ on the boundary. If instead it is located at some exterior point the discretized form of the boundary integral equation is

$$\sum_{j=1}^{N} \int_{\Gamma_j} u^{(j)} \frac{\partial u^*}{\partial n} d\Gamma = \sum_{j=1}^{N} \int_{\Gamma_j} \frac{\partial u^{(j)}}{\partial n} u^* \, d\Gamma \tag{6.15}$$

Now the integrals are no longer singular and higher order kernel functions may be used.

On expressing the field functions explicitly, as in eqn. (6.12), the regular boundary integral equation becomes

$$\sum_{j=1}^{N} \int_{\Gamma_j} [F_1 \; F_2] \frac{\partial u^*}{\partial n} d\Gamma \begin{bmatrix} u_1 \\ u_2 \end{bmatrix} = \sum_{j=1}^{N} \int_{\Gamma_j} [F_1 \; F_2] \, u^* \, d\Gamma \begin{bmatrix} \frac{\partial u_1}{\partial n} \\ \frac{\partial u_2}{\partial n} \end{bmatrix} \tag{6.16}$$

or equivalently

$$\sum_{j=1}^{N} (h_{i,j-1} + h_{i,j}) \, u_j = \sum_{j=1}^{N} (g_{i,j-1} + g_{i,j}) \frac{\partial u}{\partial n} j \tag{6.17}$$

where

$$\begin{aligned} h_{i,j-1} &= \int_{\Gamma_{j-1}} F_1 \frac{\partial u^*}{\partial n} d\Gamma; \quad & h_{i,j} &= \int_{\Gamma_j} F_2 \frac{\partial u^*}{\partial n} d\Gamma \\ g_{i,j-1} &= \int_{\Gamma_{j-1}} F_1 u^* \, d\Gamma; \quad & g_{i,j} &= \int_{\Gamma_j} F_2 u^* \, d\Gamma, \end{aligned} \tag{6.18}$$

a global numbering system has been applied to the nodal freedoms and the index '$i$' labels the kernel function. In equation (6.17) $N$ of the freedom variables are prescribed by the boundary conditions, one at each node, so that the equation is a non-homogeneous linear algebraic equation in the remaining $N$ unknowns. On employing $N$ linearly independent kernel functions a determinate linear algebraic system is obtained for the unknown paramters.

The procedure outlined appropriate to continuous elements is basically very simple but it encounters difficulties as discussed in the next section.

### *Modelling Difficulties*

When continuous elements are employed freedom nodes are located at the element edge. Usually this poses no special problem but the element edge may coincide with a discontinuity in the problem definition resulting in modelling ambiguities.

Thus in a two dimensional problem at a corner the normal to the boundary is not defined; nevertheless in a Laplacian problem the freedoms are the field

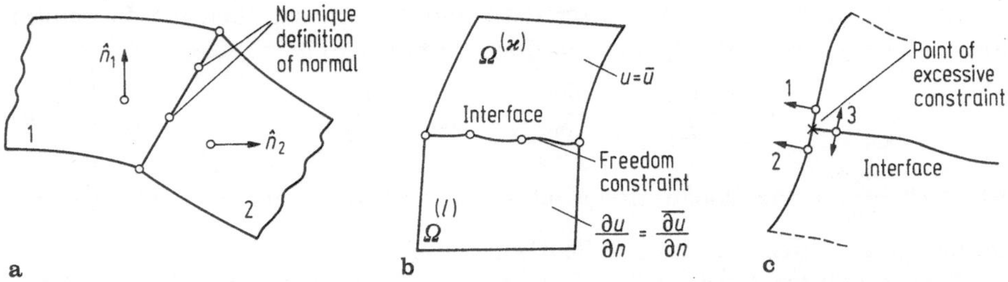


**Fig. 6.2.** **a** Non-smooth surface. **b** Freedom constraint at an interface where boundary condition type changes. **c** Freedom constraint at a point where several surfaces meet

function $u$ and its normal derivative $\frac{\partial u}{\partial n}$. Clearly the normal derivative is not defined at the corner. Other elliptic problems exhibit the same feature since the boundary equations result from integrations by parts, for instance in an elastostatic problem the surface traction is not defined at a corner. In practice two approaches have been adopted. First the problem has been ignored in that $\frac{\partial u}{\partial n}$ has been treated as a scalar variable at the corner with resulting erroneous results (an example of this appears in ref. [11]). Second the mesh has been rendered discontinuous at the corner by not joining adjacent elements and leaving the freedom nodes close together on either side of the geometric singularity [1]. Good approximate numerical results can then be obtained at the cost of incomplete modelling – the corner region is ignored. But in the singular method equations are generated using fundamental solutions with singularities at both adjacent nodes in turn which is numerically objectionable since strong linear dependence is induced in the algebraic system because of the effective linear dependence of kernel functions. In three dimensional problems the same feature arises at a boundary edge or cusp (Fig. 6.2a).

Another problem arises at the interface between $\Gamma_1$ and $\Gamma_2$, that is where the type of boundary condition changes. A freedom node on this interface is such that for one element is known and for the other $\frac{\partial u}{\partial n}$ is known (see Fig. 6.2b).

Yet again, when the numerical model is derived using subdomains [4] a similar problem arises where several subdomains meet or where two subdomains meet the exterior boundary (see Fig. 6.2c). The latter problems are discussed by Lachat, Chaudonneret and Wardle et al., [4], [12], [13].

## 6.3 Planar Discontinuous Elements

The problematic geometric features, such as a corner, discussed in the previous section are clearly inherent but the associated modelling problems only occur because of the insistence of locating freedom nodes at element edges and thereby at the singularities of problem definition. Now there is no difficulty in defining boundary elements with only interior freedom nodes. However, such elements then present difficulties in maintaining interelement continuity. Whilst such continuity may bring advantages, in the boundary element method it is not necessary. Indeed, the viability of constant elements manifestly demonstrates this fact. Studies with discontinuous elements have shown them to be practicable without the need for special strategies to cope with the previously cited problems. An obvious further development is the introduction of partially discontinuous elements, which permit interelement continuity where deemed desirable and not otherwise. In this and the next section such elements are discussed for two and three dimensional applications.

### *Discontinuous Elements*

Here, the geometry of the problem is defined, as usual, using straight line or curved elements. Again, a hierarchy of linear or other higher order variation of the field function and its flux-like variable may be taken. However, these freedoms are not defined at the geometric nodes but at "freedom nodes" which can, in principle, be freely chosen anywhere over the element. In the present assignment these freedom nodes are chosen so that in a mesh of congruent elements these nodes are equally spaced. For simplicity, the formulation of the planar problem using a discontinuous linear element is presented here, followed by a brief description of other higher order elements.

*(a) Discontinuous Linear Element* (Fig. 6.3a). Here, the freedoms are defined not at the geometric nodes, that is at $\zeta = \pm 1$, but at $\zeta = \pm 1/2$, so that the new interpolation functions are:

$$F_1 = (\tfrac{1}{2} - \zeta); \quad F_2 = (\tfrac{1}{2} + \zeta) \tag{6.19}$$

The algebraic expression of the boundary integral equation [Eq. (6.15)] for kernel function '$i$' now reads

$$\sum_{j=1}^{N} (h_{i,2j-1}\, u_{2j-1} + h_{i,2j}\, u_{2j}) = \sum_{j=1}^{N} \left\{ g_{i,2j-1} \left(\frac{\partial u}{\partial n}\right)_{2j-1} + g_{i,2j} \left(\frac{\partial u}{\partial n}\right)_{2j} \right\} \tag{6.20}$$

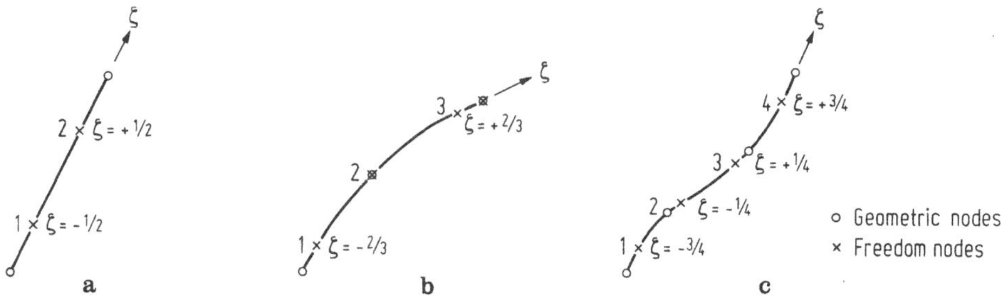


**Fig. 6.3.** **a** Discontinuous linear element. **b** Discontinuous quadratic element. **c** Discontinuous cubic element

where

$$h_{i,2j-1} = \int_{\Gamma_j} F_1 \frac{\partial u^*}{\partial n} d\Gamma; \quad h_{i,2j} = \int_{\Gamma_j} F_2 \frac{\partial u^*}{\partial n} d\Gamma$$
$$g_{i,2j-1} = \int_{\Gamma_j} F_1 u^* d\Gamma; \quad g_{i,2j} = \int_{\Gamma_j} F_2 u^* d\Gamma \tag{6.21}$$

and the summation $j$ is over each element with two nodes per element. Of the $4N$ variables $u_j$, $\left(\frac{\partial u_j}{\partial n}\right)$, $2N$ are fixed by boundary conditions, one at each freedom node. On taking $2N$ linearly independent Kernel functions a determinate non-homogeneous linear algebraic system is obtained.

*(b) Discontinuous Quadratic Elment* (Fig. 6.3b). On taking equispaced freedom nodes as before, interpolation functions are:

$$\begin{aligned} F_1 &= \zeta(9/8\,\zeta - 3/4) \\ F_2 &= (1 - 3/2\,\zeta)\,(1 + 3/2\,\zeta) \\ F_3 &= \zeta(9/8\,\zeta + 3/4) \end{aligned} \tag{6.22}$$

*(c) Discontinuous Cubic Element* (Fig. 6.3c). The interpolation functions for the four freedoms are:

$$\begin{aligned} F_1 &= (\zeta - 1/4)\,(\zeta + 1/4)\,(1 - 4/3\,\zeta) \\ F_2 &= (4\,\zeta - 1)\,(\zeta - 3/4)\,(\zeta + 3/4) \\ F_3 &= (-4\,\zeta - 1)\,(\zeta - 3/4)\,(\zeta + 3/4) \\ F_4 &= (\zeta - 1/4)\,(\zeta + 1/4)\,(1 + 4/3\,\zeta) \end{aligned} \tag{6.23}$$

The implementation of these elements and generation of a determinate algebraic system follows as for the linear element.

*Partially Discontinuous Elements*

As will be seen from the applications discussed below the discontinuous elements obviate the previous modelling difficulties at the price of introducing unnecessary freedoms where the geometry is regular. In order to avoid this shortcoming partially discontinuous elements are required which permit the use of discontinuous interpolation functions only at selected points, for instance corners. Such elements are easily developed having freedom nodes at the usual locations (including $\zeta = \pm 1$) where continuity is required, with the freeom nodes brought into the element, as for the discontinuous element, where a discontinuous interpolation is required.

*(a) Partially Discontinuous Linear Element.* Two variants are required to give discontinuity at $\zeta = +1$ or $-1$, with the freedom nodes as in Fig. 6.4a. The interpolation functions are:

| *Type 1* | *Type 2* | |
|---|---|---|
| $F_1 = 2/3\,(1/2 - \zeta)$ | $F_1 = 2/3\,(1 - \zeta)$ | |
| $F_2 = 2/3\,(1 + \zeta)$ | $F_2 = 2/3\,(1/2 + \zeta)$ | (6.24) |

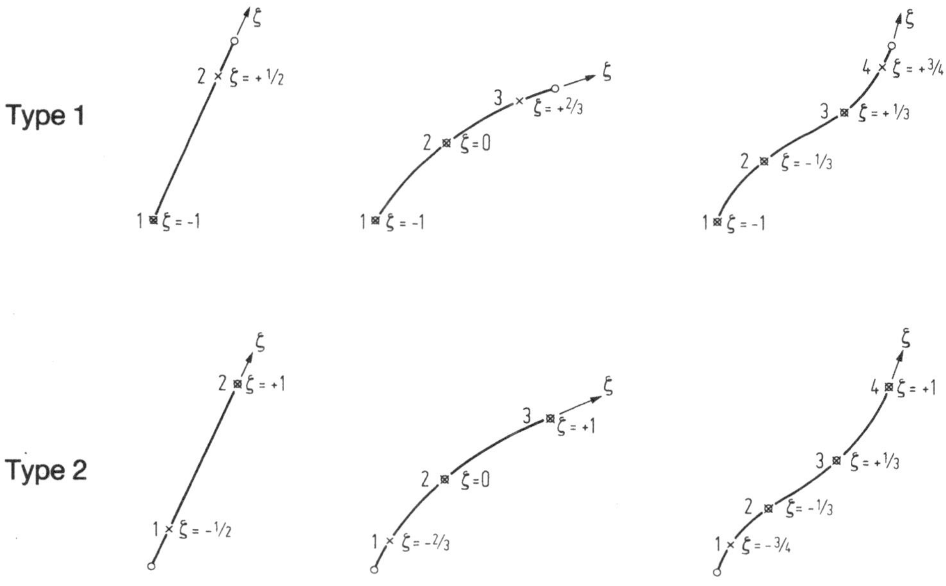


a b c

**Fig. 6.4.** **a** Partially discontinuous linear element. **b** Partially discontinuous quadratic element. **c** Partially discontinuous cubic element

*(b) Partially Discontinuous Quadratic Element.* Again there are two variants, Fig. 6.4b, with interpolation functions:

| *Type 1* | *Type 2* | |
|---|---|---|
| $F_1(\zeta) = 3/5\,\zeta(\zeta - 2/3)$ | $F_1(\zeta) = 9/10\,\zeta(\zeta - 1)$ | |
| $F_2(\zeta) = 3/2(1 + \zeta)(2/3 - \zeta)$ | $F_2(\zeta) = 3/2(1 - \zeta)(2/3 + \zeta)$ | (6.25) |
| $F_3(\zeta) = 9/10\,\zeta(\zeta + 1)$ | $F_3(\zeta) = 3/5\,\zeta(\zeta + 2/3)$ | |

*(c) Partially Discontinuous Cubic Element.* Appropriate to Fig. 6.4c, the interpolation functions are:

| *Type 1* | *Type 2* |
|---|---|
| $F_1(\zeta) = -9/14(\zeta - 3/4)(\zeta^2 - 1/9)$ | $F_1(\zeta) = -576/455(\zeta - 1)(\zeta^2 - 1/9)$ |
| $F_2(\zeta) = -27/13(\zeta - 1/3)(3/4 - \zeta)(1 + \zeta)$ | $F_2(\zeta) = -27/10(\zeta - 1/3)(3/4 + \zeta)(1 - \zeta)$ |
| $F_3(\zeta) = 27/10(\zeta + 1/3)(3/4 - \zeta)(1 + \zeta)$ | $F_3(\zeta) = 27/13(\zeta + 1/3)(3/4 + \zeta)(1 - \zeta)$ |
| $F_4(\zeta) = 576/455(1 + \zeta)(\zeta^2 - 1/9)$ | $F_4(\zeta) = 9/14(\zeta + 3/4)(\zeta^2 - 1/9)$ |

The implementation of these elements in the discretized boundary integral equations is similar to that of the continuous and discontinuous elements.

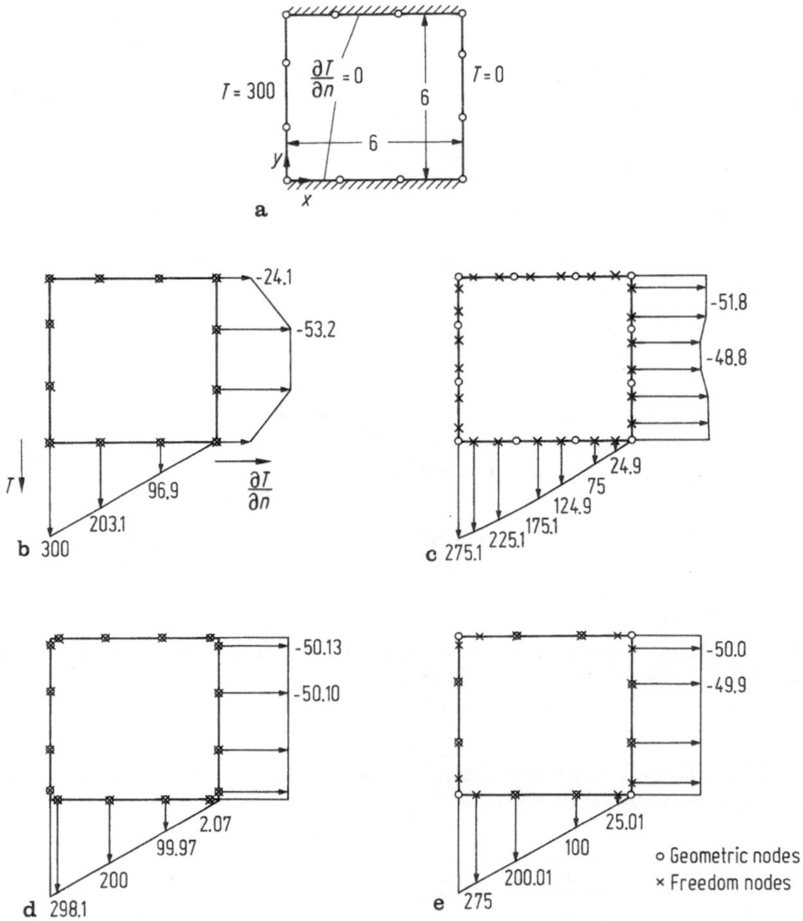


**Fig. 6.5.** **a** Numerical model for the prism problem. **b** Solution obtained using continuous linear elements. **c** Solution obtained using discontinuous linear elements. **d** Solution obtained using continuous linear elements with double nodes at corners. **e** Solution obtained using partially discontinuous elements

*Applications*

Three two-dimensional problems, thermal, fluid flow and stress, are considered which illustrate the relative merits of continuous and discontinuous elements. For each problem the solution has been obtained using, (i) continuous elements, (ii) continuous elements with double nodes at geometric singularities, (iii) discontinuous elements and (iv) partially discontinuous elements at corners and load discontinuities with continuous elements elsewhere. Two noded linear elements are used throughout. In each problem, as in the original references, cases (i) and (ii) are examined in conjunction with the conventional (singular) boundary element method while in cases (iii) and (iv) the regular boundary element method is used. All these problems subsequently have been analysed for all four cases using both methods with the result that both methods yield sensibly the same results in each case, the variations being due solely to case considered.

*(a) Steady Heat Conduction through a Rectangular Prism.* In this problem two opposite faces of the prism are insulated while the other faces are held at different constant temperatures giving a linear temperature variation between the non-insulated faces. While this problem is very simple it has, nevertheless, been of considerable use in tests [1], [11]. The numerical model employed, Fig. 6.5a, has three equal linear elements on each face. Computed results for the four cases considered are given in Figs. 6.5b–e.

*(b) Steady Inviscid Laminar Flow in a Channel Past a Disc.* A finite channel is considered with the disc located symmetrically in the channel with its normal along the channel. Only a quarter of the domain need be analysed on account of the two symmetry planes present. A graded 32 linear element mesh, Fig. 6.6a, is taken together with a stream function, $\psi$, representation. Since the flow velocity is infinite at the tip of the disc, this problem is singular [14], and so it presents a severe numerical test. Computed values of the stream velocities at the channel wall, $BC$, and over the disc face, $DO$, are given in Figs. 6.6b and 6.6c respectively.

*(c) Stress Concentration in a Rectangular Plate Having a Circular Hole.* The rectangular plate, 150 mm × 110 mm, is perforated at its centre by a circular hole of radius 15 mm and is subjected to uniform tensile loading. Again, the symmetry planes permit an analysis on the quarter geometry which is modelled using 36 graded linear elements, Fig. 6.7a. The stress concentration factor for this problem is known [15]. Computed values of the normal stress along the ligament $ED$ are given in Fig. 6.7b. The implied stress concentration factor is accurate for cases (ii), (iii) and (iv) and is 7% low for case (i).

## *Discussion*

The poor performance of the continuous elements at corners is abundantly evident in the problems considered here. Nevertheless, the effect is localized, with no general degradation being apparent so that there is no severe detriment to the interior solutions.

In the heat conduction problem, the case of continuous elements with double nodes at a singularity, case (ii), presents about 1% error in temperature at the corners, $A$, $B$, and 0.2% error away therefrom. This is accompanied by 0.1% error in flux on the faces $AD$, $CB$. The temperature values obtained with discontinuous elements, case (iii), are similar to case (ii), while the flux values show some 4% error at the corners with 1% error away therefrom. A similar result has been observed with constant elements [1]. The partially discontinuous elements, case (iv), perform best with results within 0.1% error over the whole boundary.

In the fluid flow problem, cases (ii), (iii), (iv) give the same results within 1% on the channel wall and 2% on the disc face away from its edge. At the edge of the disc, where the solution is singular, cases (iii) and (iv) give the same result with cases (i) and (ii) trailing by more than 25%. In order to achieve results with case (ii) or comparable quality to those of cases (iii) and (iv), the mesh density on the singular edge needs to be doubled.

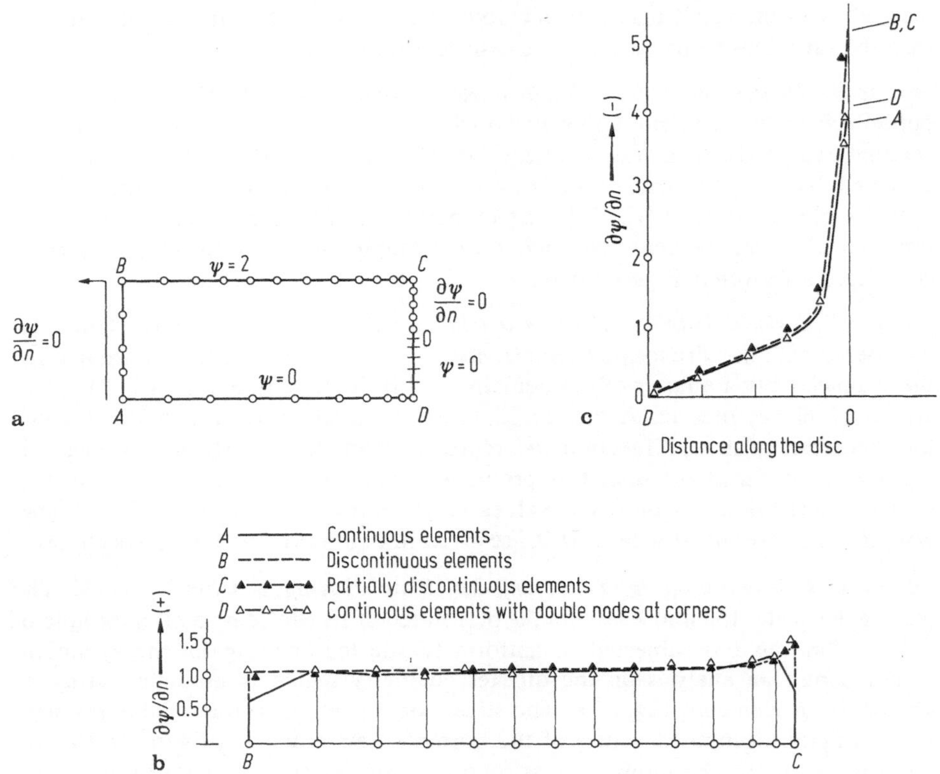


**Fig. 6.6.** **a** Boundary element discretization of flow problem. **b** Stream velocities at the channel wall, BC. **c** Stream velocities along the disc

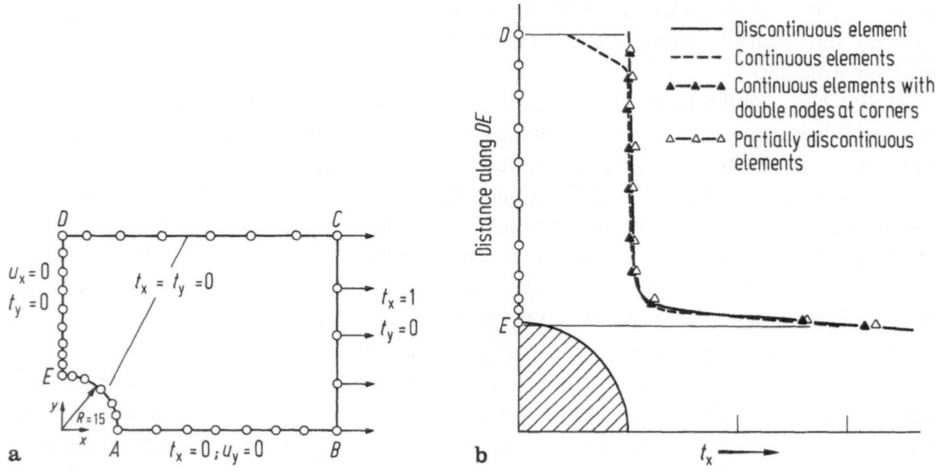


**Fig. 6.7.** **a** Boundary element discretization of plate problem. **b** Normal stress distribution along the ligament "DE" of the plate

The stress concentration factor given with cases (ii), (iii) and (iv) agrees with Peterson's value within 0.3% while in case (i) the result is 7% low.

It is clearly evident that, while the discontinuous element models have around twice the freedoms of the others, there is no consequent improvement in quality of result. This is because, despite the increase in degrees of freedom, the mesh is no more complete than a comparable mesh of continuous elements. Thus, discontinuities should only be introduced where they are required for modelling purposes.

Other case studies with singular problems [11], [16], have revealed that if the mesh is sufficiently coarse in the neighbourhood of a singularity, then a spurious oscillation can occur in the surface solution using continuous or discontinuous elements. However, this disappears on mesh refinement and is not seen in the flow problem discussed here.

Conclusions indicated are:

(i) Continuous elements perform poorly at geometric and solution singularities.
(ii) Continuous elements with double nodes at singularities give much improved results; despite mathematical objections to their use.
(iii) The discontinuous and, especially, partially discontinuous elements outperform their continuous counterparts.
(iv) Best results, overall, are obtained when using partially discontinuous elements at the problem singularities matched to continuous elements elsewhere.

## 6.4 Discontinuous Elements for Three Dimensional Analysis

Besides obviating the modelling difficulties associated with continuous elements, discontinuous and partially discontinuous elements manifest an additional advantage in three dimensions. This is that, when interelement continuity is relaxed, the element mesh may be graded freely allowing additional freedoms in mesh selection.

Any of the surface geometric elements used in finite element analysis may be used in the discretization of the surface geometry. Here, the 8 node isoparametric quadrilateral (Fig. 6.8) is assumed. The element is defined by eight geometric nodes $\mathbf{x}^\varkappa(\zeta)$, $\varkappa = 1, 2, \ldots, 8$, then the geometry is given by [9]

$$x_i(\zeta) = \sum_{\varkappa=1}^{8} M^\varkappa(\zeta)\, x_i^\varkappa, \quad i \in (1, 2, 3), \quad \zeta = (\zeta_1, \zeta_2), \quad \zeta_i \in [-1, +1] \qquad (6.27)$$

where

$$\begin{aligned} M^1(\zeta) &= 1/4(1+\zeta_1)(1+\zeta_2)(\zeta_1+\zeta_2-1) \\ M^5(\zeta) &= 1/2(1-\zeta_1^2)(1+\zeta_2) \end{aligned} \qquad (6.28)$$

and the remaining shape functions follow by induction.

As usual, having specified the geometric element, there remains a wide latitude of choice of freedom functions. Here attention is confined to 4, 8 and 12 freedom

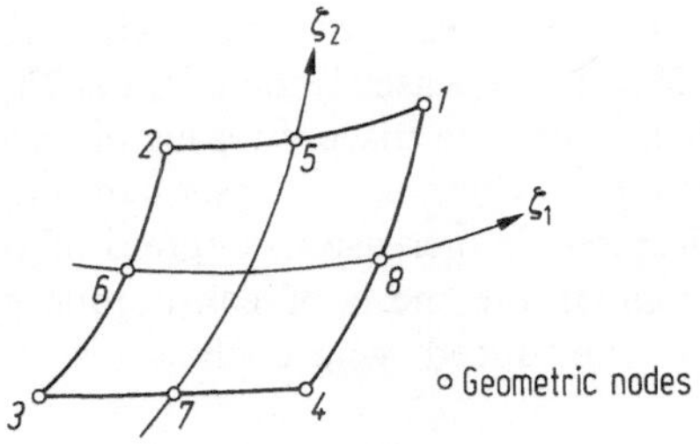


**Fig. 6.8.** 8-node isoparametric quadratic lateral element

node variants. In each case the field quantities $u$, $\frac{\partial u}{\partial n}$ are given by:

$$\begin{aligned} u(\zeta) &= \sum_{j=1}^{n} F^j(\zeta)\, u^j \\ \frac{\partial u}{\partial n}(\zeta) &= \sum_{j=1}^{n} F^j(\zeta) \frac{\partial u^j}{\partial n} \end{aligned} \tag{6.29}$$

where the element has $n$ freedom nodes, $F^j(\zeta)$ are the appropriate interpolation functions and $\zeta = (\zeta_1, \zeta_2)$; $\zeta_i \in [-1, +1]$; [17], [18].

*Discontinuous Elements*

*(a) Linear Elements.* Here, (Fig. 6.9a), the freedom nodes are taken at the four points $\zeta = (\pm\frac{1}{2}, \pm\frac{1}{2})$ and the interpolation functions are:

$$F^j(\zeta) = (\tfrac{1}{2} \pm \zeta_1)(\tfrac{1}{2} \pm \zeta_2) \tag{6.30}$$

*(b) Quadratic Element.* The eight freedom nodes are located as shown in Fig. 6.9b and typical interpolation functions are:

$$\begin{aligned} F^1(\zeta) &= 27/32\,(2/3 + \zeta_1)(2/3 + \zeta_2)(\zeta_1 + \zeta_2 - 2/3) \\ F^5(\zeta) &= 27/16\,(4/9 - \zeta_1^2)(2/3 + \zeta_2) \end{aligned} \tag{6.31}$$

the rest following by induction.

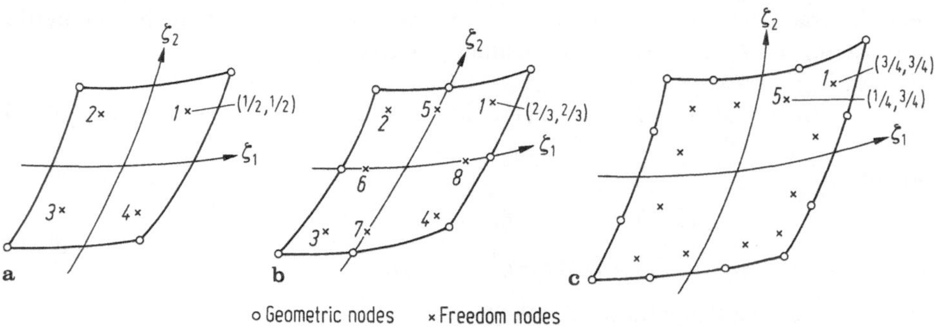


**Fig. 6.9.** **a** Discontinuous linear element. **b** Discontinuous quadratic element. **c** Discontinuous cubic element

*(c) Cubic Element.* Again, the twelve freedom nodes are located as in Fig. 6.9c, typical interpolation functions are:

$$F^1(\zeta) = 1/9\,(3/4 + \zeta_1)(3/4 + \zeta_2)(8(\zeta_1^2 + \zeta_2^2) - 5)$$
$$F^5(\zeta) = 3/2\,(1/4 + \zeta_1)(1 - 16/9\,\zeta_1^2)(3/4 + \zeta_2), \tag{6.32}$$

the rest following by induction:

*Partially Discontinuous Elements*

The formulation of partially discontinuous boundary elements is fundamentally simple. One decides at which edge, or edges, discontinuity is to occur. This dictates which freedom nodes are to be brought into the element. On locating these nodes, appropriate Lagrange shape functions may be developed with ease. In the case of the quadrilateral geometric element there are four types of partially discontinuous element, dictated by the possible incidence of discontinuous edges. Thus, there can be a single discontinuous edge, or there can be two discontinuous edges, either adjacent or non-adjacent and, finally, there can be three edges of discontinuity. The appropriate shape functions for 4, 8 and 12 noded elements have been developed by El-Sebai [18]. Here only the 4 node variant is discussed.

*a) Single Edge Discontinuity.* The freedom nodes are given in Fig. 6.10a and the interpolation functions are:

$$F^1(\zeta) = \tfrac{1}{3}(1 + \zeta_1)(1 + \zeta_2)$$
$$F^4(\zeta) = \tfrac{1}{3}(1 + \zeta_1)(\tfrac{1}{2} - \zeta_2) \tag{6.33}$$

with $F^2$, $F^2$ following by induction.

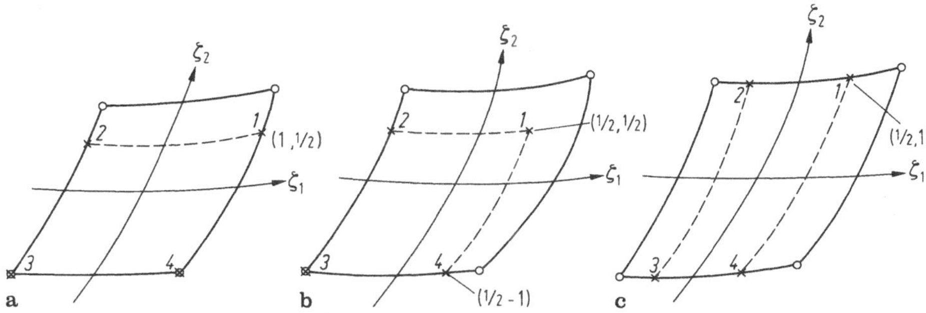


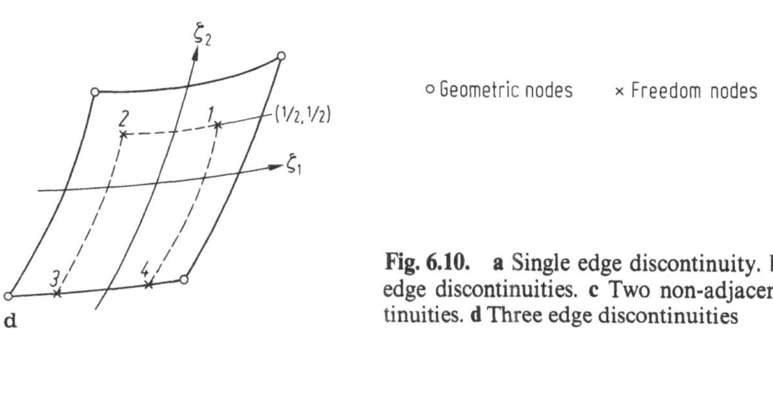


**Fig. 6.10.** **a** Single edge discontinuity. **b** Two adjacent edge discontinuities. **c** Two non-adjacent edge discontinuities. **d** Three edge discontinuities

*b) Two Adjacent Edge Discontinuities.* The freedom nodes are given in Fig. 6.10b and the interpolation functions are:

$$
\begin{aligned}
F^1(\zeta) &= \tfrac{4}{9}(1+\zeta_1)(1+\zeta_2) \\
F^2(\zeta) &= \tfrac{4}{9}(\tfrac{1}{2}-\zeta_1)(1+\zeta_2) \\
F^3(\zeta) &= \tfrac{4}{9}(\tfrac{1}{2}-\zeta_1)(\tfrac{1}{2}-\zeta_2) \\
F^4(\zeta) &= \tfrac{4}{9}(1+\zeta_1)(\tfrac{1}{2}-\zeta_2)
\end{aligned}
\tag{6.34}
$$

*c) Two Non-adjacent Edge Discontinuities.* The freedom nodes are given in Fig. 6.10c and the interpolation functions are:

$$
F^j(\zeta) = \tfrac{1}{2}(\tfrac{1}{2} \pm \zeta_1)(1 \pm \zeta_2) \tag{6.35}
$$

*d) Three Edge Discontinuities.* The freedom nodes are given in Fig. 6.10d and the interpolation functions are:

$$
\begin{aligned}
F^1(\zeta) &= \tfrac{2}{3}(\tfrac{1}{2}+\zeta_1)(1+\zeta_2) \\
F^4(\zeta) &= \tfrac{2}{3}(\tfrac{1}{2}+\zeta_1)(\tfrac{1}{2}-\zeta_2)
\end{aligned}
\tag{6.36}
$$

with $F^2$, $F^3$ following by induction.

*Applications*

As yet, there is little published information regarding the use of discontinuous, or partially discontinuous, elements in three dimensions. Discontinuous elements and their applications, for the regular boundary element method have been discussed by Patterson and El-Sebai [17], and it has been reported by Danson [19], that discontinuous elements are available in a commercial package, using the singular boundary element method, but without case study material. In reference [17] 4, 8, and 12 freedom node discontinuous elements were presented, in conjugation with a 8 node isoparametric geometric element. The accompanying case studies showed that the elements could be successfully applied, including cases where subdomains and freely graded meshes were employed. Two of the cited examples are considered here and the results are extended to include the use of partially discontinuous elements.

*a) Steady State Heat Conduction in a Bar.* A rectangular prism, $5 \times 6 \times 24$, is considered, Figure 6.11a. One end-face is held at temperature 300 and the remaining surface at temperature 0. The solution of this problem is fully three dimensional and is singular along the edges of the heated face. A fourier series solution of the problem is known [20]. The problem is modelled by partitioning the bar into two equal short bars ($5 \times 6 \times 12$). Ten elements are taken on each sub-region, one on each end face and eight on the side faces, so that each face has two equal elements which divide the long edge. The surface computed values are good, given the severely limited ability of the mesh to follow the singular surface flux. The interior solution, Figs. 6.11b, c, d, examined on lines *AB, CD, EF*, Fig. 6.11a, agrees with the fourier solution to within 0.1%.

*b) Steady State Heat Conduction in a Short Cylinder.* A cylinder of radius 3 and length 4, has its tubular face divided into halves by an axial diametrical cut. One

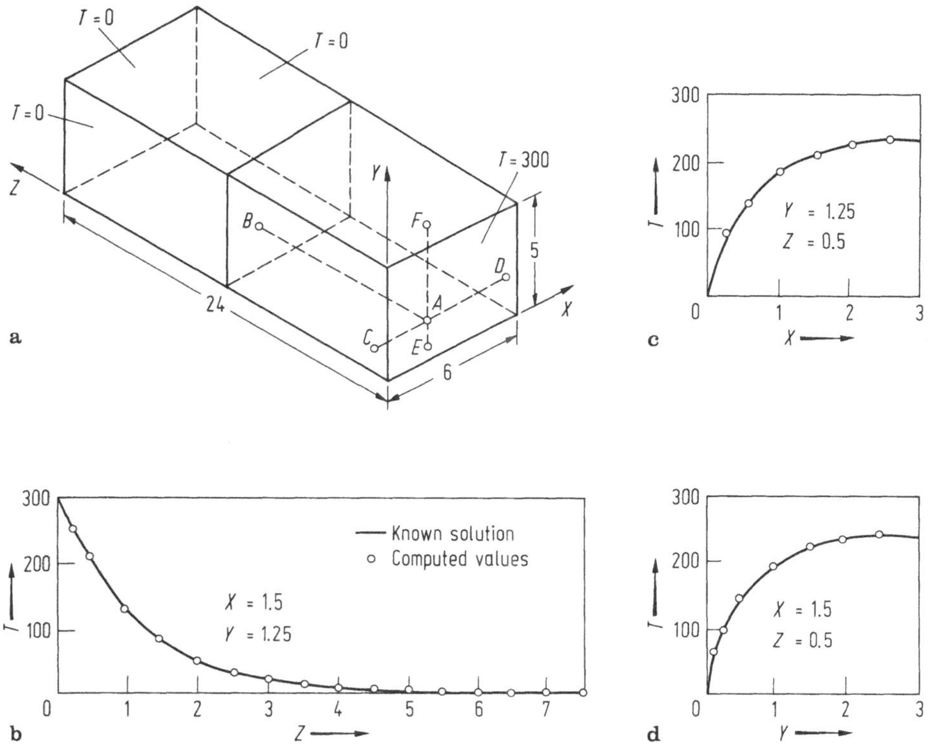


**Fig. 6.11.** **a** Bar problem – geometry and boundary conditions. **b** Temperature distribution along AB. **c** Temperature distribution along CD. **d** Temperature distribution along EF

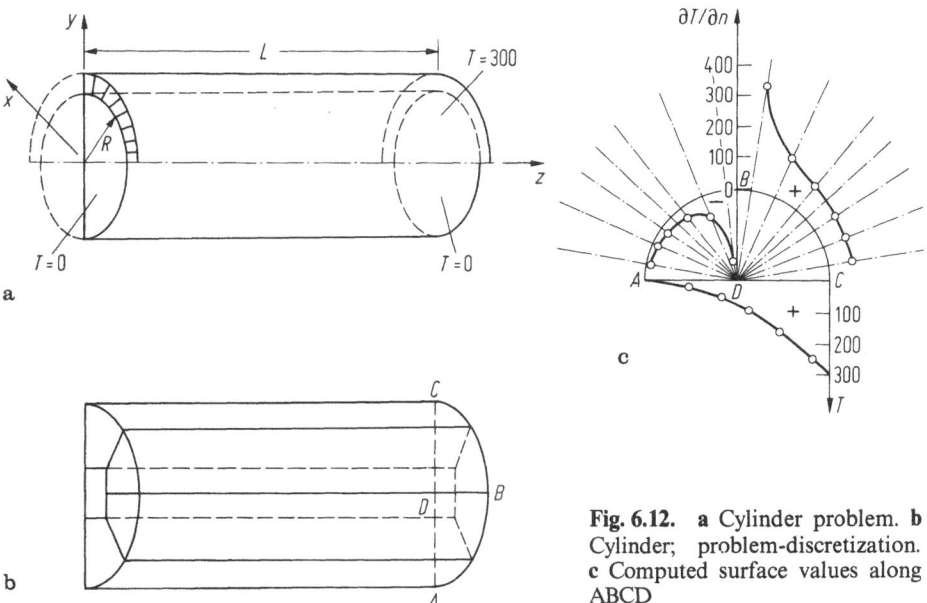

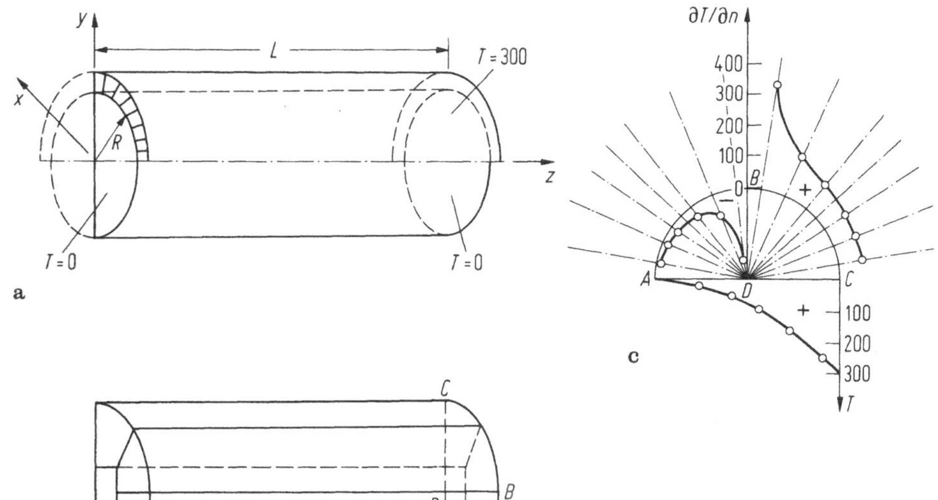

**Fig. 6.12.** **a** Cylinder problem. **b** Cylinder; problem-discretization. **c** Computed surface values along ABCD

half face is held at temperature 300 and the remaining surface of the cylinder is held at temperature 0. The solution of this problem clearly has an axial plane of symmetry, so that only half the cylinder need be modelled, Figure 6.12a. Furthermore, the solution is fully three dimensional and is singular at all edges of the heated surface. A fourier series solution is available [20]. The element mesh employs twenty mismatched elements, Figure 6.12b. Computed surface values on the end-face are given in Figure 6.12c. The internal solution, on several internal lines, was compared with the series solution and again there was less than 0.1% discrepancy.

Both of these problems have been re-analysed using partially discontinuous elements, with eight geometric and eight freedom nodes, giving results of similar quality.

In view of the manifest incompleteness of the coarse meshes employed, in their ability to describe the singular edge heat fluxes, the quality of the interior results is remarkably good for both problems. The bar problem results show that both the discontinuous and the partially discontinuous elements are well suited to subdomain modelling. Also, the cylinder problem results similarly validate the use of freely graded meshes.

## 6.5 Closure

The thesis that discontinuous elements may be employed in boundary element analysis is established experimentally by the known validity of constant elements and the similar validation of higher order discontinuous and partially discontinuous elements discussed here.

Whilst discontinuous elements may be used throughout a mesh, this is not desirable in practice because the resultant mesh is richer in freedoms, but not more complete, than a similar continuous element mesh.

Thus, the indication is that partially discontinuous elements, matched to continuous elements, should be used in practice. Discontinuities should be confined to geometric singularities (corners, etc.), to interfaces where the boundary conditions are discontinuous or change type and otherwise used in generating freely graded meshes.

Finally, discontinuous and partially discontinuous elements may be used together with subdomain models.

## References

1 Brebbia, C. A., *The Boundary Element Method for Engineers.* Pentech Press, London 1978
2 Watson, J. O., Advanced implementation of the boundary element method for two and three-dimensional elastostatics, In: *Developments in Boundary Element Methods – 1.* Eds. Banerjee, P. K. & Butterfield, R., Applied Science Publishers, London, Chapt. 3, pp. 31–63, 1979

3 Patterson, C., Sufficient conditions for convergence in the finite element method for any solution of finite energy. In: *Mathematics of Finite Elements & Applications.* Ed. Whiteman, J. R., Academic Press, London, pp. 213–224, 1973
4 Lachat, J. C., *A Further Development of the Boundary Integral Technique for Elastostatics.* Ph.D. Thesis, University of Southampton 1975
5 Patterson, C., Sheikh, M. A., A regular boundary elementmethod for fluid flow. In: Int. J. Num. Methods in Fluids, **2,** pp. 239–251, 1982
6 Patterson, C., Sheikh, M. A., Regular boundary integral equations for stress analysis. In: *Boundary Element Methods.* Ed. Brebbia, C. A., Springer-Verlag, Berlin, pp. 85–104, 1981
7 Patterson, C., Sheikh, M. A., A regular boundary element method for coupled subdomains. In: *Numerical Methods for Coupled Problems.* Eds. Lewis, R. W., Bettess, P., and Hinton, E., Pineridge Press, Swansea, pp. 115–126, 1981
8 Patterson, C., Sheikh, M. A., A higher order boundary element method for fluid flow. In: *Finite Element Flow Analysis.* Ed. Kawai, T., Tokyo University Press, Tokyo, pp. 907–913, 1982
9 Ergatoudis, J., *Isoparametric Finite Elements in Two and Three-Dimensional Analysis.* Ph.D. Thesis, University of Wales, Swansea 1968
10 Brebbia, C. A., Walker S., *Boundary Element Techniques in Engineering.* Newnes-Butterworths, London 1980
11 Patterson, C., Sheikh, M. A., Discontinuous boundary elements for heat conduction. In: *Numerical Methods in Thermal Problems – II.* Eds. Lewis, R. W., Morgan, K., and Schrefler, B. A., Pineridge Press, Swansea, pp. 25–34, 1981
12 Chaudonneret, M., On the discontinuity of the stress vector in the boundary integral equation method for elastic analysis. In: *Recent Advances in Boundary Element Methods.* Ed. Brebbia, C. A., Pentech Press, London, pp. 185–194, 1978
13 Wardle, L. J., Crotty, J. M., Two-dimensional boundary integral equation analysis for non-homogeneous mining applications. In: *Recent Advances in Boundary Element Methods.* Ed. Brebbia, C. A., Pentech Press, London, pp. 233–251, 1978
14 Milne-Thompson, L. M., *Theoretical Hydrodynamics.* Macmillan, London 1968
15 Peterson, R. E., *Stress Concentration Factors.* John Wiley & Sons, New York 1974
16 Patterson, C., Sheikh, M. A., Non-conforming boundary elements for stress analysis. In: *Boundary Element Methods.* Ed. Brebbia, C. A., Springer-Verlag, Berlin, pp. 137–152 1981
17 Patterson, C., El-Sebai, N. A. S., A regular boundary method using non-conforming elements for potential problems in three dimensions, In: *Boundary Element Methods in Engineering.* Ed. Brebbia, C. A., Springer-Verlag, Berlin, pp. 112–126, 1982
18 El-Sebai, N. A. S., *An Investigation of the Regular Boundary Element Method in Three Dimensions.* Ph.D. Thesis, University of Sheffield 1982
19 Danson, D. J., BEASY, A boundary element analysis system. In: *Boundary Element Methods in Engineering.* Ed. Brebbia, C. A., Springer-Verlag, Berlin, pp. 557–557, 1982
20 Aypaci, V. S., *Conduction Heat Transfer.* Addison-Welsey, USA 1966

## Chapter 7

# Applications in Geomechanics

*by W. S. Venturini and C. A. Brebbia*

## 7.1 Introduction

The boundary element method is already well established as an efficient numerical technique to solve a variety of continuum mechanics problems [1, 2]. Solutions exist for problems as complex as plasticity [3], viscoelasticity [4], viscoplasticity [5] and other complex time-dependent and non-linear problems. The literature is, nowadays, very extensive and the interested reader is referred to [2].

In this chapter the boundary element method is applied to solve geomechanical problems, modelling rock and soil material behaviour. The first model presented here uses the no-tension criterion and is based on the finite element work of reference [6]. The criterion was first used by Venturini and Brebbia in the context of boundary element methods and its numerical implementation is presented in detail in reference [7]. It consists of considering that the rock medium is unable to sustain any tensile stresses. The final solution is obtained using an iterative procedure in which stresses are corrected by introducing a series of initial stress fields.

Plastic and viscoplastic models are very frequently used to represent the nonlinear behaviour of soil and rock materials. They have been extensively used with finite elements since the known classical analysis by Reyes of a circular tunnel [8]. More recently finite element models have been developed using the more general Perzina's model [9] to represent viscoplasticity and these models have been extended to provide plastic and viscoplastic solutions in geomechanics [10, 11]. The model is applied in this chapter in conjunction with boundary elements following the original work carried out by Telles and Brebbia [5]. The solution is extended here by using the overlay technique [12] which allows for a better representation of material response, as shown in reference [13].

Another important problem in geomechanics is how to model the cracks or fissures discontinuities often found in the rock medium. As in finite elements [14], special relationships are used here to model the contact and slip conditions within a boundary element formulation.

Layers of material, or seams, are also frequently found in soils or rocks. As their thickness is small by comparison to other dimensions, it becomes inefficient to consider them in the same manner as other boundary element regions. A special formulation is described in this chapter which is easy to apply with boundary elements.

Several practical examples are shown to illustrate the potentialities of the boundary element formulation in geomechanics and, whenever possible, results are compared with theoretical or finite element solutions.

## 7.2 Basic Formulation

The boundary integral equation used in several geomechanical problems is given by the following expression [3], [14],

$$\begin{aligned} c_{ik}(S)\, u_k(S) + \int_\Gamma p^*_{ik}(S, Q)\, u_k(Q)\, d\Gamma(Q) &= \int_\Gamma u^*_{ik}(S, Q)\, p_k(Q)\, d\Gamma(Q) \\ &+ \int_\Omega u^*_{ik}(S, q)\, b_k(q)\, d\Omega(q) \\ &+ \int_\Omega \varepsilon^*_{imk}(S, q)\, \sigma^0_{mk}(q)\, d\Omega(q) \end{aligned} \tag{7.1}$$

In the above expression, the two domain integrals correspond to the body force and the initial stress contributions. The initial stress term, $\sigma^0_{mk}$, is necessary in order to consider effects such as temperature, shrinkage, swelling and others. The initial stress integral can also be used to model nonlinear effects such as plasticity, visco-plasticity, etc. For these cases, an iterative process needs to be carried out. The fundamental solution used in all cases is the linear Kelvin type solution [15] presented in the Appendix.

Equation (7.1) can be specialized for an interior point as follows,

$$\begin{aligned} u_i(s) = &- \int_\Gamma u_k(Q)\, p^*_{ik}(s, Q)\, d\Gamma(Q) + \int_\Gamma p_k(Q)\, u^*_{ik}(s, Q)\, d\Gamma(Q) \\ &+ \int_\Omega b_k(q)\, u^*_{ik}(s, q)\, d\Omega(q) + \int_\Omega \sigma^0_{mk}(q)\, \varepsilon^*_{imk}(s, q)\, d\Omega(q) \end{aligned} \tag{7.2}$$

After differentiating this equation and applying Hooke's law the stress at an interior point can be written as [3], [14],

$$\begin{aligned} \sigma_{ij}(s) = &- \int_\Gamma S_{ijk}(s, Q)\, u_k(Q)\, d\Gamma(Q) + \int_\Gamma D_{ijk}(s, Q)\, p_k(Q)\, d\Gamma(Q) \\ &+ \int_\Omega D_{ijk}(s, q)\, b_k(q)\, d\Omega(q) + \int_\Omega E_{ijmk}(s, q)\, \sigma^0_{mk}(q)\, d\Omega(q) \\ &- \frac{1}{8(1-\nu)} [2\, \sigma^0_{ij}(s) + (1 - 4\nu)\, \sigma^0_{ll}(s)\, \delta_{ij}] \end{aligned} \tag{7.3}$$

As it is well-known in boundary elements [1] the discretization of the boundary and domain of the body into elements and internal cells respectively allows the transformation of equation (7.1) into a system of algebraic equations as follows,

$$\mathbf{HU} = \mathbf{GP} + \mathbf{DB} + \mathbf{E}\,\boldsymbol{\sigma}^0 \tag{7.4}$$

where **H** and **G** are the usual boundary element influence matrices, the terms in **D** represent the contributions of the body forces and **E** is an influence matrix for the initial stresses.

Similarly equation (7.3) can be written in matrix form as follows,

$$\boldsymbol{\sigma} = -\,\mathbf{H}'\mathbf{U} + \mathbf{G}'\mathbf{P} + \mathbf{D}'\mathbf{B} + \mathbf{E}'\,\boldsymbol{\sigma}^0 \tag{7.5}$$

Although equation (7.3) only relates to internal points, we will assume that the boundary stresses have also been included in equation (7.5) since they are usually required in the solution of nonlinear material problems.

Equations (7.4) and (7.5) are valid for homogeneous bodies. Bodies with region with different material properties are important in practice. One can analyse them by dividing the domain into homogeneous subregions.

Consider a domain such as in Fig. 7.1 formed by several different subregions $(\Omega_1, \Omega_2 \ldots \Omega_i, \Omega_j \ldots)$. We can start by assuming that each subregion "$i$" is an independent domain and apply equation (7.4),

$$\mathbf{H}^i\mathbf{U}^i = \mathbf{G}^i\mathbf{P}^i + \mathbf{D}^i\mathbf{B}^i + \mathbf{E}^i\,\boldsymbol{\sigma}^{0i} \tag{7.6}$$

where the summation notation is not implied.

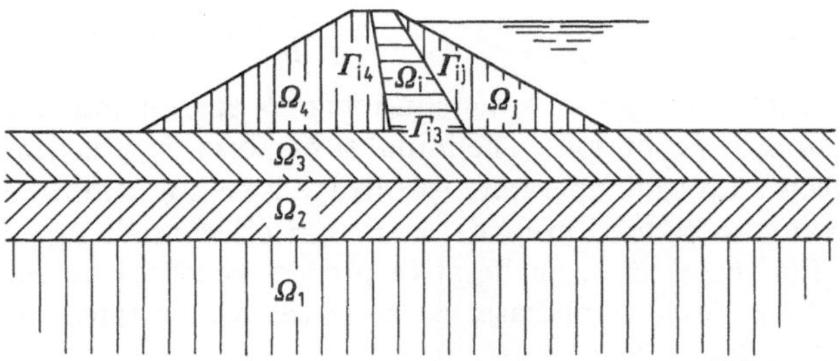


**Fig. 7.1.** Subregions

For the nodes along the interface $\Gamma_{ij}$ – common to subregions "$i$" and "$j$" – the displacement compatibility and the traction equilibrium are given by,

$$\mathbf{U}^i - \mathbf{U}^j = \mathbf{0} \tag{7.7}$$

$$\mathbf{P}^i + \mathbf{P}^j = \mathbf{0} \tag{7.8}$$

where the superscript represents the subregion to which the displacements or tractions are related.

Equation (7.6) can also be written for the subregion "$j$" and then imposing the conditions (7.7) and (7.8) the final system of equations can be obtained by assembling. This system has the same form as presented in equation (7.4) although now some of the unknowns displacements and tractions are on the interface rather than on the external boundary.

After solving the system of equations all external and interface tractions and displacements are known. Then the stresses at any point can be computed by applying equation (7.5) separately for each subregion.

The combination of subregions described above presents a particular problem for corner nodes, where the values of tractions are not uniquely defined.

*Traction Discontinuity.* In order to understand this problem, consider the two adjacent elements shown in Figure 7.2. One element is defined by the extreme points $P$ and $R$ and the other by $S$ and $T$ where the points $R$ and $S$ have the same coordinates.

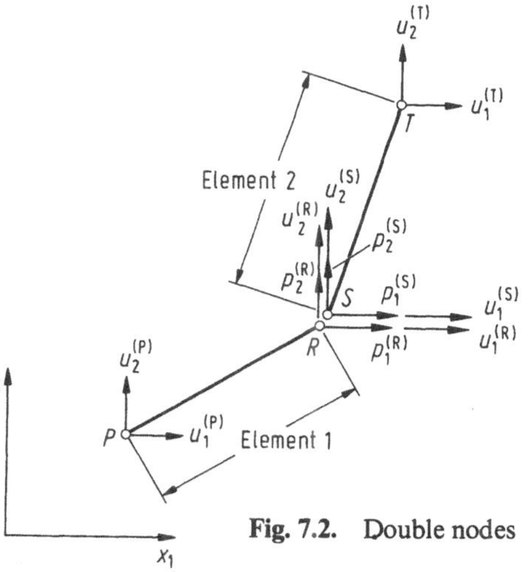


**Fig. 7.2.** Double nodes

Assuming that the stress tensor will be uniquely defined at the double nodes $R$ and $S$ and writing the traction vector for the adjacent elements as a function of the stresses, the following condition is derived for the plane case due to the symmetry of the stress tensor,

$$p_1^{(R)}\,\eta_1^{(2)} - p_1^{(S)}\,\eta_1^{(1)} = p_2^{(S)}\,\eta_2^{(1)} - p_2^{(R)}\,\eta_2^{(2)} \tag{7.9}$$

where $\eta_i^{(j)}$ is the direction cosine of element "$j$" with respect to $x_i$. Notice that the $p_i$ components have been assumed in the direction of the global system of coordinates.

Another condition for the plane case is derived by enforcing the invariance of the strain trace tensor. Using a linear interpolation function to approximate the displacements over the two adjacent elements, the invariance condition can be written,

$$\begin{aligned}
&[(u_1^{(P)} - u_1^{(R)})\,\eta_2^{(1)} + (u_2^{(R)} - u_2^{(P)})\,\eta_1^{(1)}]/l_1 \\
&\quad + [(u_1^{(T)} - u_1^{(S)})\,\eta_2^{(2)} - u_2^{(T)}\,u_2^{(S)})\,\eta_1^{(2)}]/l_2 \\
&\quad - \frac{1}{2G}[p_1^{(S)}\,\eta_1^{(2)} + p_2^{(S)}\,\eta_2^{(2)} - p_1^{(R)}\,\eta_1^{(1)} - p_2^{(R)}\,\eta_2^{(1)}] \\
&\quad = \frac{1}{2G}[\sigma_{11}^{0(S)}\,\eta_2^{(2)}\,\eta_2^{(2)} - 2\,\sigma_{12}^{0(S)}\,\eta_2^{(2)}\,\eta_1^{(2)} + \sigma_{22}^{0(S)}\,\eta_1^{(2)}\,\eta_1^{(2)} \\
&\quad - \sigma_{11}^{0(R)}\,\eta_2^{(1)}\,\eta_2^{(1)} + 2\,\sigma_{12}^{0(R)}\,\eta_1^{(1)}\,\eta_2^{(1)} - \sigma_{22}^{0(R)}\,\eta_1^{(1)}\,\eta_1^{(1)}]
\end{aligned} \tag{7.10}$$

The above two conditions can be introduced into the system of equations modifying the original equations for one of the corner nodes. This procedure removes the singularity that otherwise will appear in matrix **H** and allows different subregions to be properly combined.

*Thin Subregion.* The subregion technique presented in this section can solve systems formed of several homogeneous parts with different material properties. The technique can also be extended to solve the case of thin layers or seams which are frequently present in rock masses. However, in these cases, the interfaces defined by rock and the seam materials have to be discretized into elements small enough in order to ensure that nodes on opposite interfaces are at a sufficient distance from each other in comparison with the size of the elements, otherwise numerical inaccuracies will arise. As a result the final system of equations obtained is inconveniently large.

An alternative scheme to improve the boundary element formulation consists of finding relationships between points on different faces of the thin layer. Consider the thin layer as shown in Fig. 7.3 and apply one-dimensional stress-strain relationships for shear and compression in a finite difference manner. If the thickness of the seam element, $h$, is small by comparison with the element length, one can write,

$$\begin{aligned} \sigma_{12}^{(R)} &= (u_1^{(S)} - u_1^{(R)})\, G/h \\ \sigma_{22}^{(R)} &= (u_2^{(S)} - u_2^{(R)})\, E/h \end{aligned} \tag{7.11}$$

where $G$ and $E$ are the shear and Hooke's moduli respectively and the displacements are taken following the local system of coordinates as indicated in figure 7.3.

Using the traction stress relation one can determine the following equations relating traction and displacements,

$$\begin{aligned} p_1^{(R)} &= (u_1^{(R)} - u_1^{(S)})\, G/h\,, \\ p_2^{(R)} &= (u_2^{(R)} - u_2^{(S)})\, E/h \end{aligned} \tag{7.12}$$

Two other relationships are provided by the equilibrium conditions i.e.,

$$p_1^{(R)} + p_1^{(S)} = 0\,; \qquad p_2^{(R)} + p_2^{(S)} = 0 \tag{7.13}$$

The introduction of these equations into the global system allows for the representation of a seam.

In all cases the final system of equations is obtained by reordering the matrices in the usual boundary element manner to take into consideration the boundary conditions. This gives,

$$\mathbf{A}\mathbf{X} = \mathbf{F} + \mathbf{E}\,\boldsymbol{\sigma}^0 \tag{7.14}$$

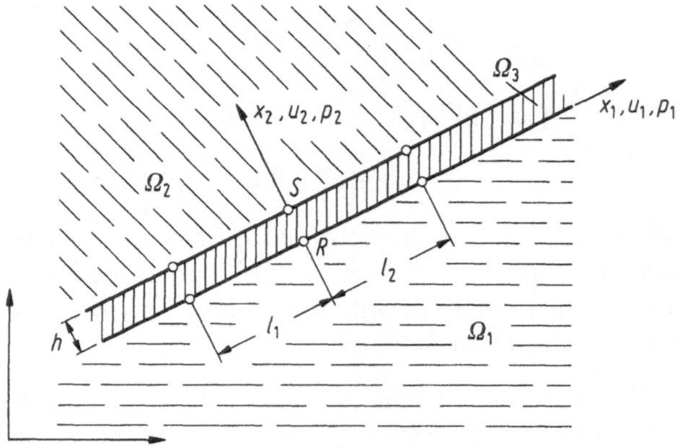


**Fig. 7.3.** Thin subregion

The vector $\mathbf{F}$ takes into consideration the applied body forces and the known displacements and tractions.

$\mathbf{A}$ in equation (7.14) is a non-symmetric matrix which may be banded or fully populated depending on which way the subregions have been considered. This matrix is usually well conditioned with dominant diagonal terms.

The solution of equation (7.14) is obtained by inverting $\mathbf{A}$ which gives,

$$\mathbf{X} = \mathbf{M} + \mathbf{R}\,\boldsymbol{\sigma}^0 \tag{7.15}$$

where

$$\mathbf{R} = \mathbf{A}^{-1}\,\mathbf{E}, \qquad \mathbf{M} = \mathbf{A}^{-1}\,\mathbf{F} \tag{7.16}$$

For the stresses (equation (7.5)) we find,

$$\boldsymbol{\sigma} = \mathbf{F}' + \mathbf{A}'\,\mathbf{X} + \mathbf{E}'\,\boldsymbol{\sigma}^0 \tag{7.17}$$

The above relationships gives the total stresses as a function of the initial stress field $\boldsymbol{\sigma}^0$. If we prefer to write it in terms of elastic stress, equation (7.17) becomes,

$$\boldsymbol{\sigma}^e = \mathbf{F}' - \mathbf{A}'\,\mathbf{X} + \mathbf{E}^*\,\boldsymbol{\sigma}^0 \tag{7.18}$$

where

$$\mathbf{E}^* = \mathbf{E}' - \mathbf{I}, \qquad \boldsymbol{\sigma}^e = \boldsymbol{\sigma} - \sigma^0 \tag{7.19}$$

Substituting (7.15) into (7.18) gives,

$$\boldsymbol{\sigma}^e = \mathbf{N} + \mathbf{S}\,\boldsymbol{\sigma}^0 \tag{7.20}$$

in which

$$\begin{aligned} \mathbf{S} &= \mathbf{E}^* - \mathbf{A}'\,\mathbf{R}, \\ \mathbf{N} &= \mathbf{F}' - \mathbf{A}'\,\mathbf{M} \end{aligned} \tag{7.21}$$

It should be noticed that the vectors $\mathbf{M}$ and $\mathbf{N}$ represent the solution of a problem without internal stresses whose effects are modelled by matrices $\mathbf{R}$ and $\mathbf{S}$.

*Application: Thin Elastic Seam Near a Circular Tunnel*

This example concerns the application of the thin layer technique. The problem has been taken from reference [16] and consists of analysing a circular tunnel excavated between two seam regions inserted in a rock medium. The solution in reference [16] was obtained using a different boundary integral formulation in which adjacent elements were taken into consideration by a particular type of singularity called quadrupoles.

For the present analysis the tunnel surface and seam interfaces have been discretized using linear elements as shown in Figure 7.4. Then three subregions are defined; two of them are constituted of sound rock with Poisson's ratio taken equal to 0.25. The third subregion, the thin layer, is formed of soft soil to which two values of its elastic modulus have been assigned for the solution of the example, $E_s/E_R = 1.0$ and $E_s/E_R = 0.01$ respectively. As can be seen in Fig. 7.5, only one quadrant of the structural system has been discretized. The load applied to the tunnel surface is one due to the removal of the initial uniaxial stress state ($\sigma_1 = -1.0$) which is assumed to exist before the excavation. The final stress distribution achieved compares very well with theoretical [17] and numerical [16] results as shown in Figure 7.6.

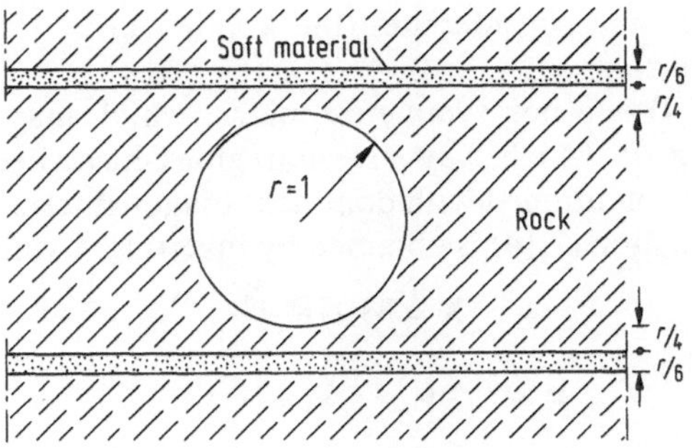


Fig. 7.4. Thin subregions in an infinite rock medium

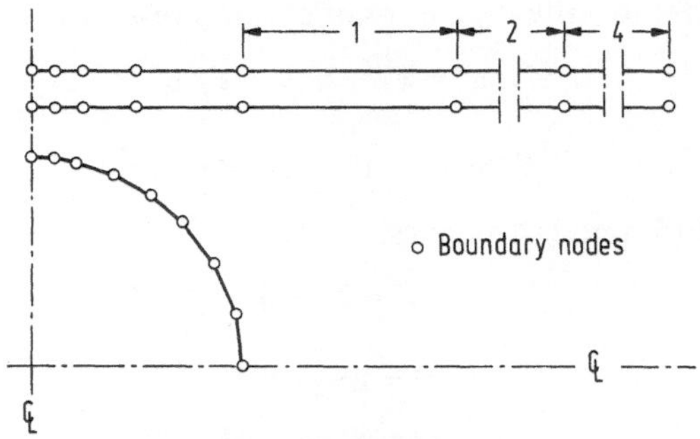


Fig. 7.5. Boundary discretization

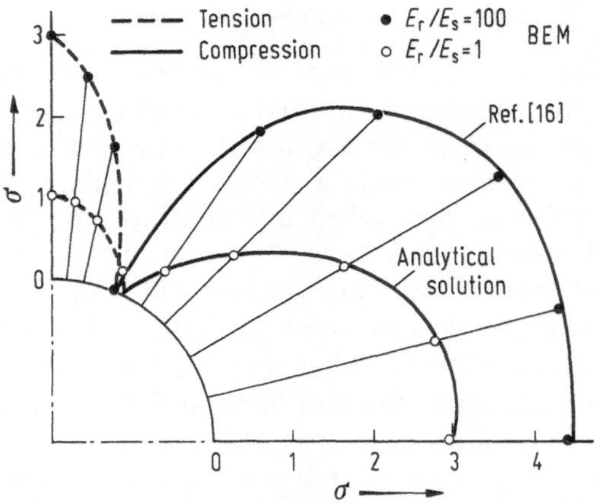


Fig. 7.6. Normal stress distribution parallel to the tunnel surface

## 7.3 No-Tension and Discontinuity Problems

For stress analysis in rock mechanics it is important to take into consideration whether the material is intact or fissured. For intact rocks the analysis is as in classical continuum mechanics as the material can be assumed to be homogeneous. When the material is extensively fissured it is also usual to carry out the same kind of consideration. In both situations the behaviour of the material when subjected to compressive or tensile stresses are quite different. In general, rock can withstand only very small tensile stresses which are usually neglected in the analysis giving rise to the so-called no-tension criterion. The compressive stress can be considered as linearly elastic. Hence the problem can be solved as in continuum mechanics assuming the rock as a linear elastic material in the direction of the compressive principal stress and at the same time no resistance to deformation in the direction of the principal tension.

In order to clarify the concept, the uniaxial relation between stress and strain for the no-tension case is presented in Figs. 7.7 and 7.8. In the first case (Fig. 7.7) the problem is assumed not to be path dependent, and complete recovery of the deformation when the load is removed can be achieved. Therefore, the final solution can be computed applying the load in one step.

For the second uniaxial stress-strain relation (Fig. 7.8), residual strains are expected on the removal of the load and the final deformation state depends on both final stress level and path stress history. Thus an incremental load process must be adopted to model the final solution, similarly as for plastic materials.

The generalization to plane strain (or stress) situations is very simple. As the third coordinate axis is principal direction, the corresponding stress is studied separately and does not affect the stresses on the plane. The only requirement needed to satisfy the criterion is that the plane principal stresses must remain less or equal to zero.

The solution of the two cases can be found by an iterative process which consists of enforcing the no-tention condition by taking into consideration an initial stress field ($\boldsymbol{\sigma}^0$).

Although the cases defined above are similar, in the non-path dependent criterion the principal stresses, which must obey the no-tension condition, are given

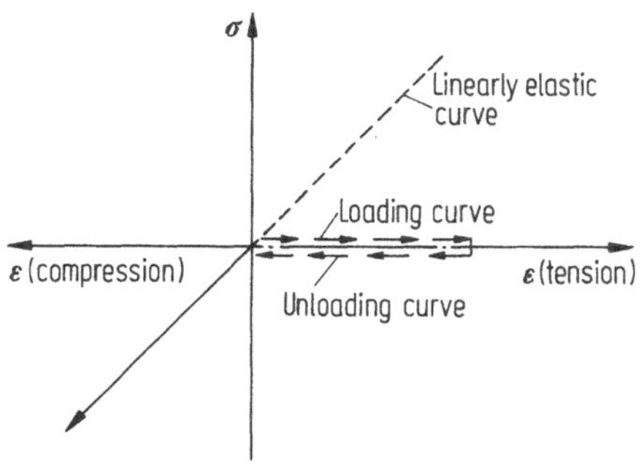


**Fig. 7.7.** Uniaxial stress-strain curve. Path independent criterion

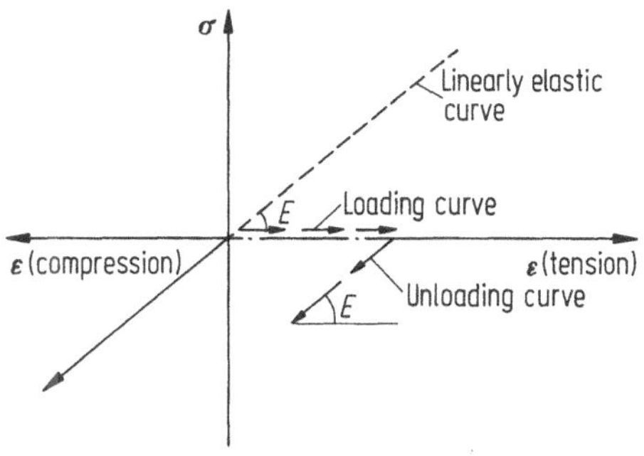


**Fig. 7.8.** Uniaxial stress-strain curve. Path dependent criterion

by the equivalent elastic stresses $\boldsymbol{\sigma}^e$ (dashed line in Fig. 7.7). For the path dependent criterion, the principal stresses are computed using the actual value of the stresses, $\boldsymbol{\sigma}$, obtained after computing the corresponding initial stress field.

Bearing in mind the difference between the two criteria for the verification of the no-tension condition, the iterative process to model the final solution is illustrated by the following steps,

(i) Compute the elastic stress increment for the iteration $n$. The increment of the elastic stress is given by (equation 7.20),

$$^{(n)}\Delta\boldsymbol{\sigma}^e = \mathbf{S}\,^{(n-1)}\boldsymbol{\sigma}^0 \tag{7.22}$$

in which $^{(n-1)}\boldsymbol{\sigma}^0$ is the tensile stress obtained in the previous iteration.

(ii) Find the stress vector $\boldsymbol{\sigma}^e$ and $\boldsymbol{\sigma}$. They are computed by the following expressions

$$\begin{aligned} ^{(n)}\boldsymbol{\sigma}^e &= {}^{(n-1)}\boldsymbol{\sigma}^e + {}^{(n)}\Delta\boldsymbol{\sigma}^e \\ ^{(n)}\boldsymbol{\sigma} &= {}^{(n-1)}\boldsymbol{\sigma}^t + {}^{(n)}\Delta\boldsymbol{\sigma}^e \end{aligned} \tag{7.23}$$

where $^{(n-1)}\boldsymbol{\sigma}^t$ represents the true stresses obtained in the last iteration.

(iii) Compute the principal stresses. Using $^{(n)}\boldsymbol{\sigma}^e$ or $^{(n)}\boldsymbol{\sigma}$ as dictated by path independent or path dependent material behaviour.

(iv) Determine the initial stress vector $^{(n)}\boldsymbol{\sigma}^0$. The initial stress field to be applied to the system is computed by taking the values of $^{(n)}\boldsymbol{\sigma}$ in the tensile principal direction obtained in (iii).

(v) Compute the true stress vector. The actual stress distribution is evaluated as follows,

$$^{(n)}\boldsymbol{\sigma}^t = {}^{(n)}\boldsymbol{\sigma} - {}^{(n)}\boldsymbol{\sigma}^0 \tag{7.24}$$

(vi) Verify the convergence. At this stage the convergence criterion must be applied. If the initial stress vectors are small with respect to the specified tolerance, the iterative process should stop. Otherwise all steps must be repeated.

For the two cases studied, the displacements can be computed after applying all loads or, if necessary at any other stage, using the accumulated initial stress values in equation (7.15).

The behaviour of rocks for the no-tension criterion described above has been assumed to be modelled within the context of continuum mechanics. However, it is know that many types of rocks are continuously disrupted by planar cracks or fissures. When the rock material is under compressive state of stress, the cracks can be shut restoring the mechanical continuity of the material, while when under tension state of stress normal to the fissures, the separation increases and quickly destroys the existing cohesion. In this case, the rock material loses its continuity and must be analysed taking into account the possible separation or slip which may occur between the two surfaces defined by the discontinuity.

In order to model this behaviour using boundary element formulation, the zones of the continuous material are assumed as subregions, while specific relations between the values of two opposite nodes along the discontinuity must be defined. The discontinuity can be interpreted as being formed by special "joint element"

(Fig. 7.9). The joint element width, $h$, represents the real gap of the discontinuity which, in general, can be filled with soft material. The width is taken equal to zero when the subregions are in contact and the coordinates of two opposite points are the same. The discontinuity has essentially no, or limited, resistance to a tension force applied in the normal direction. The slip between the two surfaces of the discontinuity is usually assumed to be governed by a linear Mohr envelope which may be also defined with limited tensile strength.

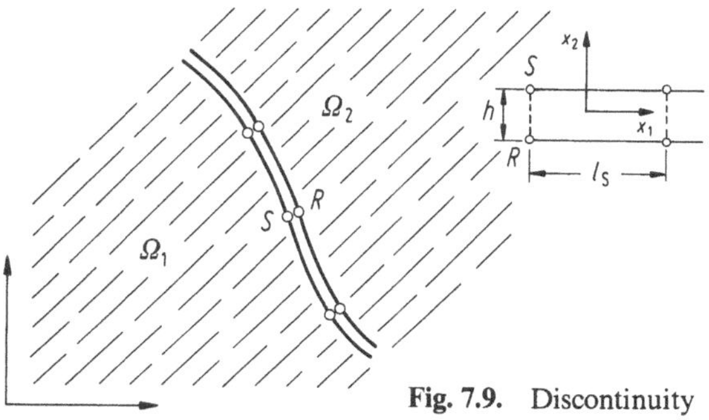


**Fig. 7.9.** Discontinuity

The equations for the joint can now be written. These equations which are always dependent on the physical characteristics of the discontinuity must be also introduced into the final system of equations. They are necessary to model any discontinuity problem and are divided in three main groups: (i) the first group represents the separation condition. This particular condition must be enforced to the problem when either a real gap between adjacent subregions exists or tensile stresses in the direction normal to the joint element exceed the admissible limit. The conditions to be enforced at each node when no soft material is filling the gap is that the surface traction is zero. (ii) The second group of equations necessary to model the discontinuity problem is formed by the contact equations,

$$\begin{aligned} u_i^{(R)} - u_i^{(S)} &= 0 \\ p_i^{(R)} + p_i^{(S)} &= 0 \end{aligned} \tag{7.25}$$

i.e. perfect continuity exists. These conditions are valid provided the surface stresses do not violate a failure criterion. (iii) The last condition to be formulated refers to the slip between the two surfaces of the discontinuity. Depending on the stress level along the discontinuity, the two surfaces can slide on each other without any relative displacement in the direction orthogonal to the discontinuity. In order to introduce such a behaviour in the system of equations, the following conditions of displacements and tractions in the direction orthogonal to the discontinuity can be written,

$$\begin{aligned} p_2^{(R)} + p_2^{(S)} &= 0 \\ u_2^{(R)} - u_2^{(S)} &= 0 \end{aligned} \tag{7.26}$$

In the direction parallel to the discontinuity, equations relating the shear and normal stresses can be written according to the assumed Coulomb's envelope. Thus for "$R$" and "$S$" the following relation is written,

$$p_1 = -p_2 \tan \varphi \tag{7.27}$$

for positive values of the shear stress, otherwise

$$p_1 = p_2 \tan \varphi \tag{7.28}$$

where $\varphi$ is the friction's angle of the material.

Equations (7.25) to (7.28) represent the new relations to be introduced into the system of equations. This can be done by changing the original boundary equations.

Discontinuity problems are markedly nonlinear; the stress level is not only a function of the displacements, but also must be related to the history of the displacements. In this case, superposition is no longer valid and an incremental procedure of loading must be followed to satisfy the path dependence of the problem.

The process of solution adopted in this work to deal with domains which show discontinuity consists of an incremental and iterative procedure in which the system of equations may be modified at each iteration according to the stress values along the discontinuity.

The sequence of steps can be summarized as follows,

(i) Assemble the system matrix: the matrix **H** is computed taking into account the conditions defined at each node over the discontinuity.
(ii) Apply boundary conditions: if it is the first iteration, the external boundary condition due to the load increment has to be applied. Otherwise traction computed in step (iii) is used.
(iii) Determine boundary traction to be applied along the discontinuity: if unadmissible tensile or shear stress were computed along the discontinuity, the traction vector necessary to re-establish the stress condition defined by the assumed envelope is computed, and the iterative process must start again at step (i). Otherwise, a new load increment has to be provided.

*No-Tension Application – Lined Tunnel*

This application consists of analysing a circular tunnel built in an infinite rock domain. The reinforced concrete lining adopted is 30 cm thick and its internal diameter is 400 cm as shown in Figure 7.10. The applied load is the internal water pressure taken equal to 100 tf/m$^2$.

The rock medium is considered to be divided into two distinct regions. A no-tension 115 cm thick ring region closer to the lining is assumed to be fully fissured, while all other points outside that region are considered capable of sustaining tensile stresses. Poisson's ratios for concrete and rock were assumed 0.15 and 0.20 respectively, while the ratio between the elastic moduli of concrete and rock is equal to 2. The residual stresses in the rock were neglected in order to allow the comparison between numerical results and the theoretical solution given in reference [18].

The problem has been solved elastically using two different boundary meshes with 24 and 48 linear elements respectively. The results (Fig. 7.10) are in total

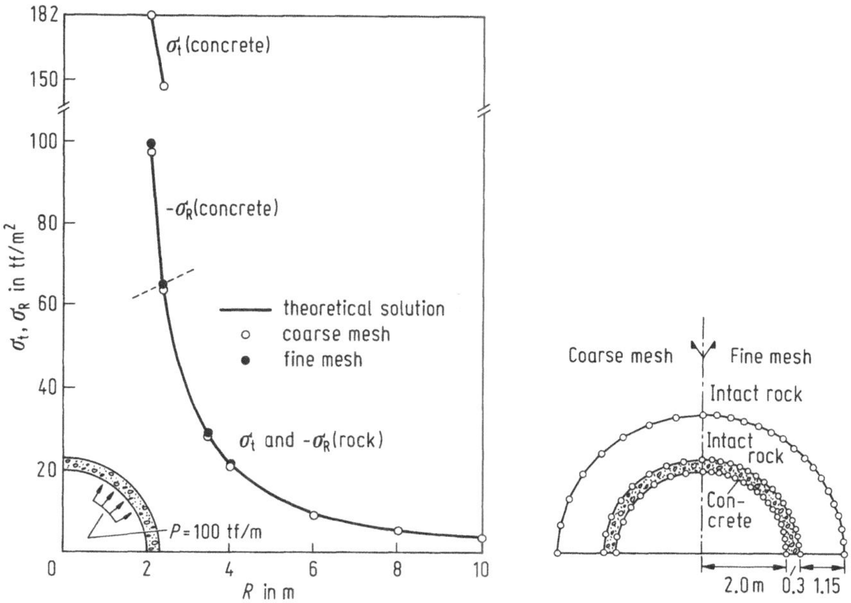


**Fig. 7.10.** Lined tunnel. Discretization and elastic results

agreement with the theoretical solution. The continuity and the accuracy of the stress profile on the two discretized interfaces shows the validity of the subregion technique employed.

The no-tension solution was first achieved by assuming the unlined case in which the load is applied directly to the rock material. Using 8 linear elements to discretize both the rock surface and the interface between intact and fissured material (Fig. 7.11), the final no-tension results are obtained.

The no-tension analysis of the lined case has been carried out using two different internal cell discretizations and a single boundary mesh consisting of 8 linear

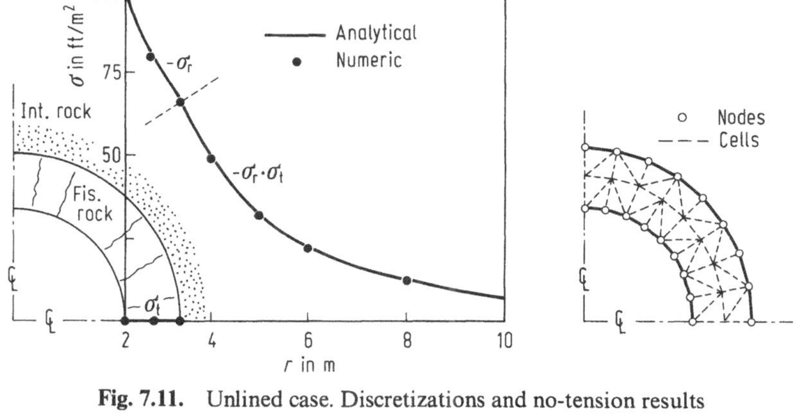


**Fig. 7.11.** Unlined case. Discretizations and no-tension results

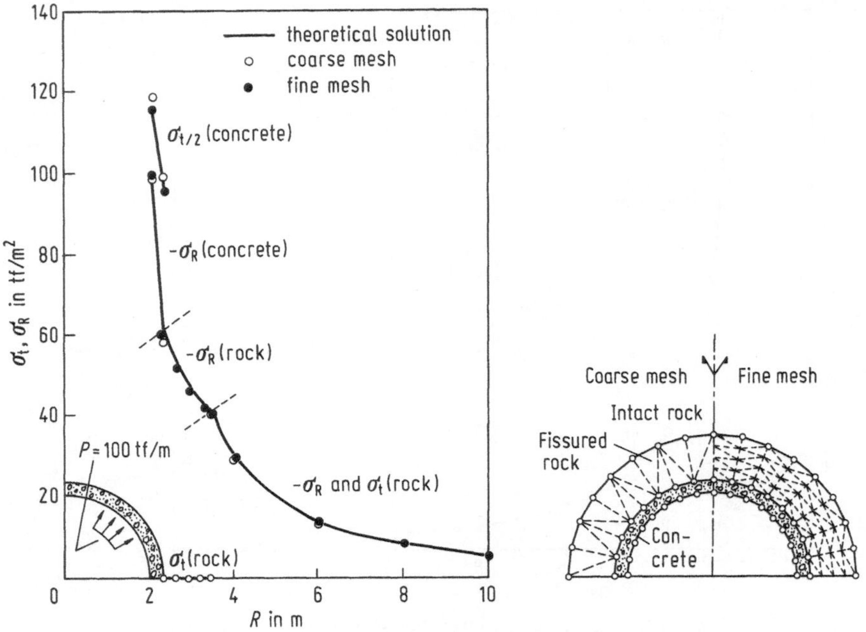


**Fig. 7.12.** Lined case. Discretizations and no-tension results

elements and 9 nodes on the internal lining surface and on the interfaces (Fig. 7.12). The final results for the coarsest mesh seem to be satisfactory; when they are compared with the theoretical solution the maximum error observed was 4.5%. For the fine mesh, better agreement between theoretical and numerical solutions was obtained; the maximum computed error is only 2%.

*Discontinuity Application – Circular Opening in a Cracked Rock Domain*

This application consists of solving a circular tunnel built in a solid rock medium with two joints parallel to the opening axis. The Poisson's ratio assumed for the rock material is 0.3. On the discontinuity the shear stresses are governed by a simple Coulomb's friction law for which the friction angle $\varphi$ and the cohesion $c$ are assumed to be equal to 30° and 0.0 respectively. The discretizations for the boundary and crack are presented in Fig. 7.13, as can be seen no relative

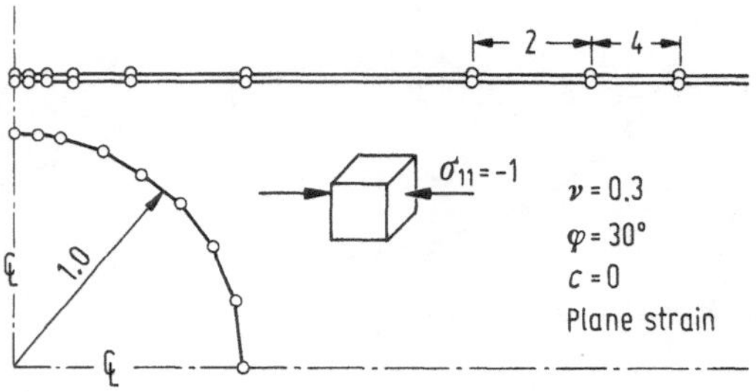


**Fig. 7.13.** Circular tunnel and discontinuity surface. Geometry and discretization

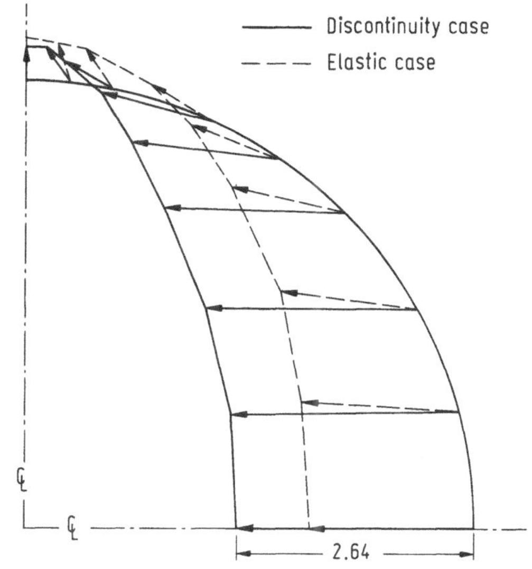


**Fig. 7.14.** Boundary displacements

displacement between the two discontinuity surfaces exists before the application of the load. Only horizontal loads are applied to the system in order to relieve the uniaxial system of residual compressive horizontal stresses ($\sigma_{11} = -1.0$). The results obtained with the boundary technique are shown in Figures 7.14 and 7.15. In the first figure the new position of the boundary nodes is presented in comparison with the corresponding displacements for non fissured case. The second diagram

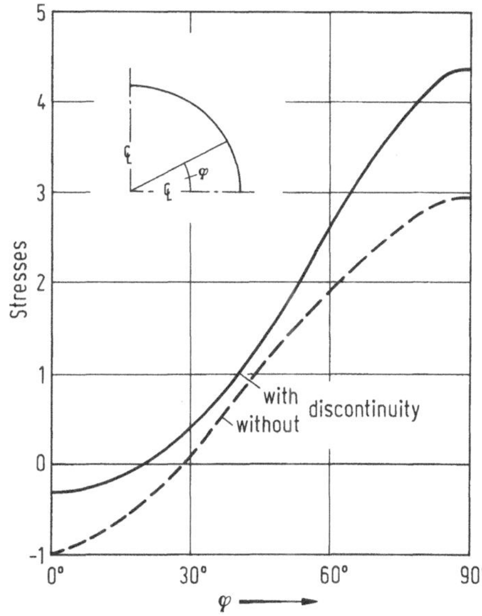


**Fig. 7.15.** Normal stress on the tunnel surface

shows the distribution of the normal stress in the tangential direction over the tunnel boundary. These results also illustrate the effects of the discontinuity when compared with the solution obtained using intact rock medium.

## 7.4 Viscoplasticity

For the analysis of stresses and displacements of any soil or rock structural system the material properties must be considered as realistically as possible in order to provide safe and stable design. The previous no-tension and plasticity theories are based on neglecting the time-dependency of the material properties. However, it is known that in many geomechanical problems the actual behaviour of the material is governed by rheological properties. For instance, plastic solutions are acceptable when the onset of the plastic deformations occurs much faster than the loading time, i.e. the concept of instantaneous development of permanent and irreversible strain is valid. On the other hand, the viscous effects become important when permanent deformations take a comparatively long time to be developed.

In order to introduce time-dependent effects into the boundary formulation, the Perzyna's model [9] is adopted here. Using this model the stress conditions which indicate the onset of viscoplastic deformations are governed by a well-known plastic yield function, i.e. irreversible viscous deformations take place when

$$F(\sigma_{ij}, k) \geqq 0 \tag{7.29}$$

where $F$ is the yield function which can also be represented by,

$$F(\sigma_{ij}, k) = f(\sigma_{ij}) - Y(k) \geqq 0 \tag{7.30}$$

In equation (7.30) $Y(k)$ stands for the yield stress, $k$ is a history dependent hardening (or softening) parameter, and $f(\sigma_{ij})$ can be interpreted as an equivalent uniaxial stress value $\sigma_e$.

In the usual manner for nonlinear problems, the total strain can be separated into elastic and inelastic parts. In this case in particular, the total strain rate, $\dot{\varepsilon}_{ij}$, can be expressed as,

$$\dot{\varepsilon}_{ij} = \dot{\varepsilon}^e_{ij} + \dot{\varepsilon}^p_{ij} \tag{7.31}$$

in which $\dot{\varepsilon}^e_{ij}$ and $\dot{\varepsilon}^p_{ij}$ stand for the elastic and viscoplastic strain rates.

Adopting the concept of plastic potential from the plasticity theory the viscoplastic strain rate in equation (7.31) can be written as,

$$\dot{\varepsilon}^{vp}_{ij} = \gamma \langle \Phi\,(F/F_0) \rangle \frac{\partial g\,(\sigma_{ij})}{\partial \sigma_{ij}} \tag{7.32}$$

where $\gamma$ is the viscosity parameter of the material and $F_0$ denotes any convenient reference value of $F$, usually $Y(k)$, for the dimensionless representation of $\Phi$.

The function $g(\sigma_{ij})$ is also a scalar function of the stress values and equal to $f(\sigma_{ij})$ for associative viscoplasticity cases.

In equation (7.41) the notation $\langle\,\rangle$ implies

$$\begin{aligned} \langle \Phi\,(F/F_0) \rangle &= \Phi\,(F/F_0) \qquad \text{if } F > 0 \\ \langle \Phi\,(F/F_0) \rangle &= 0 \qquad\qquad\quad \text{if } F \leqq 0 \end{aligned} \tag{7.33}$$

The function $\Phi$ is defined from experimental tests and may assume different forms, as suggested in reference [9] in which the following representations have been proposed

$$\begin{aligned} \Phi\,(F/F_0) &= F/F_0 \\ \Phi\,(F/F_0) &= (F/F_0)^n \\ \Phi\,(F/F_0) &= \exp\,(F/F_0) - 1 \end{aligned} \tag{7.34}$$

which correspond to linear, power and exponential types respectively.

After multiplying both sides of equation (7.32) by $\sigma_{ij}$ gives,

$$\sigma_{ij}\,\dot{\varepsilon}_{ij}^{p} = \gamma \langle \Phi\,(F/F_0) \rangle\, \sigma_{ij} \frac{\partial g\,(\sigma_{ij})}{\partial \sigma_{ij}} \tag{7.35}$$

Assuming that $g\,(\sigma_{ij})$ is homogeneous of degree one and applying Euler's theorem, one obtains,

$$\sigma_{ij}\,\dot{\varepsilon}_{ij}^{p} = \gamma \langle \Phi\,(F/F_0) \rangle\, g\,(\sigma_{ij}) \tag{7.36}$$

Using the definition of work hardening as in reference [19] equation (7.36) becomes

$$\dot{\varepsilon}_{e}^{p} = \gamma \langle \Phi\,(F/F_0) \rangle \tag{7.37}$$

in which $\dot{\varepsilon}_{e}^{p}$ stands for the equivalent viscoplastic strain rate.

*Overlay Model.* As it is known, for soils and rocks the use of overlay models [12] can be useful for a better representation of the material behaviour. The overlay technique is based on assumptions that the body is formed by several layers or overlays for which different material properties and a weighting parameter (overlay thickness) are assigned. By enforcing the same strain pattern in each overlay, different stress responses are obtained which are employed to compute the final stress values as follows,

$$\sigma_{ij} = {}^{(m)}\sigma_{ij}\, h_m \qquad (m = 1, \ldots, M) \tag{7.38}$$

where ${}^{(m)}\sigma_{ij}$ stands for the stress level at each overlay, $M$ is the total number of overlays adopted $h_m$ represents the overlay thickness, and,

$$\sum_{m=1}^{M} h_m = 1 \tag{7.39}$$

Using the general form of equation (7.31) for a continuum problem, the total strain rate tensor of a particular overlay can be represented by,

$${}^{(m)}\dot{\varepsilon}_{ij} = {}^{(m)}\dot{\varepsilon}_{ij}^{e} + {}^{(m)}\dot{\varepsilon}_{ij}^{p} \tag{7.40}$$

where the total strain rate tensor must be the same in all overlays.

Multiplying equation (7.40) by the elastic compliances gives,

$${}^{(m)}\dot{\sigma}_{ij} = {}^{(m)}\dot{\sigma}_{ij}^{e} - {}^{(m)}\dot{\sigma}_{ij}^{p} \tag{7.41}$$

For practical purposes the incremental form of equation (7.41) obtained by integrating it over an increment of time $\Delta t$ is usually adopted as follows,

$$^{(m)}\Delta\sigma_{ij} = {}^{(m)}\Delta\sigma^{e}_{ij} - {}^{(m)}\Delta\sigma^{p}_{ij} \tag{7.42}$$

which can also be written as a function of the total values,

$$\Delta\sigma_{ij} = \Delta\sigma^{e}_{ij} - \Delta\sigma^{p}_{ij}\,. \tag{7.43}$$

As the same strain pattern is enforced in all overlays the viscoplastic stress increment tensor, $\Delta\sigma^{p}_{ij}$, is also computed following equation (7.37), i.e.,

$$\Delta\sigma^{p}_{ij} = {}^{(m)}\Delta\sigma^{p}_{ij}\, h_m \qquad (m = 1, 2, 3, \ldots, M) \tag{7.44}$$

in which $M$ is the overlay number.

The procedure adopted for the viscoplastic solution is similar to those used to model no-tension responses presented in the previous section, and consists of the following steps,

(i) Starting from known values of $^{(m)}\boldsymbol{\sigma}^n$ and $^{(m)}F^n$, which represent the total stress vector and the yield function $F$ for the overlay "$m$" at time $t^n$, the strain rate vector, $\dot{\boldsymbol{\varepsilon}}^p$, is computed by equation (7.32); then the corresponding stress rate vector, $^{(m)}\dot{\boldsymbol{\sigma}}^p$ is also evaluated.

(ii) Choosing a time step $\Delta t$ and adopting a simpler Euler time integration scheme, the viscoplastic stress increment vector of one overlay is calculated as follows,

$$^{(m)}\Delta\boldsymbol{\sigma}^p = \int_t^{t+\Delta t} {}^{(m)}\dot{\boldsymbol{\varepsilon}}^p\, dt = {}^{(m)}\dot{\boldsymbol{\varepsilon}}^p\, \Delta t \tag{7.45}$$

and the weighted value computed by,

$$\Delta\boldsymbol{\sigma}^p = {}^{(m)}\Delta\boldsymbol{\sigma}^p\, h_m \qquad (m = 1, 2, \ldots, M) \tag{7.46}$$

(iii) Applying the weighted viscoplastic stress increments, $\Delta\boldsymbol{\sigma}^p$, as an initial stress field, gives the total elastic stress increment vector,

$$\Delta\boldsymbol{\sigma}^e = \mathbf{S}\,\Delta\boldsymbol{\sigma}^p \tag{7.47}$$

or the corresponding elastic value for each overlay,

$$^{(m)}\Delta\boldsymbol{\sigma}^e = {}^{(m)}\mathbf{S}\,\Delta\boldsymbol{\sigma}^p \tag{7.48}$$

(iv) The total stress vector for one overlay "$m$" can now be obtained at time $t^{n+1} = t^n + \Delta t$ by accumulating the elastic and viscoplastic stress increment components, i.e.,

$$^{(m)}\boldsymbol{\sigma}^{n+1} = {}^{(m)}\boldsymbol{\sigma}^n + {}^{(m)}\Delta\boldsymbol{\sigma}^e - {}^{(m)}\Delta\boldsymbol{\sigma}^p \tag{7.49}$$

then the total stress vector for the whole body is computed once more using the weighted parameter,

$$\boldsymbol{\sigma}^{n+1} = {}^{(m)}\boldsymbol{\sigma}^{n+1}\, h_m \qquad (m = 1, 2, \ldots, M) \tag{7.50}$$

(v) Before computing the next time step, the new value of the equivalent plastic strain vector can be determined as follows,

$$\varepsilon_e^{p(n+1)} = \varepsilon_e^{p(n)} + \Delta\varepsilon_e^p \tag{7.51}$$

in which $\Delta\varepsilon_e^p$ is computed employing the equivalent viscoplastic strain rate given in equation (7.37).

It is important to notice that the steps shown above only model the time behaviour starting from any state of stress of the body under consideration. In order to consider the application of any load, whether at the beginning or during the process, its elastic effects must be instantaneously added to both elastic and total stress vectors.

The success of the incremental technique shown above for solving elasto/viscoplastic problems is directly dependent upon the appropriate time step selection. Then, in order to regulate the time step length, two procedures borrowed from finite element formulation are adopted here.

The first corresponds to the application of a limit time step formulated for creep problems [20], and frequently adopted in geomechanics. The time step length is chosen in order to achieve accuracy in the results and is calculated as a function of the relation between the equivalent viscoplastic strain rate and the accumulated equivalent strain.

The technique recommended above cannot avoid instability problems due to error propagations. Therefore, the time step limit proposed by Cormeau [21] must also be followed. In this case, the time step length is evaluated based on the material properties, the adopted yield function and the viscoplastic constitutive law.

In the examples which follow this section, the viscoplastic algorithm is not only employed to model time-dependent behaviour but can also be used for plasticity. As has been known for some time, the plastic behaviour can be obtained by applying the load in increments and allowing the solution to progress until stationary conditions are attained. After applying an increment of load the steady state solution which corresponds to the condition $F \leqq 0$ is approached asymptotically with the viscoplastic strain rate diminishing as the yield surface is approached. As for plastic solution, the condition $F \leqq 0$ must be attained during the whole loading process; the accuracy of such an approach is directly dependent on the increment lengths adopted.

*Plastic Application – Strip Footing*

In this example the ultimate capacity of a 10 ft wide flexible strip footing resting on the surface of elastic perfectly plastic soil is analysed. The material was assumed to have an elastic modulus of $4.82 \times 10^6$ lb/ft$^2$, a Poisson's ratio equal to 0.3, and to be isotropic and homogeneous. The plastic deformations were assumed to be governed by the Mohr-Coulomb criterion with associative flow rule for which the following material parameters were used,

$$c = 1440 \text{ lb/ft}^2$$

$$\varphi = 20^\circ$$

The final plastic solution was obtained by using the viscoplastic algorithm. The load was applied in small increments and the convergence was assumed when the stationary conditions were achieved. The time step and the viscosity parameter have been chosen in order to obey the stability conditions given in reference [21].

*a) Weightless Soil Case.* We shall start the analysis of the strip footing by neglecting the soil weight. The boundary and internal discretizations employed to run the problem are shown in Fig. 7.16 together with the boundary conditions prescribed in this case. In common with finite element analysis for this example (see reference [10]) a closed domain has been chosen. This kind of restriction for which the displacements are enforced to be zero at points not very distant from the applied load is a normal procedure in finite element techniques and has proved to be acceptable for the bearing capacity determinations.

The collapse load predicted by Prandtl was simulated with 1% of error. The load increment for which the convergence was verified gives the total amount of 21 600 lb/ft$^2$, while the theoretical solution is 21 370 lb/ft$^2$. The problem has also been solved with the associated Drucker-Prager yield criterion and the same bearing capacity was obtained.

The load displacement curve for both criteria are shown in Fig. 7.17 where one can see that the displacements for the Drucker-Prager criterion are larger than the corresponding values computed using the Mohr-Coulomb surface.

The growth of the plastic zone defined by the Mohr-Coulomb criterion is shown in Figure 7.18. It began at points underneath the load and only spread all over the domain when almost all increments of load had been applied.

The determination of the limit load has also been carried out by analysing the elasto/viscoplastic response. The load is applied in one increment and the final

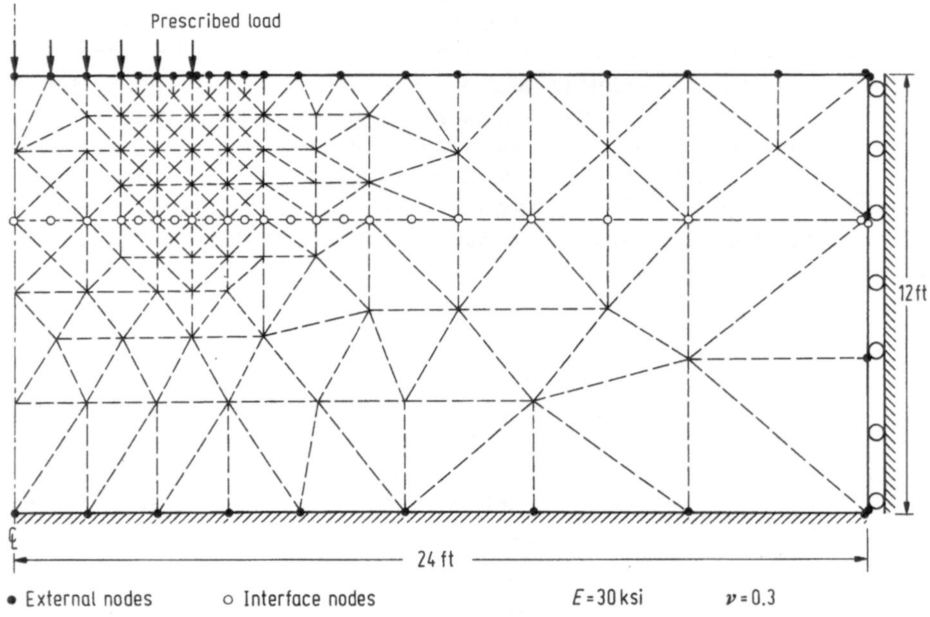


**Fig. 7.16.** Strip footing. Problem definition and discretizations

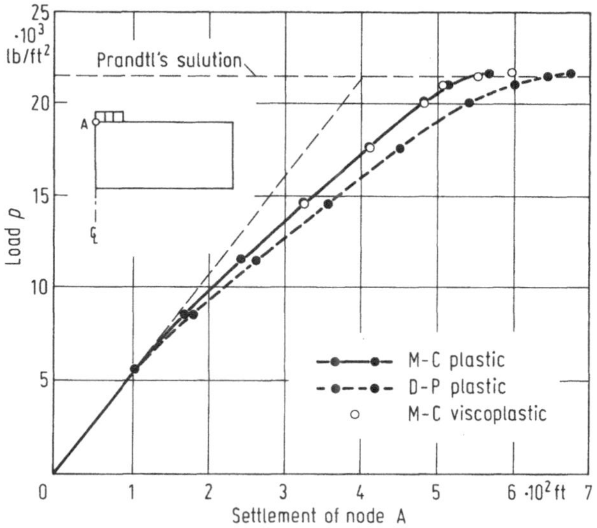


**Fig. 7.17.** Load displacement curve

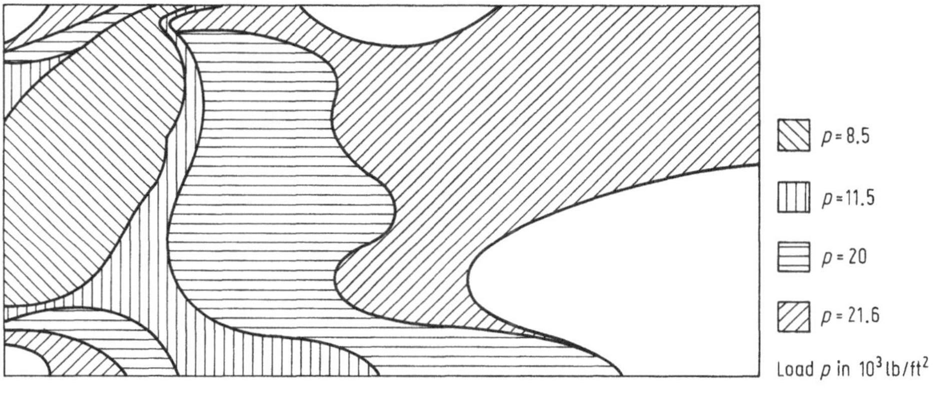


**Fig. 7.18.** Plastic zones

solution is obtained when the stationary conditions are achieved. By solving the problem several times it was found that a load equal to 21 600 lb/ft$^2$ was the largest value or which the steady solution was obtained. For any load greater than this value, a continuous growth of the displacements with respect to time was observed. In spite of being a non incremental procedure, the displacements computed with this procedure are similar to those obtained following the incremental scheme. Significant differences have been verified only near the limit load, as shown in Figure 7.17.

*b) Bearing Capacity Considering the Weight of the Soil.* This example represents a more realistic situation if compared with the previous case. For instance, it can

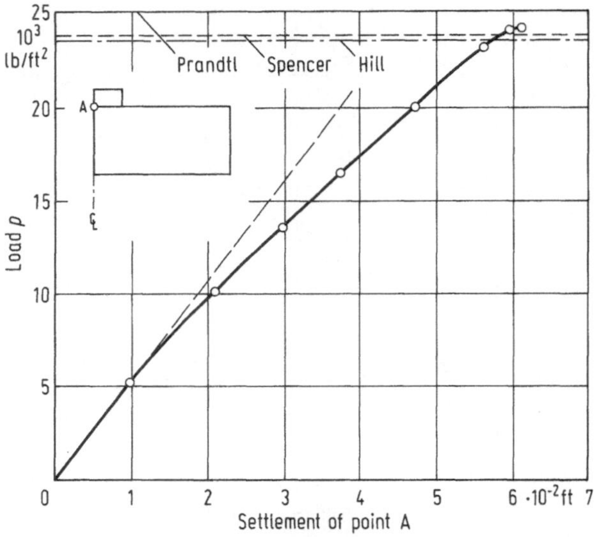


**Fig. 7.19.** Load displacement curve considering the weight of soil

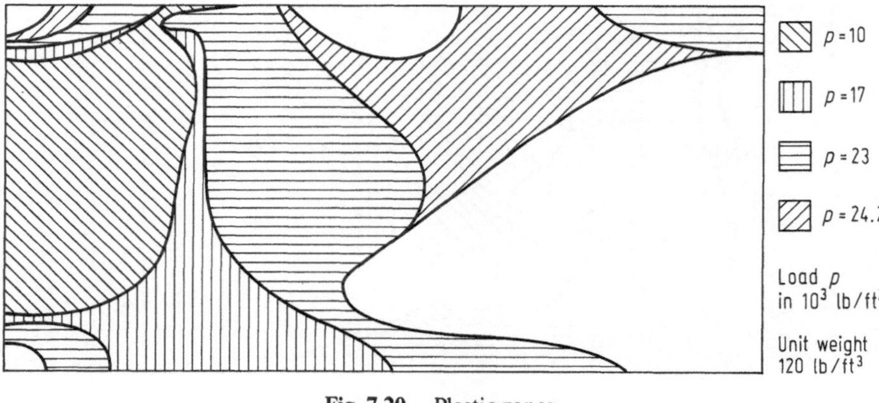


**Fig. 7.20.** Plastic zones

represent any foundation resting on the ground surface. The self-weight of the soil $\bar{\gamma} = 120$ lb/ft$^3$ is taken into consideration by assuming a pre-existing stress field in the material. The vertical stresses are assumed to be equal to the soil weight while the horizontal values are computed considering the earth pressure coefficient at-rest, $k_0$, which is given in this case by $\nu/(1-\nu)$.

The same analysis carried out for the weightless case has been repeated here. A limit load equal to 24200 lb/ft$^2$ has been achieved and is much smaller than the collapse load computed using the Prandtl's mechanism, which is 25080 lb/ft$^2$. However, the collapse load assuming Prandtl's mechanism is rigorously an upper bound. In this case, the ultimate capacity achieved by adopting the Hill mechanism [22] and assuming valid the Terzaghi superposition [23] is 23450 lb/ft$^2$, only 3% smaller than the numerical result and can be considered a more realistic solution. Spencer [24] has also achieved a collapse load for this problem by using a perturbation technique, and his solution, 23800 lb/ft$^2$, is less than 2% smaller than the numerical limit obtained. Assuming the Drucker-Prager criterion as for case "*a*", the limit load obtained was virtually the same.

The load displacement curve for Mohr-Coulomb yield criterion is shown in Fig. 7.19 where the theoretical limit loads mentioned above are also presented. The plastic zone developed during the loading is presented in Figure 7.20. By comparison with the first case, the final yield zone is smaller, which can be justified by the compressive state of stress assumed to exist before the loading.

*Viscoplastic Application – Tunnelling Problem*

The example presented in this section consists of analysing the displacements around a circular opening and in the concrete lining which was inserted some time after the excavation. This problem has been taken from reference [11] where the corresponding finite element solution is presented.

For all cases analysed, the rock material was assumed to have a constant residual stress field before the tunnel is cut. The vertical stresses have been taken equal to 80000 lb/ft$^2$, while the horizontal stresses were computed according to the earth pressure coefficient at-rest, $k_0$, assumed to 0.5. For the rock and concrete, the following values of the elastic parameter have been adopted.

**Table 7.1.** Circular Tunnel. Elastic values

| Parameter | Material | |
|---|---|---|
| | Rock | Concrete |
| Elastic modulus (lb/ft$^2$) | $7.2 \times 10^7$ | $4.32 \times 10^8$ |
| Poisson's ratio | 0.15 | 0.20 |

Before the lining insertion, the opening has an internal diameter equal to 11 ft. After the application of the concrete support the internal diameter is reduced to 10 ft. The boundary and internal discretizations, together with the geometry of the problem, are presented in Figure 7.21. As can be seen, only a quarter of the domain needed to be discretized due to the symmetry. Several loadings are here assumed to represent different stages of the tunnel construction and they will be illustrated separately in this section.

*a) Unlined Case.* This is a situation which often ocurs in practical tunnelling problems, either due to the deliberate omission of the structural lining or due to a prolonged delay before the insertion of the support.

Depending on both the viscous properties of the rock and the speed of the excavation, elastoplastic or visco/elastoplastic solutions are acceptable to model the stress and displacement distribution around the cavity. Elastoplastic behaviour is assumed when the time-dependent displacements occur very quickly in com-

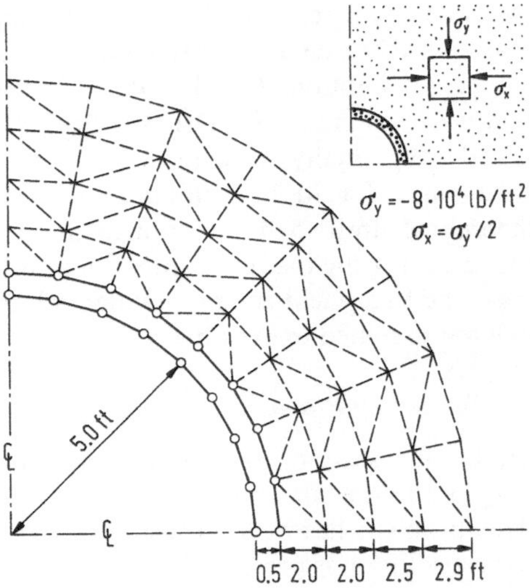


**Fig. 7.21.** Circular tunnel discretizations

parison with the face advancing. However, the case showing long-time displacements in which inelastic deformations are noticed far behind the face and for quite a long time is the most common situation.

In order to solve the tunnel excavation assuming visco/elastoplastic behaviour, Perzyna's model was employed together with an associative Mohr-Coulomb flow rule. The linear function $\Phi$ (see equation 7.34) was applied. The time was defined in units of $1/\gamma$ ($\gamma$ is the viscosity parameter) and the incremental stepping procedure has been taken according to the Cormeau's time step. The cohesion and the friction angle were taken to be $1.44 \times 10^4$ lb/ft$^2$ and 30° respectively. At this stage, the only load applied is due to the removal of the residual stress on the opening surface.

Assuming that the load is applied in one single increment, the final elasto/viscoplastic solution is obtained and the displacements over the tunnel surface are shown in Fig. 7.22, in which displacements after time equivalent at $5\Delta t$ are also presented.

*b) Lined Case.* In this case, the insertion of the lining is considered in order to restrain the viscous deformations. The loads, equivalent to the removal of the core, are applied in one increment and then the lining is placed at a practical distance from the face. As time and viscous parameter have been hypothetically assumed to solve this example, the distance from the application of the lining to the face is considered to be equivalent to a time interval equal to $5\Delta t$. Therefore, part of the viscous displacements have already occurred before the application of the support (see Fig. 7.22).

The lining is assumed to be of concrete material and will be displaced from the initial position only due to the viscous deformation of the rock. Figure 7.23 shows the final steady displacements for the lining, which compare well with the original finite element solution given in reference [11].

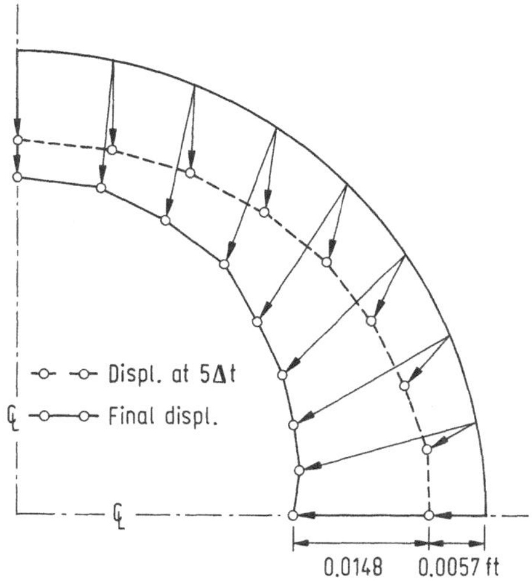


Fig. 7.22. Displacements of the rock without lining

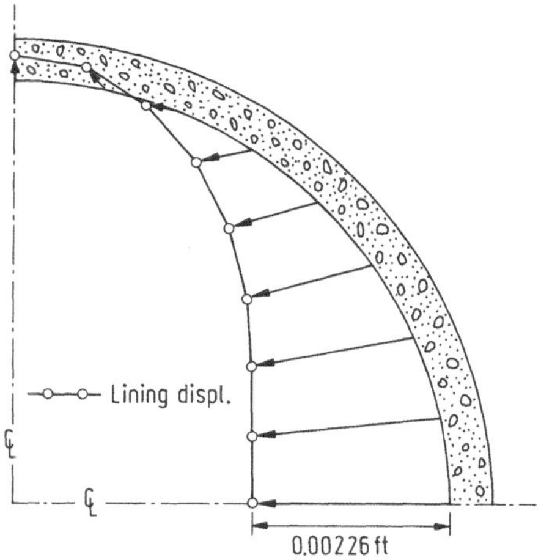


Fig. 7.23. Final lining displacements

*c) Alternative Case.* The transference of the stress from the rock to the lining support obtained in the last example is only due to the time-dependent deformations which took place in a small region (viscoplastic zone) in the vicinity of the interface between concrete and rock materials. However, in many tunnel designs it is convenient to assume that viscoelastic or creep effects occur for any change in the rock stress level. Using the overlay concept we can define a model to give both viscoplastic and viscoelastic responses. Here, the viscoplastic effects are modelled assuming the same conditions adopted in cases "*a*" and "*b*", and the viscoelastic deformations are governed by an equivalent Kelvin-Voight unit. Figure 7.24 shows the rheological model obtained by representing the rock material using two overlay models. Also in Fig. 7.24 the parameters adopted to simulate the viscoelastic and viscoplastic behaviours are presented. The internal discretization for this example has to be sufficiently increased in order to take into account all significant viscous deformations.

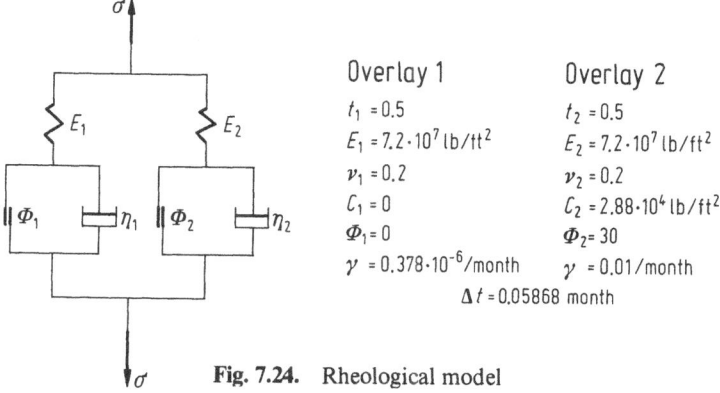

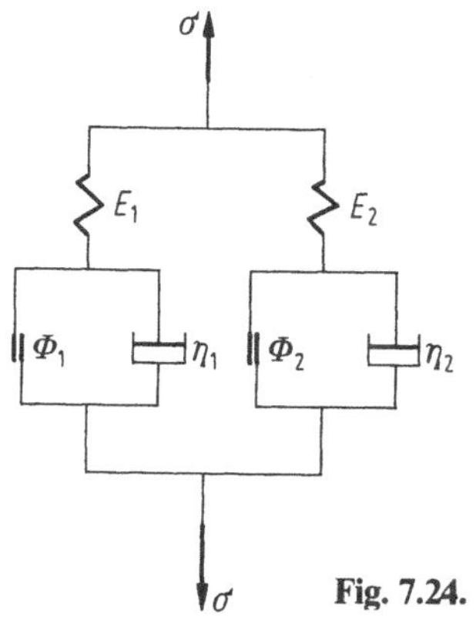


| Overlay 1 | Overlay 2 |
|---|---|
| $t_1 = 0.5$ | $t_2 = 0.5$ |
| $E_1 = 7.2 \cdot 10^7$ lb/ft$^2$ | $E_2 = 7.2 \cdot 10^7$ lb/ft$^2$ |
| $\nu_1 = 0.2$ | $\nu_2 = 0.2$ |
| $C_1 = 0$ | $C_2 = 2.88 \cdot 10^4$ lb/ft$^2$ |
| $\Phi_1 = 0$ | $\Phi_2 = 30$ |
| $\gamma = 0.378 \cdot 10^{-6}$/month | $\gamma = 0.01$/month |

$\Delta t = 0.05868$ month

Fig. 7.24. Rheological model

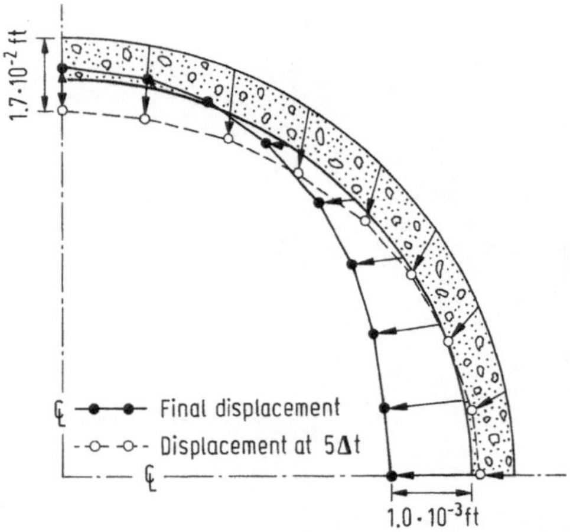


**Fig. 7.25.** Rock displacements after 5 $\Delta t$ and final lining displacements

As was expected with the assumptions of viscous effects everywhere, considerable changes in the final solution were verified. Figure 7.25 presents the boundary displacements after five time steps which also represent the time of the lining application. In this case, the displacements due to viscous deformations occur faster than in the previous case. As a consequence, the final steady solution (Fig. 7.25) shows smaller displacements in the lining, and the final forces acting on the interface between concrete and rock are smaller in comparison with the corresponding values obtained in "*b*" (Fig. 7.26).

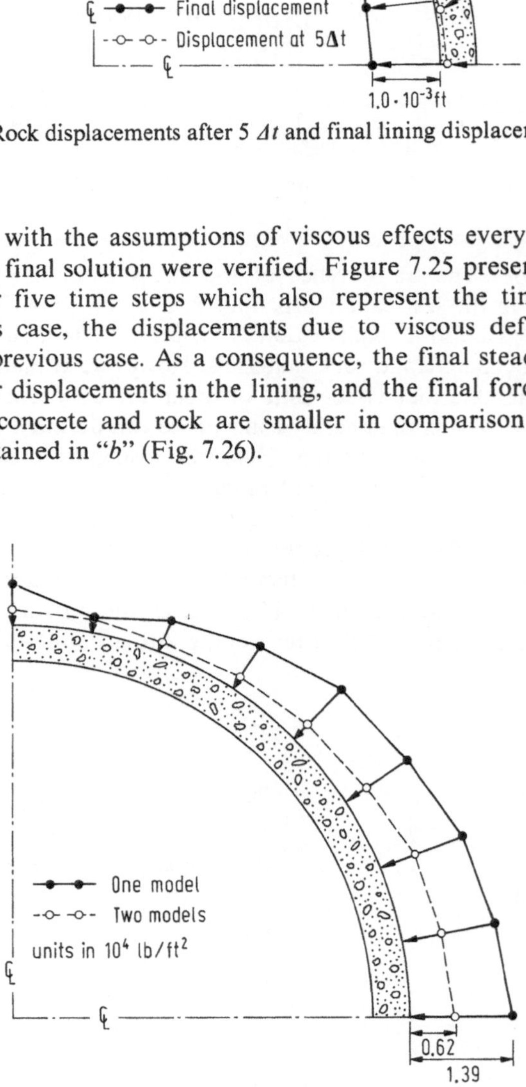


**Fig. 7.26.** Reaction on the lining surface

## 7.5 Conclusions

In this chapter the formulation of the boundary element method to deal with the no-tension and discontinuity processes of geomechanics has been presented. The examples presented here show the accuracy of the solutions. Boundary elements is attractive for solving geomechanical problems because of the considerable reductions in the data required to run a problem and the possibility of modelling properly boundaries extending to infinity.

The chapter explains how to formulate the nonlinear algorithm required for the solution of no-tension materials. They have also been extended to deal with viscoplastic problems using the overlay technique and Perzyna's model. The procedure can be used as well for the solution of time independent plasticity as a particular case.

## Appendix: Fundamental Solutions and Other Tensor Forms

In the section 2 of this chapter the boundary integral equations for displacement and stress determinations have been introduced. They involved the, so called, Kelvin's fundamental solution and other tensor forms. All these tensors correspond to responses at a source point "*s*" due to a unit load applied at a field point "*q*" in an infinite plane domain.

For completeness we resume here all of these forms,

fundamental solution for displacements:

$$u_{ij}^*(s,q) = \frac{-1}{8\pi(1-\nu)G}\{(3-4\nu)\ln r\,\delta_{ij} - r_{,i}\,r_{,j}\} \tag{A.1}$$

fundamental solution for tractions:

$$p_{ij}^*(s,q) = \frac{1}{4\pi(1-\nu)r}\left\{[(1-2\nu)\,\delta_{ij} + 2r_{,i}\,r_{,j}]\frac{\partial r}{\partial n} - (1-2\nu)(r_{,i}\,\eta_j - r_{,j}\,\eta_i)\right\} \tag{A.2}$$

Other tensor forms:

$$\varepsilon_{imk}^*(s,q) = \frac{1}{8\pi(1-\nu)Gr}\{(1-2)(r_{,k}\,\delta_{im} + r_{,m}\,\delta_{ik}) - r_{,i}\,\delta_{mk} + 2r_{,i}\,r_{,m}\,r_{,k}\} \tag{A.3}$$

$$\begin{aligned} S_{ijk}(s,q) = {} & \frac{2G}{4\pi(1-\nu)r^2}\{2r_{,n}[(1-2\nu)\,\delta_{ij}\,r_{,k} + \nu(\delta_{ik}\,r_{,j} + \delta_{jk}\,r_{,i}) - 4r_{,i}\,r_{,j}\,r_{,k}] \\ & + 2\nu(\eta_i\,r_{,j}\,r_{,k} + \eta_j\,r_{,i}\,r_{,k}) + (1-2\nu)(2\eta_k\,r_{,i}\,r_{,j} + \\ & + \eta_j\,\delta_{ik} + \eta_i\,\delta_{jk}) - (1-4\nu)\,\eta_k\,\delta_{ij}\} \end{aligned} \tag{A.4}$$

$$D_{ijk}(s,q) = \frac{1}{4\pi(1-\nu)r}\{(1-2\nu)[\delta_{ik}\,r_{,j} + \delta_{jk}\,r_{,i} - \delta_{ij}\,r_{,k}] + 25_{,i}\,r_{,j}\,r_{,k}\} \tag{A.5}$$

$$E_{ijmk}(s, q) = \frac{1}{4\pi(1-\nu)\, r^2} \{(1-2\nu)\,[\delta_{ik}\,\delta_{jm} + \delta_{jk}\,\delta_{im} - \delta_{ij}\,\delta_{mk} + 2\,\delta_{ij}\, r_{,m}\, r_{,k}]$$
$$+ 2\nu\,[\delta_{im}\, r_{,j}\, r_{,k} + \delta_{jk}\, r_{,i}\, r_{,m} + \delta_{ik}\, r_{,j}\, r_{,m} + \delta_{jm}\, r_{,i}\, r_{,k}]$$
$$+ 2\,\delta_{mk}\, r_{,i}\, r_{,j} - 8\, r_{,i}\, r_{,j}\, r_{,m}\, r_{,k}\} \tag{A.6}$$

In all the above expressions the $r_{,i}$ represents the derivatives of the distance between "$s$" and "$a$" with respect to the $x_i$ Cartesian coordinate of "$q$", i.e.,

$$r_{,i} = \frac{\partial r(s, q)}{\partial x_i(q)} = \frac{x_i(q) - x_i(s)}{r(s, q)}$$

where

$$r(s, q) = [(x_1^{(q)} - x_1^{(s)}]^2 + [x_2^{(q)} - x_2^{(s)})^2]^{1/2}$$

# References


1 Brebbia, C. A., *The Boundary Element Method for Engineers.* Pentech Press, London 1978

2 Brebbia, C. A., Telles, J. C. F., Wrobel, L. C., *Boundary Elements Techniques. Theory and Applications in Engineering.* Springer-Verlag, 1984

3 Telles, J. C. F., Brebbia, C. A., Plasticity. In: *Progress in Boundary Element Methods.* Ed. by Brebbia, C. A., Chapter 5, **2,** Springer-Verlag, 1981

4 Rizzo, F. J., Shippy, D. J., An application of the correspondence principles of linear viscoelastic theory. SIAM, J. Appl. Math., **21,** pp. 321–330, 1971

5 Telles, J. C. F., Brebbia, C. A., Elastic/viscoplastic problems using boundary elements. Int. J. Mech. Sci., Vol. 24, pp. 605–618, 1982

6 Zienkiewicz, O. C., Villiappan, S., King. I. P., Stress analysis of rock as a no-tension material. Geotechnique, **18,** No. 1, pp. 56–66, 1968

7 Venturini, W. S., Brebbia, C. A., Boundary element formulation to solve no-tension problems in geomechanics. In: *Numerical Methods in Geomechanics.* Ed. by Martins, J. B., NATO ASI Series, REIDEL, 1982

8 Reyes, S. F., Deere, D. U., Elastic-plastic analysis of underground openings by the finite element method. Proc. 1st Congress of the Int. Soc. of Rock Mechanics, **2,** pp. 477–483, Lisbon 1966

9 Perzyna, P., Fundamental problems in viscoplasticity. Advances in Applied Mechanics, **9,** pp. 243–377, Academic Press, New York 1966

10 Zienkiewicz, O. C., Humpheson, C., Lewis, R. W., Associated and non-associated visco-plasticity and plasticity in soil mechanics. Gèotechnique, **25,** 671–689, 1975

11 Zienkiewicz, O. C., Humpheson, C., Viscoplasticity: A generalized model for description of soil behaviour. In: *Numerical Methods in Geotechnical Engineering.* Ed. Desai, S. et al., McGraw-Hill 1977

12 Owen, D. R. J., Prakash, A., Zienkiewicz, O. C., Finite element analysis of non-linear composite materials by use of overlay systems. Computers and structures, **4,** pp. 1251–1267, 1974

13 Venturini, W. S., Brebbia, C. A., The Boundary Element Method for the Solution of No-tension Materials. In: *Boundary Element Methods.* Ed. Brebbia, C. A., Springer-Verlag, Berlin Heidelberg New York 1981

14 Goodman, R. E., Taylor, R. L., Brekke, T. L., A model for the mechanics of jointed rock. J. Soil Mech. Found. Div., ASCE, **94,** pp. 637–657, 1968

15 Love, A. E. H., *Treatise on the Mathematical Theory of Elasticity.* Dover 1944

16 Hocking, G., Stress analysis of Underground excavations incorporating slip and separation along discontinuities. In: *Recent Advances in Boundary Element Methods.* Ed. Brebbia, C. A., Pentech Press, 1978

17 Jaeger, J. C., Cook, N. G. W., *Fundamentals of Rock Mechanics.* John Wiley and Sons, Inc. New York 1976

18 Camargo, W. M., Projeto de Tuneis em Maçico Rochose sob Pressaõ Hidrostática Interna. Ph.D. Thesis, University of Sao Paulo 1968
19 Mendelson, A., *Plasticity: Theory and Applications.* Macmillan, New York 1968
20 Sutherland, W. H., AXICRIP; Finite element computer code for creep analysis of plane strain and axisymmetric bodies. Nucl. Eng. and Design, **11**, pp. 769–285, 1970
21 Cormeau, I. C., Numerical stability in quasi-static elasto/viscoplasticity. Int. J. Num. Meth. Engng., **9**, pp. 109–127, 1975
22 Hill, R., *The Mathematical Theory of Plasticity.* Clarendon Press, Oxford 1950
23 Terzaghi, K., *Theoretical Soil Mechanics.* Wiley, New York 1983
24 Spencer, A. J. M., Perturbation methods in plasticity. III, Plane strain of ideal soils and plastic solids with body forces. J. Mech. Phys. Solids, **10**, pp. 165–177, 1962

## Chapter 8

# Applications in Mining

*by G. Beer and J. L. Meek*

## 8.1 Introduction

Mining is the term used for the extraction of mineral from the earth's crust. The problems of analysis which arise are thus almost exclusively concerned with the excavation of ore from the "host material" and the stresses and displacements induced by the consequent mining excavations. The determination of these stresses and displacements then allows predictions to be made about the stability of the mine working and may affect the design of the mine layout. There are basically two ways in which ore is deposited, namely,

1. Tabular: In seams, lenses or strata of large extent in two dimensions but with relatively small thickness in the third dimension.
2. Massive: A three dimensional body of complicated shape.

For tabular orebodies the terms "roof" and "floor" or "hanging wall" and "footwall" are used for the surrounding rock mass surfaces adjacent to the seam. Figure 8.1 shows a cross-section of the Mount Isa Mine, Queensland, in Australia with both tabular and massive orebodies. One method of excavation is to create openings (or stopes) separated by ribs of the ore material to form pillars. These pillars provide support and prevent roof or floor collapse. The design or dimensioning of these pillars is critical to the efficiency of the mining operation.

In a second approach used mainly in coal mining, the roof is kept stable only in the working area. Collapse is permitted in mined areas away from the current excavation zone. The design of the support systems in the working area is vital to the success of this type of operation.

For massive orebodies, or steeply inclined tabular deposits, ore can be extracted by a method called "open stoping" in which large voids (or stopes) are formed by blasting. The dimensions of these "stopes" can be very large (in the order of 100 m). After blasting, the fragmented ore is drawn from the base of the stope. The stability of the walls of the stope (hanging wall and footwall) is of concern mainly because the collapse of host rock with no mineralisation into the stope results in dilution of the ore causing inefficiencies in the mineral processing operations. In many cases, cable dowels or rock bolts are installed to attempt to prevent hanging wall collapse. In the "open stoping", a certain pattern is followed leaving enough material in the pillars to ensure regional stability (Fig. 8.2). After the ore has been withdrawn, the stopes are backfilled with material which may be a mixture of

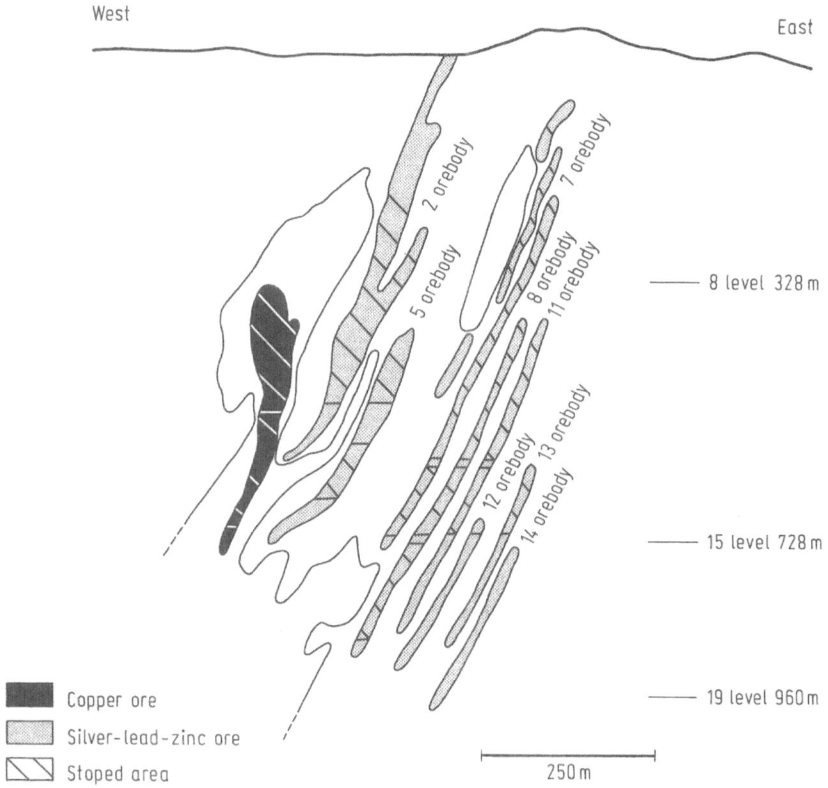


**Fig. 8.1.** Cross section of Northern Mine area, Mt. Isa, Queensland, Australia

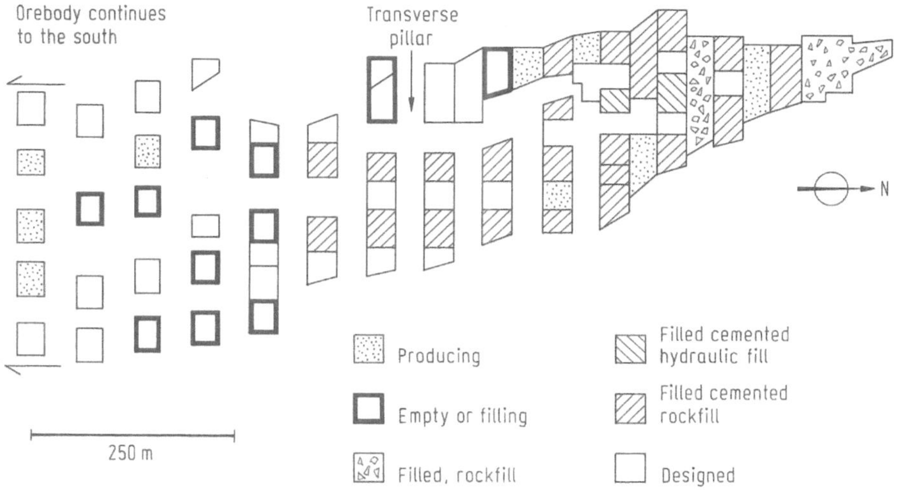


**Fig. 8.2.** Plan of open stoping in the 1100 orebody at Mount Isa

milltailings, cement and possibly rock fragments (rock fill). Extraction of the pillar ore follows. The objective is to achieve maximum extraction of ore with a minimum dilution and at the same time maintaining regional stability.

The above examples of mining methods are given to highlight some of the problems encountered in a mine operation. It can be seen that the analysis problems associated with a mine are large, complex and three-dimensional (see for example the large number of stopes and pillars in Figure 8.2). To contain the analysis effort to reasonable proportions, simplifications have to be introduced and these have influenced the development and application of Boundary Solution methods.

## 8.2 Review of the Development and Application of Boundary Element Formulations in Mining

The papers by Salamon [1–4] seem to be the earliest references on the application of Boundary Solution procedures to mining problems. Salamon concentrated on a solution for the problem when the thickness of the reef or seam is negligible compared with its overall dimensions. A good approximation to this problem is to assume that the thickness of the excavation is zero but that the displacements of the roof are different from those of the floor. A fundamental solution for a displacement discontinuity or dislocation in an infinite medium is derived and a "Face Element" approach used to obtain the solution for the actual excavation problem. The "Face Element" method is, in fact, identical to the Boundary Element method with constant Boundary Elements. The approach by Salamon was generalised and adapted to digital computers by Starfield [5] and Crouch [6]. The method has become increasingly popular in recent years and many workers have contributed to the further developments which have resulted in improvements in capability and economy [7–10]. The popularity of the method with mining engineers stems from the fact that the data input is extremely simple. Many problems which are too complex and too large to analyse by a full 3-D analysis can be solved.

Deist and co-workers [11–12] were the first to propose a method for analysing excavations in massive ore bodies. In this method, the excavation surface is discretised into "Face Elements" and a distribution of singularities is determined which reduces the traction at the center of each element to zero. The analysis is fully three-dimensional and the Kelvin solution is used. The method is essentially an indirect Boundary Element method (with constant Boundary Elements) because the displacements are not determined directly but rather from the distribution of singularities. In the implementation on the computer, Deist has been particularly concerned with efficient and economic solution of the system of equations. Complex mine configurations in some instances required discretisations of up to 4800 Face Elements with 14400 degrees of freedom. The solution of such large systems of equations is a formidable task especially when it is realized that the coefficient matrix is fully populated and non-symmetric. Deist uses an accelerated successive over-relaxation technique and "lumping" to solve the large system minimising the use of computer resources. The "lumping" method, which will be discussed in more detail later, gives coefficient matrices which are sparsely populated and a substantial decrease in the numerical effort required to solve the equations. The direct Boundary Element Method [13] was used by Cruse [14] to analyse isolated rock pillars in three dimensions.

The analyses discussed so far allow consideration only of homogeneous, linear elastic material. The extension of the Boundary Element method to piecewise homogeneous material has been presented by Lachat and Watson [15, 16] for the direct method and Banerjee [17] for the indirect method. Lachat and Watson seem to have been the first to use curved Boundary Elements with a higher order functional to obtain higher accuracy with fewer Boundary Elements.

Brady [18, 19] reformulated the Displacement Discontinuity method for analysis of openings in tabular orebodies. Instead of using displacement discontinuity solutions quadruple and hexapole solutions are used (these are fundamental solutions for four or six point forces acting opposite each other across a small gap). To simulate the opening, a set of forces is determined which reduces the total normal and shear stress components at the centres of Boundary Elements to zero. Agreement of solutions from this technique with analytical solutions appears to be good. As yet it does not seem to have been applied to practical problems in mining.

The first Boundary Element conference in Southampton in 1978 is significant for the presentation for the first time of practical applications of the B.E.M. in elasto-plasticity, although the theory was published earlier [20]. At this conference, Hocking [21] presented an algorithm for analysing slip and separation along discontinuities using a combination of the Displacement Discontinuity and Boundary Element method and Wardle and Crotty [22] dealt with practical applications of the piecewise homogeneous Boundary Element method in mining. There has been rapid development in the application of the B.E.M. to elasto-plasticity since that time [23]. More recently. efficient ways to combine the Boundary and Finite Element method have been investigated. Application of the coupled-method to mining problems has been reported by Beer [24, 25].

This chapter begins with a discussion of the different formulations used for the analysis of openings in the initial stress field. The formulations used are (1) the Displacement Discontinuity method, (2) the indirect B.E.M. and (3) the direct B.E.M.

In the literature, there has been a tendency for vastly different notations to be used in the different formulations. Herein the following notation has been applied rigorously

1) *Subscripts* denote *direction* ($i = 1, 2, 3; j = 1, 2, 3$). For example, $T_{ij}$ is the traction component in $x_i$ direction due to a source in $x_j$ direction. The usual summation rules apply.
2) *Values in parenthesis* denote *location.* For example, $T_{ij}(x, y)$ is the traction at location $x$ having the co-ordinates $x_i$ due to a source at location $y$, co-ordinates $y_i$. Alternatively, for constant Boundary Elements, $a_{ij}(m, m)$ is the influence coefficient at element $m$ due to a source at Element $n$.

3) *Superscripts* denote *region.* For example, $u_i^k$ is the displacement component in region $k$. Alternatively, $u_i^+$ is the displacement at the roof and $u_i^-$ the displacement at the floor of an excavation.

## 8.3 Displacement Discontinuity Formulation

In this formulation, the excavation is treated as having negligible thickness. Therefore, the roof and floor of the excavation in Fig. 8.3 is assumed to be at the same location. The displacements are now defined in mining terminology

$$\begin{aligned} d_1 &= u_1^- - u_1^+ \quad \text{Ride} \\ d_2 &= u_2^- - u_2^+ \quad \text{Ride} \\ d_3 &= u_3^- - u_3^+ \quad \text{Closure} \end{aligned} \tag{8.1}$$

There are only 3 stress components considered to act on the seam. That is, $\sigma_{33}$, the stress normal to the plane of the seam and the shear stresses $\sigma_{13}$ and $\sigma_{23}$ in the plane.

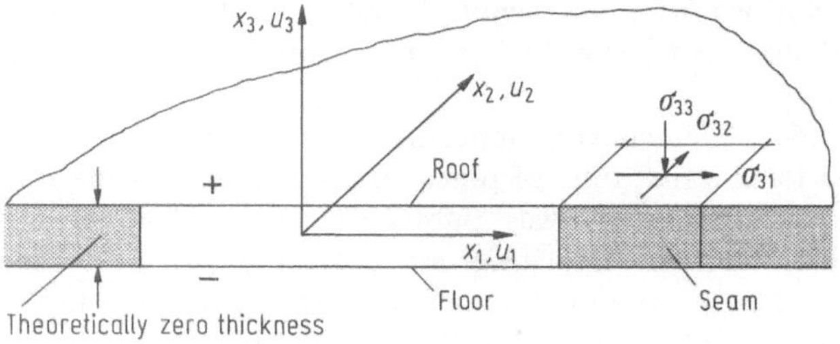


**Fig. 8.3.** Displacement discontinuity in an infinite medium

For the mined portion of the seam all components of stress must be reduced to zero. That is, the induced stresses at the excavation surfaces must exactly balance the premining stress components $p_{13}, p_{23}, p_{33}$. Thus,

$$\sigma_{i3} = -p_{i3} \tag{8.2}$$

In the unmined portions the following must hold if the seam material is infinitely stiff:

$$d_i = 0 \tag{8.3}$$

For an elastic seam the following relationship between stress and displacement exists

$$\sigma_{33} = E/t\, d_{33} \tag{8.4}$$

$$\sigma_{i3} = G/t\, d_{i3} \quad (i \neq 3) \tag{8.5}$$

where $E$ and $G$ are the compression and shear moduli of elasticity of the seam and $t$ is a fictitious thickness.

Solutions for the excavation in seams with any geometry can be found by subdividing the region of interest into a large number of small square zones or Boundary Elements of area $a$. The displacement discontinuities and stresses are assumed to be constant over each element and a central point assumed to be representative of the element. The stresses in element "$n$" can then be expressed as

$$\sigma_{33}(n) = \sum_{m=1}^{M} a_{33}(n, m)\, d_3(m) \tag{8.5}$$

and

$$\sigma_{i3}(n) = \sum_{m=1}^{M} a_{ij}(n, m)\, d_j(m) \quad (i, j \neq 3)$$

In the above, $d_i(m)$ is the average dislocation at element $m$ and the summation is for all elements $M$. The coefficients $a_{ij}(n, m)$ are fundamental solutions for influences of unit dislocations. The fundamental solutions have been derived using harmonic functions by Salomon [1] and Starfield [5]. The solutions can also be obtained by using the Somigliana identity and the Kelvin solution (see Appendix B).

In the early displacement discontinuity programs, a regular matrix of elements was assumed and an element defined by its row and column number. In this regular matrix excavated elements were flagged within the grid and any excavation pattern could be reproduced. The use of a regular grid allows the influence coefficients to be a function of row and column numbers (or more specifically differences in row and column numbers) thus making their evaluation more efficient. In addition, the data input is simplified to a great extent. For mine size problems however, the number of unknowns tends to become large and in the computer programs developed to date an iterative solution method with "lumping" has been used.

The explanation of the method herein has been restricted to the excavation of a single seam in an infinite domain. The formulation can be extended to excavations of more than one seam in proximity to each other in either infinite or semi-infinite media. These extensions are discussed in detail by Sinha [26]. When an iterative solution of the equations is used it is also possible to enforce Mohr-Coulomb and no-tension conditions in the unmined portions. Here the yield conditions are checked every few iterations and the stresses adjusted as required. This process seems to work [7] although results from non-linear analyses have not been tested rigorously.

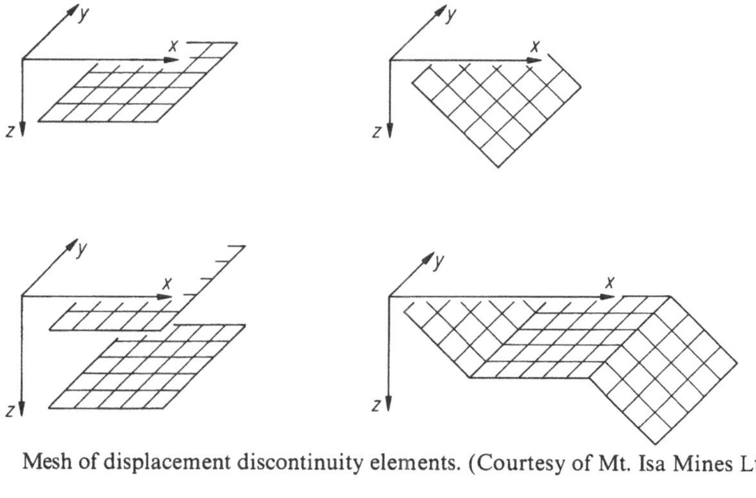


**Fig. 8.4.** Mesh of displacement discontinuity elements. (Courtesy of Mt. Isa Mines Ltd.)

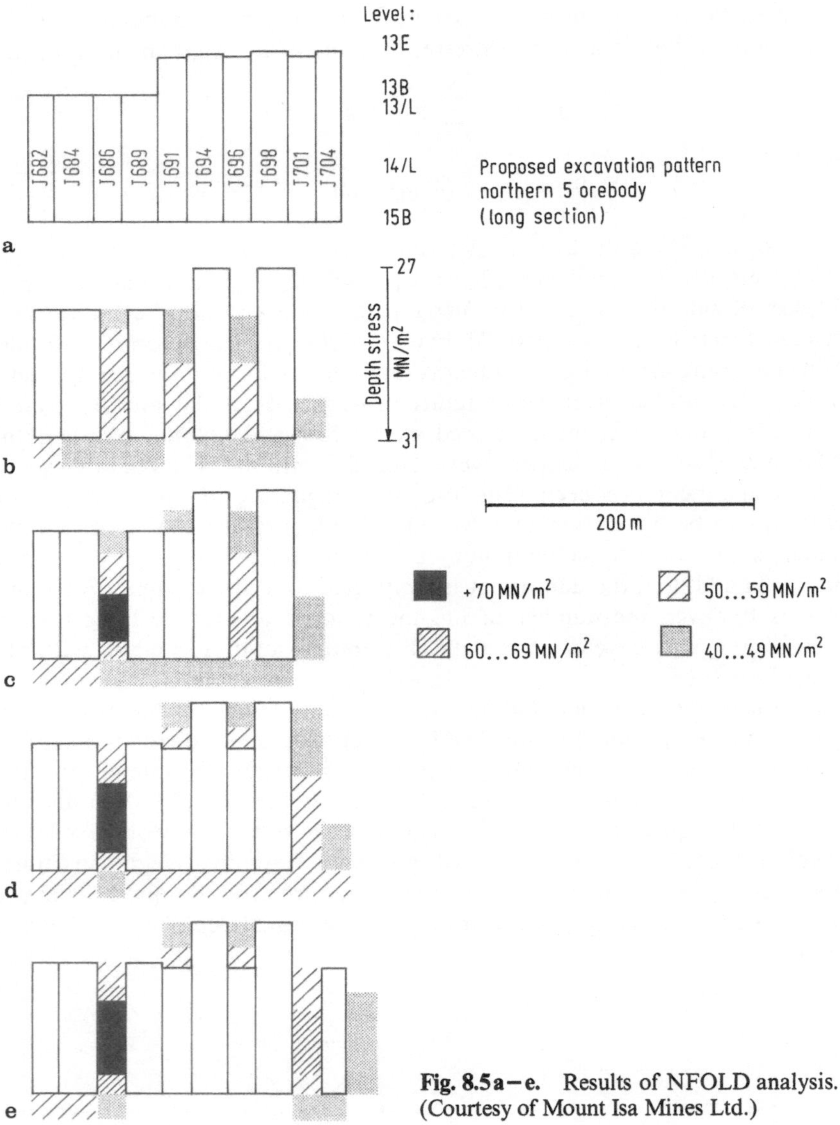


**Fig. 8.5 a–e.** Results of NFOLD analysis. (Courtesy of Mount Isa Mines Ltd.)

Figure 8.4 shows typical grid patterns used for the analysis of tabular seams. The pillar stresses for particular excavation stages at Northern 5 orebody at Mount Isa [68] are shown in Figure 8.5. The values of pillar stresses are found to be in good agreement with measurements. While the program gives good results for average pillar stresses, the detailed stress components in the vicinity of the edges of the excavation are found to be unreliable. Better results can be obtained by assuming a linear or parabolic variation of dislocation and stresses instead of constant or average values. Development in this direction has been carried out by Diering [27] and more recently by Crawford [10]. Displacement discontinuity programs have been extensively used by mining engineers mainly in South Africa, Australia and the USA. The reason for their popularity is the ease in data input which is oriented towards the methodology used in mining. Also, the solution of thin tabular excavations could prove to be difficult with classical Boundary Element methods because of the proximity of the roof and floor surfaces, a situation which can lead to ill-conditioning of the coefficient matrix.

## 8.4 The Boundary Element Formulation

*Indirect Boundary Element Formulation*

An indirect Boundary Element formulation was first used by Deist [12] in 1973 to analyse the excavation of massive orebodies. The method is called "indirect" because no direct relation is derived between the displacements and tractions at the Boundary. Instead, a distribution of fictitious forces is determined which gives a traction free excavation surface. From this distribution, the displacements are determined on the excavation surface and the induced stresses in the rock mass.

In the indirect method by Deist, the excavation surface is divided into a number of plane Elements as shown in Figure 8.6. The vector normal to the Element $k$ is $n(k)$ and the stress at the center of Element $k$ before the excavation is made is is $\sigma_{ij}^{0}(k)$. Subsequent to the excavation the traction

$$t_i(k) = n_j(k)\,\sigma_{ij}^{0}(k) \tag{8.8}$$

has to be reduced to zero.

If this traction is assumed to be constant over Element "$k$" the stress resultant is:

$$F_i(k) = t_i(k)\,S(k) \tag{8.9}$$

where $S(k)$ is the area of element "$k$".

A distribution of fictitious forces (sources) $P_i(l)$ is considered at the centres of all elements $L$, which gives forces $F_i(k)$, thence:

$$F_i(k) = \sum_{l=1}^{L} \Delta T_{ij}(k, l)\,P_j(l) \tag{8.10}$$

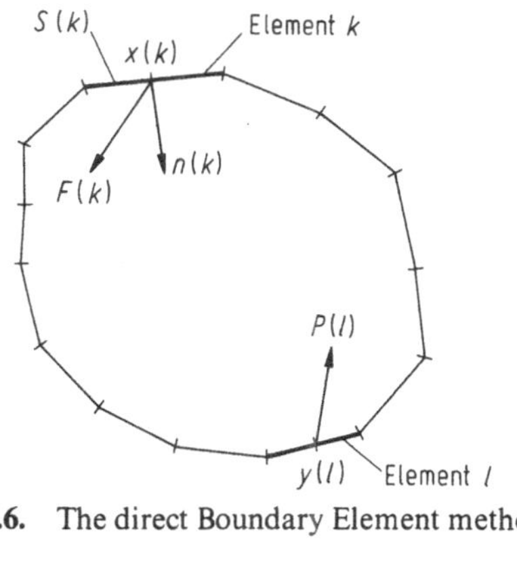


**Fig. 8.6.** The direct Boundary Element method

In equation (8.10), $\Delta T_{ij}$ is defined by,

$$\Delta T_{ij}(k, l) = \int_{S(k)} T_{ij}(x, y(l))\, dS(x) \tag{8.11}$$

In equation (8.11), $T_{ij}(x, y)$ is the fundamental solution for the tractions at $x$ due to a source at $y$ and $y(l)$ is the centre of element $l$. For $l = k$ i.e. when the source is at Element $k$ the fundamental solution tends to infinity and the integral does not exist. For smooth surfaces Deist [12] has shown that the coefficient can be computed from:

$$\Delta T_{ij}(k, l) = \tfrac{1}{2}\, \delta_{ij} \tag{8.12}$$

where $\delta_{ij}$ is the Kronecker Delta.

Equation (8.10) can be solved for the fictitious forces $P_j(l)$. The displacements at point $x$ inside the continuum can be computed from:

$$u_i(x) = \sum_{l=1}^{L} \Delta U_{ij}(x, l)\, P_j(l) \tag{8.13}$$

where

$$\Delta U_{ij}(x, l) = U_{ij}(x, y(l)) \tag{8.14}$$

$U_{ij}(x, y)$ is the fundamental solution for the displacement at $x$ due to a source at $y$.

The stresses at $x$ are

$$\sigma_{ik}(x) = \lambda\, \delta_{ik}\, \varepsilon_{mm} + 2G\, \varepsilon_{ik} \tag{8.15}$$

where $\lambda$ and $G$ are Lamè Constants of elasticity and

$$\varepsilon_{mn} = \frac{1}{2}\left(\frac{\partial u_m}{\partial x_n} + \frac{\partial u_n}{\partial x_m}\right) \tag{8.16}$$

Finally

$$\sigma_{ik}(x) = \sum_{l=1}^{L} \Delta D_{ikj}(x, l)\, P_j(l) \tag{8.17}$$

In equation (8.17), $\Delta D_{ikj}$ is a third order tensor obtained by differentiation of (8.13) and substitution into (8.15). Other indirect formulations have been used by Banerjee [29], Hocking [21], Brady and Bray [30] and Hoek and Brown [31].

The indirect method seems to be simpler than the direct method and easier to explain. It was applied in Civil and Mining Engineering before the direct method (e.g. Massonet [32], Oliviera [33]). An explanation of the method without the use of a single equation is given in [31] and this may appeal in particular to practising mining engineers. The method seems to give an accuracy similar to the direct method for examples involving smooth boundaries. Some doubts have been expressed recently about the accuracy of the indirect method when analysing problems with sharp corners [34].

An application of the indirect method to a large mining problem is shown in Fig. 8.7 which depicts a typical cross-section of the mine analysed. Using a "lumping procedure" to increase the sparsity of the system of equations, Deist and co-workers [12] claim to have solved the problem which comprises 4800 Elements and 14400 degrees of freedom in 3 hours on an IBM 370/145 computer. Unfortunately, any results of the analysis and the full mesh used have never been published.

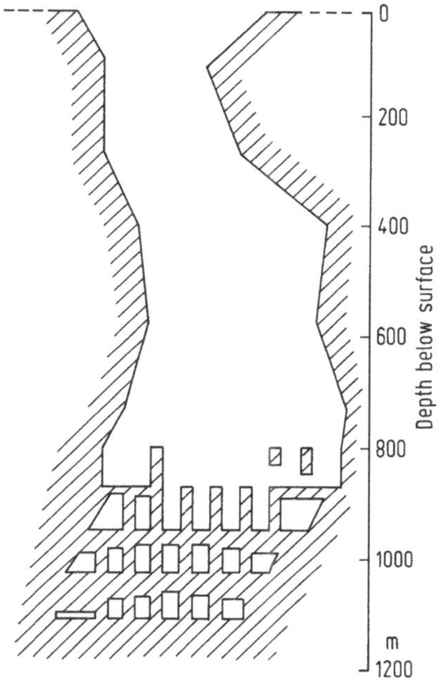


**Fig. 8.7.** Cross section of a mine involving multiple pillars

*The Direct Boundary Element Formulation*

A direct Boundary Element formulation was first proposed by Rizzo [13] in 1966. In this formulation Betti's reciprocal theorem is used to obtain an integral equation on the boundary in terms of displacement ($u_i$) and traction ($t_i$) values. This leads to:

$$c_{ij}(x)\, u_j(x) + \int_S T_{ij}(x, y)\, u_j(y)\, dS = \int_S U_{ij}(x, y)\, t_j(y)\, dS \tag{8.18}$$

where $T_{ij}$ and $U_{ij}$ are fundamental solutions for the tractions and displacements and $c_{ij}(x)$ is $\frac{1}{2}\delta_{ij}$ for smooth boundaries. The integral equation may be solved by approximating the boundary variables $u_i$, $t_i$ as piecewise constant over a finite area. Thus the surface of the problem to be analysed is divided into $M$ Boundary Elements and (8.18) rewritten in algebraic form as:

$$c_{ij}\, u_i(n) + \sum_{m=1}^{M} \Delta T_{ij}(n, m)\, u_j(m) = \sum_{m=1}^{M} \Delta U_{ij}(n, m)\, t_j(m) \tag{8.19}$$

where

$$\Delta T_{ij}(n, m) = \int_{S(m)} T_{ij}(x(n), y)\, dS(y) \tag{8.20}$$

and

$$\Delta U_{ij}(n, m) = \int_{S(m)} U_{ij}(x(n), y)\, dS(y) \tag{8.21}$$

where $x(n)$ is the location at the centre of Element $n$ and $S(m)$ is the surface of element $m$. In three dimensions the number of equations in (8.19) is $3M$. For an excavation problem the tractions on the excavation boundary have to be reduced to zero and the system of equation can be written as

$$[A]\{a\} = \{F\} \tag{8.22}$$

where $\{a\}$ is a vector containing the displacements at the centres of boundary elements and the coefficients of $\{F\}$ are defined as

$$\begin{aligned} f_1 &= \sum_{m=1}^{M} \Delta U_{1j}(n, m)\, t_j^0(m) \\ f_2 &= \sum_{m=1}^{M} \Delta U_{2j}(n, m)\, t_j^0(m) \quad \text{etc.} \end{aligned} \tag{8.23}$$

where

$$t_i^0 = -n_j \sigma_{ij}^0 \tag{8.24}$$

and $\sigma_{ij}^0$ is the premining stress tensor.

The coefficient matrix $[A]$ is fully populated and unsymmetric. The displacements at the boundary are determined by solution of (8.22). The displacement at an interior point $x$ is

$$u_i(x) = \sum_{m=1}^{M} \Delta U_{ij}(x, m)\, t_j(m) - \sum_{m=1}^{M} \Delta T_{ij}(x, m)\, u_j(m) \tag{8.25}$$

The stresses at $x$ can be determined using expression (8.15), that is,

$$\sigma_{ik}(x) = \sum_{m=1}^{M} \Delta D_{ikj}(x, m)\, t_j(m) - \sum_{m=1}^{M} \Delta S_{ikj}(x, m)\, u_j(m) \tag{8.26}$$

where the third order tensors $\Delta D_{ikj}$ and $\Delta S_{ikj}$ are determined from differentiation of (8.25).

The first application of the direct formulation to mining problems was reported by Cruse [14], who analysed a room and pillar structure representative of the excavation geometry of a coal mine. A more recent application of the direct B.E.M. with the displacements and tractions assumed constant over each Boundary Element has been presented by Diering [35]. Figure 8.8 shows the geometry used to model the interaction between an open pit and underground excavations. A plane of symmetry was assumed thus reducing the discretisation effort by one half. For this problem 350 Boundary Elements were used and it is remarkable that the solution was achieved on a 16 bit Data General mini-computer. This was made possible by using a lumping procedure and solving the system of simultaneous equations iteratively. The results of the analysis high-lighted areas of high tension or high shear on specified planes.

Ricardella [36, 37] appears to be the first to suggest a higher order variation of displacements and tractions across the Boundary Elements. He implemented a linear variation in a two-dimensional Boundary Element program. This program has been developed further and applied to the two-dimensional analysis of mining problems by Crotty and Wardle [38]. A 3-D analysis program using triangular Boundary Elements and linear variation of the unknowns as developed by Cruse and modified by Wardle to handle excavation problems but has not yet been applied to large scale mining problems [39]. Lachat and Watson [15, 16, 34, 40, 41] have written programs which use curved isoparametric elements with parabolic

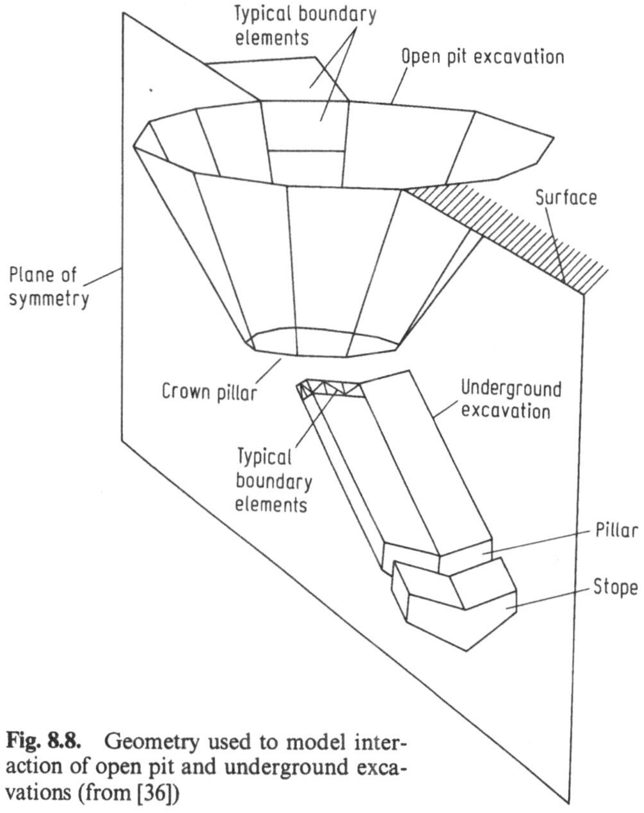


**Fig. 8.8.** Geometry used to model interaction of open pit and underground excavations (from [36])

shape and parabolic variation of the unknowns in an attempt to further increase the solution accuracy and decrease solution costs. Using parabolic Boundary Elements for the analysis of a spherical opening, Watson has shown [34] that if 4 Elements are used for one eight of the sphere an accuracy of about 1% can be obtained.

There is no doubt in the authors mind that the use of higher order shape functions make the analysis more economic and accurate. Using curved isoparametric Boundary Elements fewer are needed to describe complex boundary shapes and the use of higher order functions should improve the representation of tractions and displacements.

Watson [42] has recently developed this idea further by using Hermitian cubic Boundary Elements. These Elements have not only a cubic variation of shape and functions but also $C^1$ continuity on smooth surfaces. Thus, in three-dimensions, each node has three displacements plus derivatives in each of two tangential directions. Since there are no midside nodes the ratio of elements to nodes is high (i.e. a node is shared by several Elements). On the other hand, the use of better shape functions should give a significant improvement in accuracy. For Boundary Elements near corners special functions can be used which represent the singularities and stress concentration factors calculated directly. The implementation of

Hermitian cubic Boundary Elements requires considerable programming skill and the integration of the Kernel function products is quite complicated. At the present time, only a 2-D program with Hermitian cubic Elements has been implemented and tested. It remains to be seen if such a level of sophistication is desirable for the efficient analysis of mine size problems.

### *Subregions and Interfaces*

In the formulations discussed so far, the continuum has been assumed to be homogeneous and linearly elastic, a condition rarely attained in nature. Usually the ore and the host rock will have different properties and the host rock properties may change from one region to another. In addition, discontinuities may develop either at the interfaces of different materials or at pre-existing fractures. To model these properties, the rock mass can be divided into subregions which are connected along interfaces. Now, not only do the boundaries of the problem have to be discretised, but also the interfaces between subregions.

The Boundary Element equation (8.22) is set up for each subregion and at the interfaces the following conditions may exist:

a) no slipping or separation of adjacent regions with continuity of normal and shear stresses (rigid interface).
b) elastic slip or separation of adjacent regions (elastic interface).
c) inelastic slip or separation (rigid-plastic interface).
d) elastic and inelastic slip or separation (elasto-plastic interface).

For a rigid interface, the conditions to be satisfied between regions $k$ and $l$ are:

$$u_j^k = u_j^l \tag{8.27}$$

$$t_j^k = -t_j^l \tag{8.28}$$

For an elastic interface or joint, the tractions on the interface are related to the displacements by:

$$t_n^l = t_n^k = K_n(u_n^k - u_n^l) \tag{8.29}$$

$$t_s^l = t_s = K_s(u_s^k - u_s^l) \tag{8.30}$$

where $t_n$ and $t_s$ are the traction normal and tangential to the interface and $K_n$ and $K_s$ are the normal and shear stiffnesses.

For elasto-plastic and rigid-plastic interfaces two yield conditions must be checked for each node point.

*Yield condition 1* (no-tension)

$$F_T = t_n - T < 0 \tag{8.31}$$

*Yield condition 2* (Mohr-Coulomb friction law)

$$F_S = |t_s| + t_n \tan \varphi - c < 0 \tag{8.32}$$

When the yield conditions are violated the excess shear or tension has to be redistributed.

For a node where $F_T > 0$ separation occurs and shear can no longer be supported. The excess tractions $t_j^*$ are

$$t_s^* = -t_s \quad \text{and} \quad t_n^* = -F_T \tag{8.33}$$

For a node where $F_S > 0$ inelastic slip occurs at the interface. The excess traction parallel to the interface is

$$t_s^* = -\operatorname{sign}(t_s) \cdot F_S \tag{8.34}$$

The non-linear problem may be solved iteratively by redistributing excess tractions at each step. The iteration continues until the yield conditions are satisfied at all points of the interfaces. The procedure has been described for 2-D problems here but extension to 3-D involves only one additional tangential traction component. The yield and flow conditions can be written to include peak and residual strengths, joint dilatancy and other features which give a more realistic description of rock fractures. The treatment of non-linear material properties is identical to the one used for Finite Element and there is ample literature on the subject [43].

For rigid interfaces, the coefficient matrices $[A]$ for each subregion are assembled into a global matrix $[A]^G$ as shown in Fig. 8.9 for a problem consisting of two subregions. It can be seen that the matrix is no longer fully populated. Thus the effort in establishing and solving the system of equations can be significantly reduced when subregions are used. The use of subregions, therefore, is also advisable in cases where the material is homogeneous. Watson has used this technique extensively for three dimensional stress analysis problems [34]. Special solvers for the solution of such sparse matrices have been developed [44]. An application of the multi-region Boundary Element method to a mining problem was undertaken by Hocking [44]. He analysed the extraction of a massive orebody at Mount Isa Mines (Queensland, Australia) using a highly idealised model (Fig. 8.10). Here each pillar (shaded) is treated as a subregion and symmetry conditions have been applied in the $x-y$, $x-z$ and $y-z$ planes. The mesh consists of

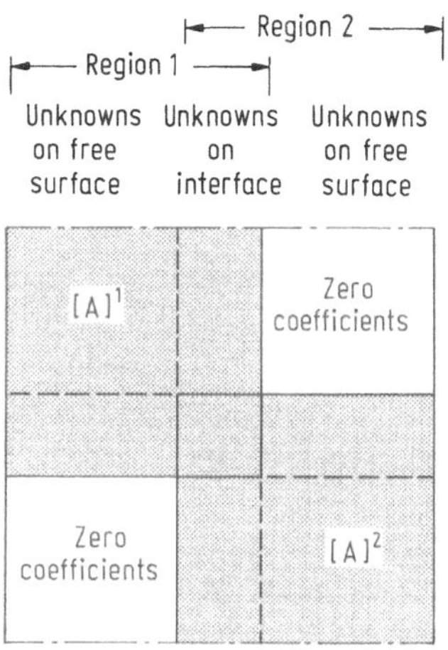


**Fig. 8.9.** Matrix $[A]$ for multi-region problem

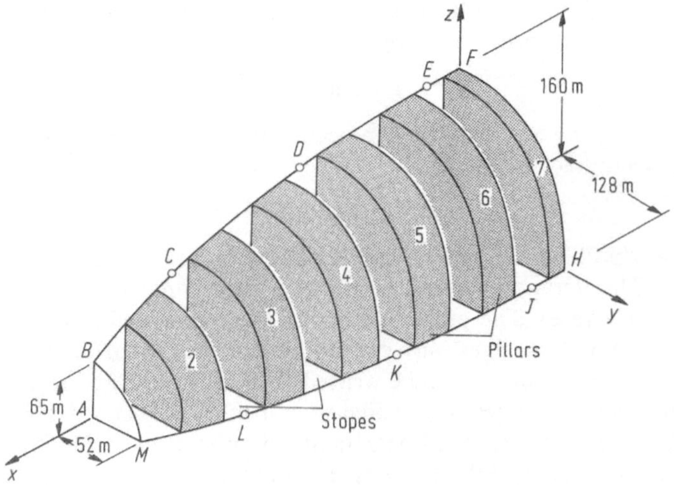


**Fig. 8.10.** Idealization used for the 1100 orebody at Mount Isa (from [46])

seven regions with over 100 isoparametric Boundary Elements of parabolic shape. The main result obtained by the study were the pillar stresses at different stages of excavation. However, these are of little practical use since the actual problem is much more complicated than the simple idealisation and the boundaries used. One obvious limitation must have been the number of Boundary Elements needed to discretise each of the 30 excavations in more detail. In three-dimensional stress analysis by the Boundary Element method, the costs of constructing and solving the system of equations escalates rapidly with the number of excavations making the solution of problems involving the thousands of elements impossible unless special techniques are used. This is discussed in more detail later.

When the infinite medium is divided into subregions the interfaces theoretically extend to infinity and an infinite number of Boundary Elements are needed to discretise the interface. Such problems can be solved by truncating the interface discretisation at points which are sufficiently far away from the region of interest. This method has been applied successfully by Crotty [46] and Bolteus [47]. A better and more accurate method might be to use infinite Boundary Elements as suggested by Watson [34].

Rigid-plastic interfaces can be analysed by redistributing excess tractions iteratively. This process may be included in the iterative solution of the system of equations. An example of an analysis of underground excavations in faulted rock is shown in Figures 8.11 a and 8.11 b. The figures show results of a program BITEMJ [46] which has been recently released by CSIRO, Division of Applied Geomechanics.

Another way of analysing excavations in faulted rock is by a combination of Boundary and Finite Element methods. In this approach two or more Boundary Element regions are connected with each other by joint Elements. The coupled analysis procedure will be discussed later.

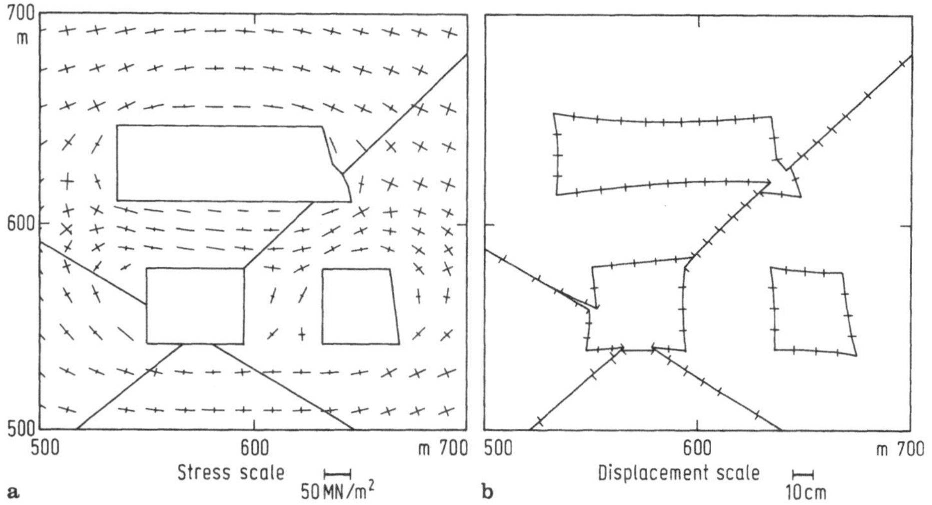


**Fig. 8.11.** Results of BITEMJ analysis of multiple excavation in jointed rock. (Courtesy of CSIRO, Div. of Geomechanics). **a** Displacements vectors. **b** Displaced boundaries

## 8.5 Elasto-plastic Material Behaviour

In addition to major faults, pre-existing fractures and fissures are present in the rock mass. These are classified under the term "structural discontinuities" and in most cases, their frequency and continuity is such that it is uneconomic to model them individually.

One approach has been to consider the combined or overall effect of these structures i.e. consider the rock mass as an elasto-plastic material. Mohr-Coulomb or Drucker-Prager yield conditions have been applied to materials which are essentially isotropic, i.e. where the structural discontinuities are distributed in a random manner. In cases where fractures have a predominant direction, a multi-laminate model [48] can be used. In this model, Mohr-Coulomb and no tension conditions are checked on certain planes only with stress redistribution occuring in each plane. This model is in many respects similar to the joint model discussed earlier except that the fracture is not modelled explicitly. Instead, the stresses at a number of interior points are checked. For a point at which the failure criteria are exceeded, the excess stresses are treated as initial stresses and redistributed. This initial stress approach is well known to workers in the Finite Element method [43].

It has been shown recently that these techniques can also be used with the Boundary Element method but require a discretisation of the continuum in the zone where elasto-plastic straining occurs. This discretisation is needed because the redistribution of the excess stresses involves the computation of a volume integral. Equation (8.18) is rewritten for elastoplastic problem as:

$$c_{ij}(x)\, u_j(x) + \int_S T_{ij}(x,y)\, u_j(y)\, dS = \int_S U_{ij}(x,y)\, t_j(y)\, dS + \int_V E_{ijk}(x,y)\, \sigma^0_{jk}(y)\, dV$$

where $E_{ijk}$ is the fundamental solution for the strain at $x$ [49]. To compute the volume integral the interior is divided into Finite Elements or cells. In the simplest case, rectangular cells are used and the stresses assumed to be constant over the cell. The discretised form of (8.35) is

$$c_j u_i(n) + \sum_{m=1}^{M} \Delta T_{ij}(n, m)\, u_j(m) = \sum_{m=1}^{M} \Delta U_{ij}(n, m)\, t_j(m) + \sum_{l=1}^{L} \Delta E_{ijk}(n, l)\, \sigma_{jk}^{0}(l) \tag{8.36}$$

where $L$ is the number of cells and

$$\Delta E_{ijk}(n, l) = \int_{V(l)} E_{ijk}(x(n), y)\, dV(y) \tag{8.37}$$

with the other variables defined previously.

The elasto-plastic analysis now proceeds as follows:

1. An elastic analysis is performed with specified values of tractions and zero initial stress $\sigma_{jk}^{0}$.
2. The stress is calculated at the centre of each cell and the yield condition checked.
3. For a point when the yield condition is violated, the excess stresses are computed and applied as initial stresses.
4. The problem is reanalysed with only the initial stresses applied and yield conditions rechecked for the superimposed results.

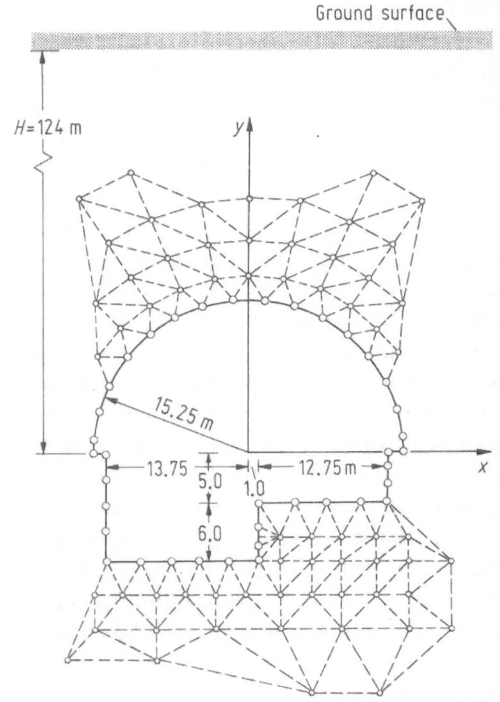


**Fig. 8.12.** Boundary discretisation and internal cells of cavern. (Courtesy of C. A. Brebbia [50])

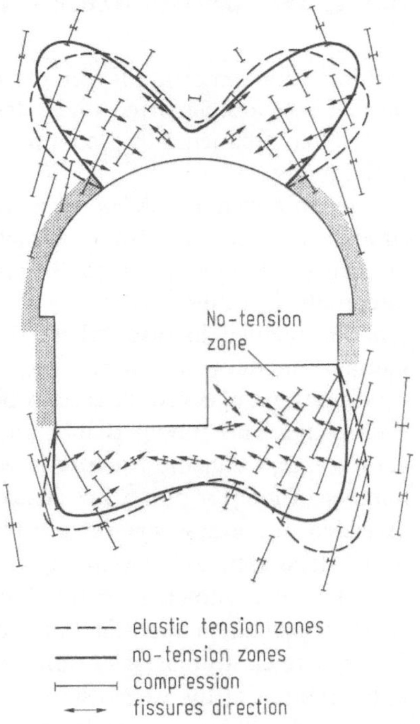


**Fig. 8.13.** Elastic and no-tension results. (Courtesy of C. A. Brebbia [50])

The process continues until the yield conditions are satisfied at all points. The elasto-plastic analysis thus involves an interior stress computation for each cell centroid, a volume integration and a solution for each iteration.

Examples of non-linear analysis in geomechanics are given by Venturini and Brebbia [50] and in Chap. 7 of this volume. In Fig. 8.12, the discretisation of an underground rock cavern into Boundary Elements and internal cells is shown. The result of the analysis assuming the rock to be a no-tension material is shown in Figure 8.13. There has been an increasing amount of research into the application in plasticity. However, elasto-plastic Boundary Element analyses have, to the authors' best knowledge, not been applied to mining problems.

## 8.6 Combination of BEM with Other Techniques

*Combination with the Finite Element Method*

In the Finite Element method a discretisation is necessary, not only on the boundary of the openings, but also through the volume of the continuum. This has obvious disadvantages when unbounded problems are analysed as the mesh has to be truncated or infinite elements used. Both approaches introduce errors in the solution. Also, because in most cases the displacement functions are defined locally, compatibility and equilibrium may be violated in the continuum at certain locations. While the Boundary Element method seems to be well suited for linear elastic, isotropic, homogeneous problems, the Finite Element method possesses certain advantages for non-linear and anisotropic problems. In such cases, a volume discretisation is also needed with the Boundary Element method. In addition, in the Finite Element method, the system of equations is banded and symmetric thus reducing the solution effort considerably.

For mining problems which involve non-linear and non-homogeneous material, the use of a combination of the Finite and Boundary Element methods becomes attractive. In its application, the best features of each method can then be used.

The idea of a "marriage a la mode" was presented by Zienkiewicz [52] some years ago. The coupling has been applied to problems in fluid mechanics and potential flow. Applications in elasticity have appeared more recently [52, 24, 25, 53]. A coupled Boundary Element – Finite Element computer program has been developed at the Department of Civil Engineering, University of Queensland [55] and applied to mining problems. The implementation of Boundary Elements into an existing Finite Element program required additional subroutines for the integration of Boundary Element contributions for the assembly into the coefficient matrix and for the solution and processing of results. Details of the efficient implementation of a coupled analysis capability is discussed elsewhere [53]. Here only the main features of the coupled approach are discussed. For a coupled analysis, the possibility exists to discretise parts of both the volume and the boundary. An example of this is shown in Fig. 8.14 where the boundary of a square opening is discretised into Boundary Elements and the pillar into Finite Elements. Both meshes are connected at the interface nodes. The vectors of

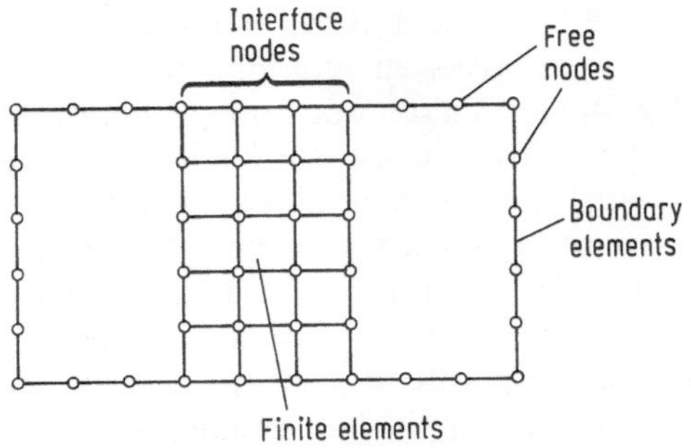


**Fig. 8.14.** Coupled finite element-boundary element mesh

displacements and tractions at the interface are $\{a\}_{II}$ and $\{b\}_{II}$ and the corresponding values at the free surface are $\{a\}_I$ and $\{b\}_I$. The Boundary Element equations can be written as an extension of equation (8.22).

$$[A]\begin{Bmatrix}\{a\}_I\\ \{a\}_{II}\end{Bmatrix} = [B]\begin{Bmatrix}\{b\}_I\\ \{b\}_{II}\end{Bmatrix} \tag{8.38}$$

Now only the tractions on the free nodes $\{b\}_I$ are known. The matrices $[A]$ and $[B]$ are partitioned and reordered thus:

$$[[A]_I\,]B]_{II}]\begin{Bmatrix}\{a\}_I\\ -\{b\}_{II}\end{Bmatrix} = \{f\} - [A]_{II}\,\{a\}_{II} \tag{8.39}$$

with,

$$\{f\} = [B]_I\,\{b\}_I \tag{8.40}$$

Equation (8.39) can be solved to give

$$\begin{Bmatrix}\{a\}_I\\ -\{b\}_{II}\end{Bmatrix} = \begin{Bmatrix}\{\hat{a}\}_I\\ \{\hat{b}\}_{II}\end{Bmatrix} - \begin{bmatrix}[Q]_I\\ [Q]_{II}\end{bmatrix}\{a\}_{II} \tag{8.41}$$

where $\{\hat{a}\}_I$ and $\{\hat{b}\}_{II}$ are the solutions with the interface nodes fixed and the columns of $[Q]_I$ and $[Q]_{II}$ are solutions for unit values of displacements at interface nodes. The tractions at the interface can now be expressed as a function of the displacements at the interface by

$$\{b\}_{II} = [Q]_{II}\,\{a\}_{II} - \{\hat{b}\}_{II} \tag{8.42}$$

At the interface, the conditions of equilibrium and compatibility have to be satisfied. If the nodes at the interface are assigned the same displacement degrees of freedom and the shape function of the Finite Elements and Boundary Elements are identical, compatibility is satisfied. Equilibrium is satisfied in an average sense when the variation of the potential energy with respect to the displacements at the interface nodes is zero. The potential energy at the Boundary Element interface is:

$$\pi = \tfrac{1}{2}\int_S u_i\,\tilde{t}_i\,dS - \int_S u_i\,\hat{t}_i\,dS \tag{8.43}$$

where $\tilde{t}$ are tractions due to displacements and $\hat{t}$ other tractions. For the interface the coupled equation is obtained from this (see ref. [53]):

$$([K]_{\text{F.E.}} + [K]_{\text{B.E.}})\,\{a\}_{\text{II}} + \{F\} = 0 \tag{8.44}$$

$[K]_{\text{B.E.}}$ is the stiffness matrix for the interface nodes. This matrix is symmetric and for two-dimensional problems has the size of $2M \times 2M$ where $M$ is the number of interface nodes. The vector $\{F\}$ is the nodal force vector. After the stiffness matrix and the load vector have been obtained for the interface, the analysis proceeds using standard Finite Element software. The Boundary Element region is treated as a Super Finite Element and assembled in the usual manner.

When non-linear problems are analysed, the iterations now only involve degrees of freedom of the Finite Element nodes including the interface. In the non-linear analysis by the Boundary Element method, the degrees of freedom of all the boundary nodes have to be considered. Also, in the Finite Element analysis only the nodes of each Finite Element contribute to the stress computation at points inside it. This is in contrast to the non-linear solution by the Boundary Element method where each interior stress computation involves all the Boundary Element nodes. For problems where the number of interface nodes is small compared with the total number of boundary nodes a coupled analysis is thus expected to be more efficient. Some typical non-linear mining problems analysed using the coupled Finite-Boundary Element program BEFE [69] are shown herein.

**Example I.** *Mine Pillar Problem* [24]. In Fig. 8.15 a problem is shown where excavations have been made in virgin ground. As a result, the material between the excavations, termed a "mine pillar", is subjected to high stress. This situation occurs frequently in mining especially in coal mining although in particular applications the dimensions will vary and several pillars may be present. Because of the high stress, the pillar is expected to behave non-linearly and thus it has been discretised into Finite Elements. In the example, the pillar was assumed to have planes of weakness inclined 30° to the horizontal. These planes were assumed to have an angle of friction of 10° and a cohesion of 1 MPa. These properties were modelled using a viscoplastic multi-laminate model [48]. The vertical stresses in the pillar for the elastic analysis are shown in Fig. 8.16, together with those for the first 3 time steps in the visco-plastic analysis. The vertical pre-mining stress was taken as 30 MPa.

**Example II.** *Crown Pillar Analysis* [54]. This is similar to the previous example except that the excavation is inclined to the horizontal and the shape of the material between the excavations is more complex (see Figure 8.17). Because of the

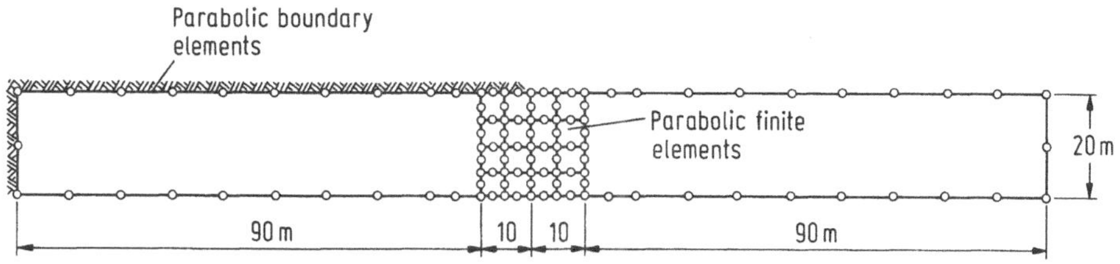


**Fig. 8.15.** Mine pillar problem: mesh and dimensions

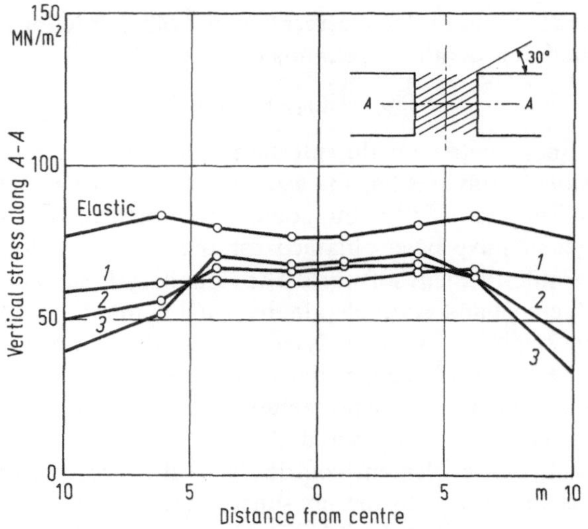


**Fig. 8.16.** Pillar stresses for the first 3 time steps for pillar with planes weakness at 30°

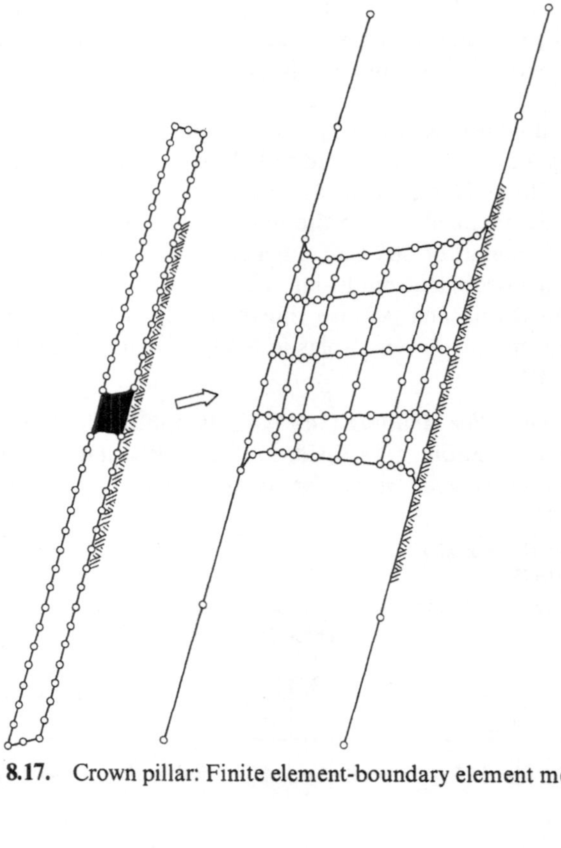

**Fig. 8.17.** Crown pillar: Finite element-boundary element mesh

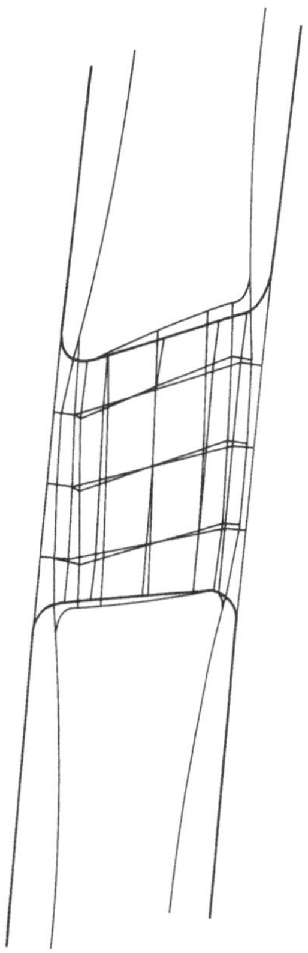

Fig. 8.18. Crown pillar problem: displaced shape

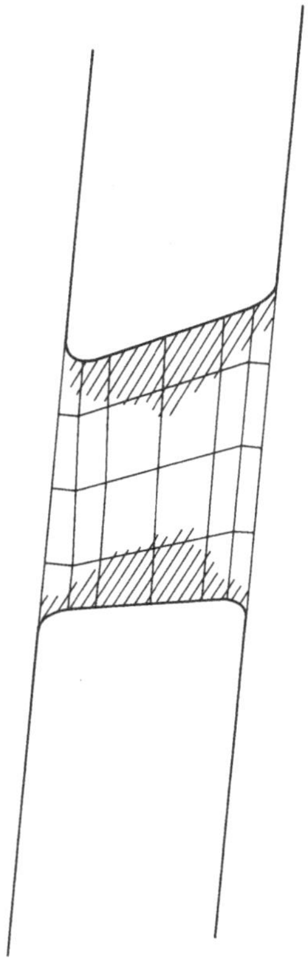

Fig. 8.19. Crown pillar problem: visco-plastic zones

orientation of the stopes, it is termed a "crown pillar". The coupled Boundary Element – Finite Element mesh is shown in Figure 8.17. In this case, the pillar material is assumed to be isotropic and to follow a Drucker Prager yield condition with an angle of friction of 50° and a cohesion of 20 MPa. The intial stress is 34 MPa at right angles to the long axis of excavation. The resulting displacements are shown in Figure 8.18. The zones where visco-plastic straining has taken place are shown in Figure 8.19.

**Example III.** *Hanging Wall Problem* [56]. This is a problem of particular interest in the mining industry and has been the subject of an extensive study [56]. Figure 8.20 shows a cross-section of an excavation where the over-hanging rock has been discretised into Finite Elements. The boundary associated with it is called a hanging wall. If there are pre-existing fractures in the material, the hanging wall has the tendency to collapse and cause ore dilution. In the present analysis, it has been assumed that the rock mass has a distinct bedding parallel to the opening with particularly weak bedding planes occuring at regular intervals. Along these weak bedding planes, discontinuous behaviour will take place and to model this Contact Elements with Mohr-Coulomb and no-tension properties have been used between

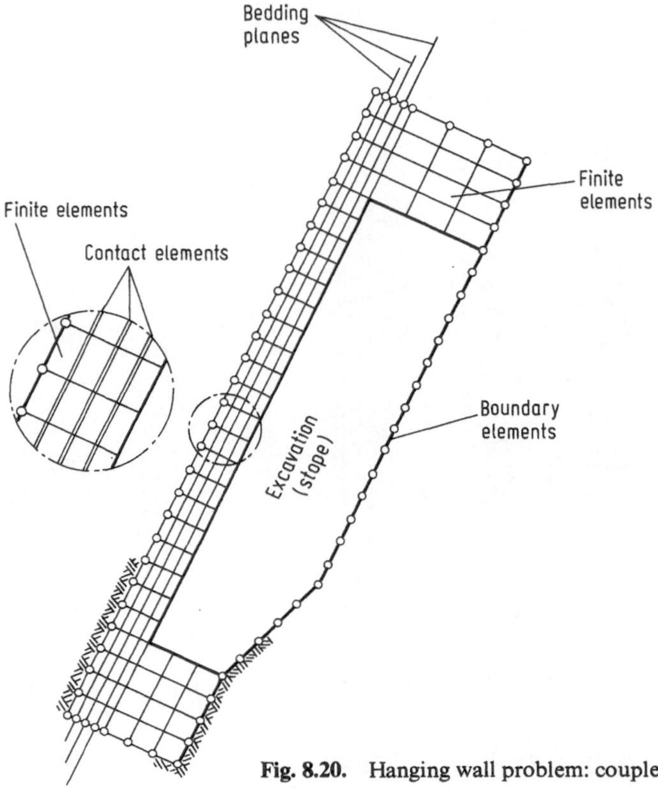


**Fig. 8.20.** Hanging wall problem: coupled mesh

the layers of Finite Elements. The angle of friction has been assumed to be 10° with no cohesion. This value corresponds to particularly smooth and graphitic bedding planes. Figure 8.21 shows the displacements which occur in the hanging wall after the excavation has been made in the initially stressed rock mass. It can be seen that there is significant differential slip occuring near the abutment and, associated with this, an opening of fractures. This is an example of the coupled analysis where the problem could not have been analysed as efficiently by either Finite or Boundary Element methods.

*Combination of Displacement Discontinuity and Boundary Element Formulations*

It has been shown that the different formulations of the Boundary Element method have applications for specific mining problems. The displacement discontinuity formulation is suitable for analysis of tabular excavation problems in which the distance between roof and floor is small compared with the overall dimensions. That is, for the excavation of seams or tabular orebodies. For obvious reasons this formulation can not be used for the analysis of massive orebodies where the classical Boundary Element formulations (direct and indirect) can be applied. However, their application to slit type excavations is limited because of the excessive amount of discretisation required to obtain a stable solution (as a general

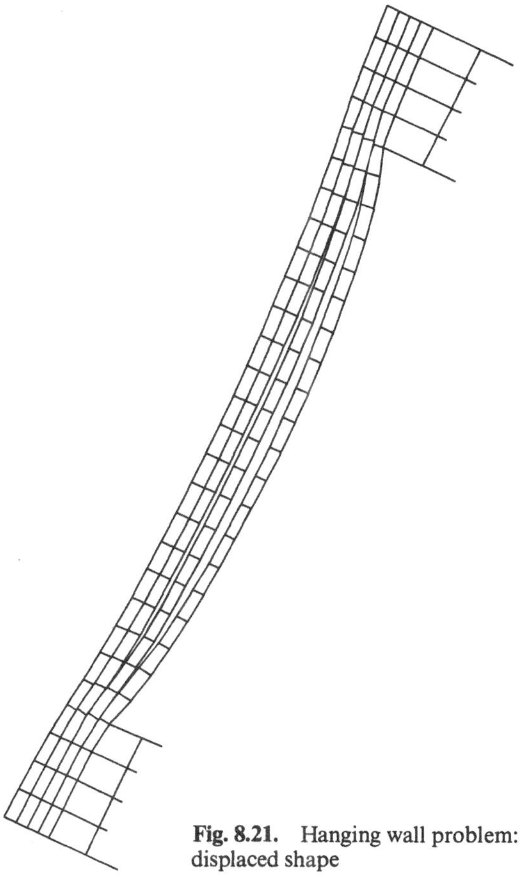

**Fig. 8.21.** Hanging wall problem: displaced shape

rule, two Boundary Elements should not be closer to each other than half of their length otherwise the solution can become unstable).

There are some problems in mining that involve both massive and seam type excavations and there is obviously some benefit in having both formulations in one computer program. The two methods are in fact very similar except that instead of displacements, the closure and rides are the variables in the displacement discontinuity method. In the Boundary Element equations, the influence of the displacement discontinuity element has to be included i.e., Eq. (8.19) is rewritten:

$$c_{ij}\,u_j(n) + \sum_{m=1}^{M} \Delta T_{ij}(n, m)\,u_j(m) + \sum_{l=1}^{L} \Delta T_{ij}(n, l)\,d_j(l) = \sum_{m=1}^{M} \Delta U_{ij}(n, m)\,t_n(m) \quad (8.45)$$

In the above, $M$ is the number of Boundary Elements and $L$ the number of Displacement Discontinuity Elements. For the Displacement Discontinuity elements the following relationship may be written for the stress,

$$\sigma_{i3}(n) = \sum_{l=1}^{L} a_{ij}(n, l)\,d_j(l) + \sum_{m=1}^{M} \Delta S_{ijk}(n, m)\,u_j(m) - \sum_{m=1}^{M} \Delta D_{ijk}(n, m)\,t_j(m) \quad (8.46)$$

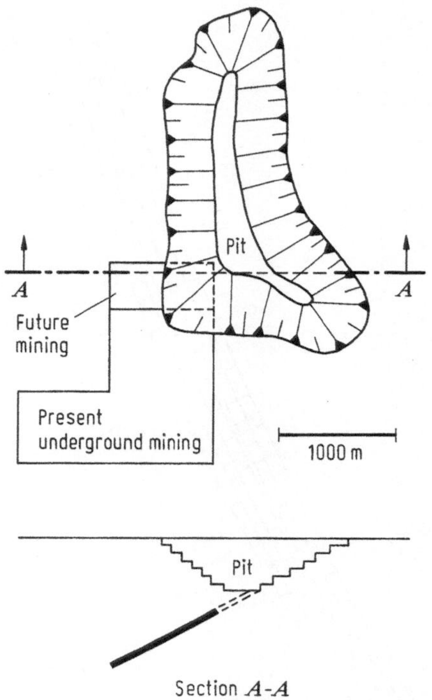


**Fig. 8.22.** Schematic representation of open pit and tabular underground excavations (from [36])

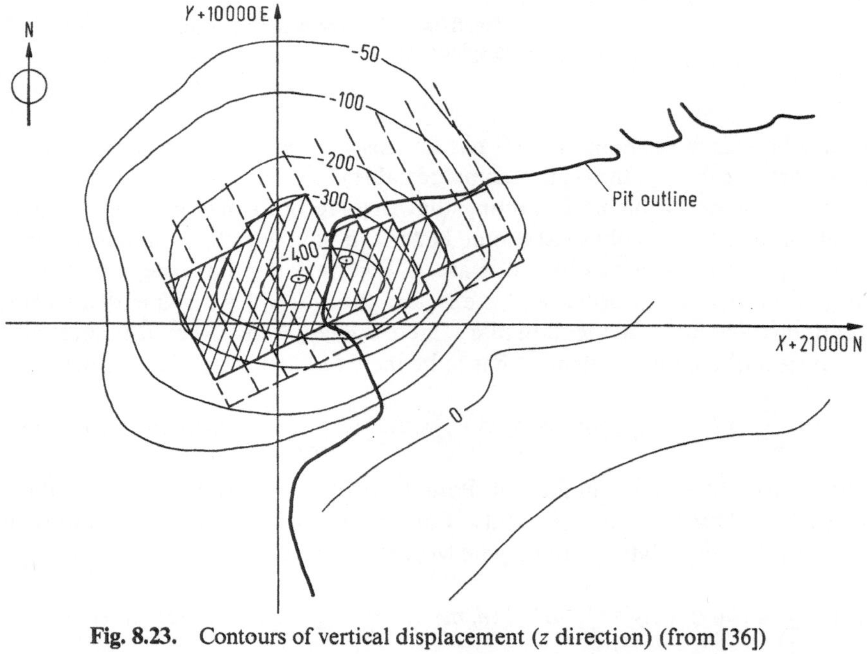


**Fig. 8.23.** Contours of vertical displacement ($z$ direction) (from [36])

The equations (8.45) and (8.46) are then solved for the values of displacements and displacement discontinuities. For a seam inclined to the global axes, the tractions and displacements have to be converted to a local coordinate system. Details are given by Diering [35].

Figure 8.22 shows a typical application of the mixed method where a pit excavation interacts with a tabular excavation. This problem was analysed by Diering [35] who discretised the Pit surface into Boundary Elements and the seam into Displacement Discontinuity elements. A typical result is shown in Fig. 8.23 which depicts the contours of displacement of the ground surface. Similar mixed procedures have been used by Hocking [21] to analyse excavations with fault zones.

## 8.7 Summary and Future Outlook

There appears to be no doubt that the Boundary Element method is a major analytical tool in the stress analysis associated with mine opening problems. Its ability to model the far field accurately is ideal for many problems. In three dimensions, it leads to the possibility of solutions to problems beyond the scope of the finite element method. Attention should be focused on the Mindlin solution as well as the Kelvin solution so that the free surface can be included without discretisation.

Unfortunately, problems in rock mechanics do not in general lend themselves to simple interpretation using continuum mechanics theory alone. The presence of joints, bedding planes and material variability leads to the conclusion that a combination of the continnum and discrete element type models is necessary. In this way joint release and non-conservative joint slip can be accomodated. In the writers' opinion, this can best be achieved by a combination of not only finite element and boundary element techniques, but also the inclusion of some ideas of the discrete element models. The "marriage a la mode" of Professor Zienkiewicz then contains three partners, Boundary Elements, Finite Elements and Discrete Elements.

### Acknowledgements

Some of the work presented in this chapter has been sponsored by Mount Isa Mines. Ltd., Mount Isa, Australia. The permission to publish mine cross-sections of the Mount Isa mine and the results of a mining problem analysed by the displacement discontinuity program NFOLD is gratefully acknowledged. Thanks must go to T. Diering for permission to include some figures from his thesis here. The permission of the CSIRO, Division of Applied Geomechanics to include results from program BITEMJ is acknowledged. Finally, the authors would like to thank C. A. Brebbia for making available plots of the no tension analysis.

# Appendix A: Equation Solvers

All Boundary Element formulations discussed here lead to large systems of simultaneous equations. These coefficient matrices are fully populated and non-symmetric. This contrasts with the sparsely populated and symmetric matrices of the Finite Element method. Thus more storage is required and more time used in the solution of a given number of equations. On the other hand, however, the number of unknowns is much less in a Boundary Element discretisation.

Basically, two different approaches can be used for the solution, that is, iterative and direct methods. In addition, hybrid schemes are possible. One quite popular iterative solution technique is the Successive Overrelaxation technique.

The solution of

$$[A]\{a\} = \{b\} \tag{8.47}$$

is approximated after iteration $n$ by:

$$\{a\}^{n+1} = \{a\}^n + \omega [D_A]^{-1} (\{b\} - [A]\{a\}^n) \tag{8.48}$$

where $\{a\}^n$ is the displacement vector after $n$ iterations $[D_A]$ is a matrix containing the diagonal coefficients of $[A]$ and $\omega$ is an over-relaxation factor which is usually in the range of 1 to 2. The number of multiplications and divisions $M^i$ for each iteration is then,

$$M^i = N(N+3) \tag{8.49}$$

where $N$ is the dimension of matrix $[A]$. The iteration process continues until the vector of the displacements is identical (within a given tolerance) for two successive iterations. The number of iterations required depends on the problem size and the condition of $[A]$. As a general rule, the larger the problem the more iterations will be needed for the solution. Also, the number of iterations will be reduced if the diagonal coefficients of matrix $[A]$ are strongly dominant. An untried scheme which has proved to be invaluable in Finite Element analyses is the Wilson Energy Balance for accelerated convergence (see ref. [64]). In situations for which the standard Gauss-Seidel Iteration method has proved to have convergence problems, this energy balance procedure has given excellent results.

The method is as follows. After a given number of cycles, let $\{a\}_1$ denote the calculated displacements and suppose that a better approximation, $\{a\}$ is related to $\{a\}_1$ by a factor $\delta$, such that,

$$\{a\} = \delta \{a\}_1 . \tag{8.50}$$

The total strain energy may be calculated from the load vector $\{b\}$ and also as a quadratic function of displacements. That is,

$$U = \tfrac{1}{2}\{b\}^T\{a\} = \tfrac{1}{2}\{a\}^T[A]\{a\} . \tag{8.51}$$

Since $\{a\}_1$ is known this affords a mean to calculate $\delta$. Substitution for $\{a\}$ gives,

$$\{b\}^T \delta \{a\}_1 = \delta \{a\}_1^T [A]\{a\}_1 . \tag{8.52}$$

That is,

$$\delta = \frac{\{b\}^T\{a\}_1}{\{a\}_1^T[A]\{a\}_1} . \tag{8.53}$$

The factor $\delta$ may be calculated every 10 to 20 cycles. The degree of improvement will be problem orientated and a certain amount of experience is needed in its use.

The Direct Method, using Gaussian elimination involves three steps:

1. Triangular decomposition of matrix $[A]$
2. Forward substitution } (re-)solution
3. Back-substitution }

The triangular decomposition phase requires the largest number of operations. The number of multiplications and divisions is approximately

$$M^d = \frac{N^3}{3}. \tag{8.54}$$

The advantage of this method is that once the matrix $[A]$ has been decomposed solutions for any number of right hand side vectors can be obtained at relatively small cost. This is particularly important for the coupled B.E.M.−F.E.M. solutions discussed previously. For direct methods, the solution time is not influenced by the problem geometry and condition of $[A]$.

A comparison between Gauss elimination and successive over-relaxation (S. O. R.) is made in Table 8.1 for a problem with 3000 degrees of freedom, assuming it takes 100 iterations for the S. O. R. to converge. It can be seen that for both methods the numerical work is quite considerable. It may well be that for very large problems an economic limit is reached.

**Table 8.1**

| | Gauss | S. O. R. |
|---|---|---|
| No. of coefficients | $9 \times 10^6$ | $9 \times 10^6$ |
| No. of operations | $9 \times 10^9$ | $0.9 \times 10^9$ |

An inspection of the matrix $[A]$ shows that for most problems many of the off-diagonal coefficients are very small compared with the diagonal terms. This is because the Kernel functions decay rapidly with distance, particularly for the three-dimensional formulation.

One way of exploiting this property of $[A]$ has been proposed recently by Bettess [57]. In this method, the smaller off-diagonal terms are replaced by zero and a sparsely populated matrix obtained. This matrix is then solved using a skyline solver which eliminates most of the operations on zero coefficients. The results are then multiplied with the full matrix $[A]$ and compared with the original right hand side. The difference between the results obtained is then taken as a new right hand side and the system resolved using the factorised sparse coefficient matrix. The process is a combination of both the direct and iterative method. For the problems considered in [58], the solver reduced the execution time by a factor of about two. This factor could increase significantly for very large problems, but no further details are available.

Another method, perhaps more physically obvious, has been developed by Deist and used by others. Called the "lumping" method, it takes advantage of the fact that inter-element influences decay rapidly with distance. However, these remote influences can not be neglected because their combined effect is still significant. For constant Boundary Elements, the lumping procedure is as follows: If an Element is sufficiently far away from a group of Elements, its effect can be evaluated on a central point within the group. Similarly, the effects of a group of elements on any remote element can be represented by the influence of a "composite" element located at a central point of the group. It can be seen that this method will introduce sparseness into $[A]$ because for distant elements only one point in the lump contributes to the coefficient matrix. The bigger the lumps the sparser the matrix will become. In addition, fewer coefficients will have to be calculated. To make the "lumping" procedure fully automatic is not straight forward and details of the method are discussed by Deist [12]. Diering [35] has recently adopted a much simpler approach where lumping information is input by the user. That is, the user defines "lumps" of elements which are used to evaluate coefficients for distant elements. In the "lumping" procedure, the matrix $[A]$ will become sparsely populated but not banded. The iterative solution techniques are not affected much by the lack of banding, but for the direct solution, efficient solvers for unbanded sparse matrices are not freely available. For this reason, the "lumping" has only been used to date in conjunction with iterative solvers. However, there are no technical reasons why lumping can not be implemented into direct solvers.

## Appendix B: Fundamental Solutions

Fundamental solutions are functions which satisfy the differential equations of equilibrium and compatibility. The most widely used is the Kelvin solution for a point load in an infinite medium. The displacements in direction $i$ at $x$ due to unit sources in direction $j$ at $y$ is:

$$U_{ij}(x, y) = \frac{C_1}{2R}(C_2\,\delta_{ij} + r_i\,r_j) \tag{8.55}$$

where referring to Fig. 8.24

$$r_i = \frac{R_i}{R} = \frac{x_i - y_i}{R} \tag{8.56}$$

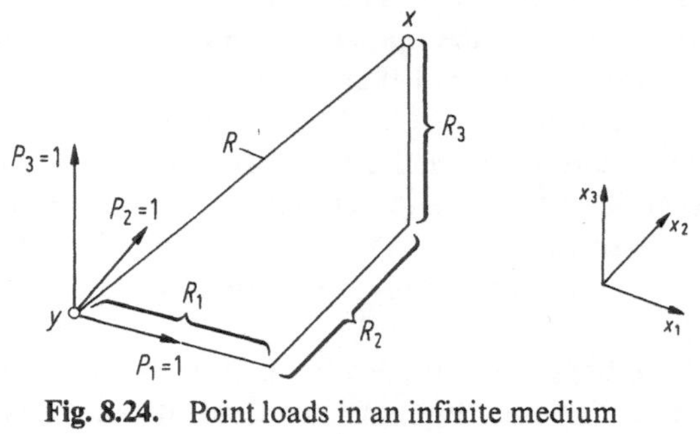


**Fig. 8.24.** Point loads in an infinite medium

and

$$C_1 = \frac{(1+\nu)}{4\pi E(1-\nu)}$$
$$C_2 = 3 - 4\nu\,. \tag{8.57}$$

In equations (8.52) and (8.53) the elastic constants are $E$, Young's Modulus and $\nu$ Poisson's ratio. The stress at $x$ is given by

$$\sigma_{ik} = \lambda\,\delta_{ik}\,\varepsilon_{mm} + 2G\,\varepsilon_{ik} \tag{8.58}$$

with

$$\varepsilon_{mm} = \frac{1}{2}\left(\frac{\partial U_m}{\partial x_n} + \frac{\partial U_n}{\partial x_m}\right). \tag{8.59}$$

In the above, $\lambda$ and $G$ are the Lamè constants. Differentiation of equation (8.55) and substitution in equations (8.59) and (8.58) gives the stress components at $x$ due to a unit source in direction $j$ at $y$.

$$D_{ijk}(x,y) = \frac{C_3}{2R^2}\left[C_4(\delta_{ij}r_k + \delta_{kj}r_i - \delta_{ik}r_j) + 3r_ir_jr_k\right]$$
$$C_3 = \frac{1}{4\pi(1-\nu)} \tag{8.60}$$
$$C_4 = (1-2\nu)\,.$$

The tractions on a surface through $x$ with outward normal $n_i$ can be calculated from:

$$T_{ij}(x,y) = D_{ikj}(x,y)\,n_k(x) \tag{8.61}$$

which gives

$$T_{ij}(x,y) = \frac{C_3}{2R^2}\left(C_4(n_jr_i - n_ir_j) + [3r_ir_j + C_4\delta_{ij}]\,r_ln_l\right). \tag{8.62}$$

Solutions for plane strain are presented in [58].

Another useful solution which can be used is by Mindlin, for sources in the interior of a semi-infinite medium. This solution is particularly suitable for shallow mining excavation because the traction free surface does not have to be discretised. Solutions for the displacements and stresses have been given by Mindlin [59]. Application of the solution to the plane strain analysis of shallow tunnel excavations have been presented recently by Telles [60].

The fundamental solutions discussed so far assume that the solid is isotropic and homogeneous. Fundamental solutions for homogeneous transversely isotropic media have been given by Salamon [1] and Wardle [61], [62] and Cruse [28]. In addition, Salamon also presented fundamental solutions for multi-membrane material, i.e. a material which has no shear resistance.

Another type of fundamental solution is that of a displacement discontinuity in infinite and semi-infinite media. This solution has been derived by Salamon [1] and Starfield [5] using harmonic functions. The derivation presented here uses the Somigliana identity and the Kelvin solution.

Figure 8.25 shows a displacement discontinuity in an infinite medium. The thickness $D$ is assumed to be zero and therefore the surfaces $S_+$ and $S_-$ are at identical locations. It is assumed that the tractions $t_i$ and displacements $u_i$ are

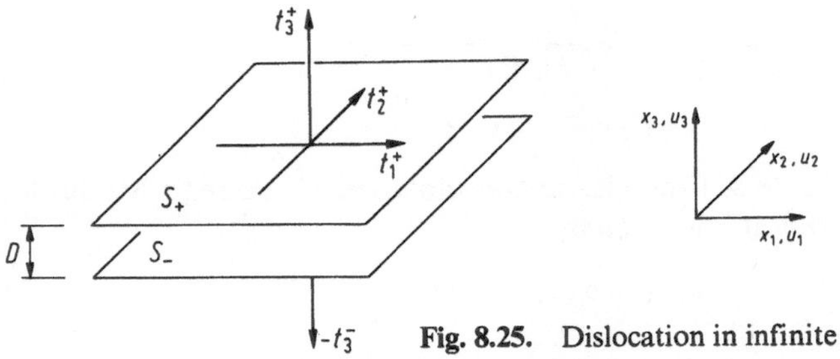


**Fig. 8.25.** Dislocation in infinite medium

constant over a small area $S$. The displacements at any point $x$ due to displacements and tractions at surface $S_+$ are given by:

$$u_i^+(x) = \Delta U_{ij}^+(x, y_c)\, t_j^+ + \Delta T_{ij}^+(x, y_c)\, u_j^+ \tag{8.63}$$

where

$$\Delta U_{ij}(x, y_c) = \int_S U_{ij}(x, y)\, \mathrm{d}S(y) \tag{8.64}$$

and

$$\Delta T_{ij}(x, y_c) = \int_S T_{ij}(x, y)\, dS(y) \tag{8.65}$$

At the dislocation we have,

$$t_i^+ = -\, t_i^- \tag{8.66}$$

Also, since $S_+$ and $S_-$ are at identical location but have opposite outward normals

$$\Delta U_{ij}^+ = \Delta U_{ij}^- \quad \text{and} \quad \Delta T_{ij}^+ = -\,\Delta T_{ij}^- \tag{8.67}$$

Therefore the displacements at $x$ due to the displacements and tractions at both $S_+$ and $S_-$ are

$$u_i(x) = \Delta U_{ij}^+\, t_j^+ + \Delta U_{ij}^-\, t_j^- + \Delta T_{ij}^+\, u_j^+ + \Delta T_{ij}^-\, u_j^- \tag{8.68}$$

Substitution of (8.67) gives:

$$u_i(x) = -\,\Delta T_{ij}^-\,(u_j^+ - u_j^-) = \Delta T_{ij}^+\, d_j \tag{8.69}$$

where

$$d_j = u_j^+ - u_j^- \tag{8.70}$$

Using equation (8.58) the stresses are given by

$$\sigma_{ik}(x) = \Delta S_{ikj}\, d_k \tag{8.71}$$

where

$$S_{ijk} = \frac{C_5}{r^3}\{3\,[C_4\,\delta_{ij}\, r_k + \nu\,(\delta_{ik}\, r_j + \delta_{jk}\, r_i) - 5\, r_i\, r_j\, r_k]\, n_l\, r_l$$
$$+\, 3\nu\,(n_i\, r_j\, r_k + n_j\, r_i\, r_k) + C_4\,(3\, n_k\, r_i\, r_j + n_j\,\delta_{jk} + n_i\,\delta_{jk}) - C_6\, n_k\,\delta_{ij}\} \tag{8.72}$$

$$C_5 = \frac{E}{8\pi\,(1-\nu^2)}; \qquad C_6 = 1 - 4\nu\,.$$

Diering [35] has shown that the solutions obtained here are in fact identical to the ones by Starfield [5]. The derivations presented here assume that the displacement discontinuity is parallel to the $x_1 - x_2$ plane. If this is not the case, the displacements and tractions have to be transformed into local directions. Details are given by Diering [35].

# References


1 Salamon, M. D. G., Elastic analysis of displacements and stresses induced by the Mining of seam or reef deposits. *Part I Jnl. of S. Afr. Inst. Min and Metall,* **64** (4), pp. 129–149, 1963

2 Salamon, M. D. G., Elastic analysis of displacements and stresses induced by the mining of seam or reef deposits. Part II, *Jnl. S. Afr. Inst. Min and Metall,* **64** (6), pp. 197–218, 1964

3 Salamon, M. D. G., Elastic analysis of displacements and stresses induced by the mining of seam or reef deposits. *Part III, Jnl. S. Afr. Inst. Min and Metall,* **64** (10), pp. 468–500, 1964

4 Salamon, M. D. G., Analysis of displacements and stresses induced by the mining of seam of reef deposits. *Part IV, Jnl. S. Afr. Inst. Min and Metall,* **65** (5), pp. 319–338, 1964

5 Starfield, A. M., Crouch, S. L., Elastic analysis of single seam extraction. In: *New Horizons in Rock Mechanics.* (Eds. Hardy, H. R., Jr. and Stefhanko, R.) ASCE, N.Y., pp. 421–439, 1973

6 Crouch, S. L., Solution of plane elasticity problems by the displacement discontinuity method. *Int. J. Numer. Meth. in Eng.,* **10,** pp. 301–343, 1976

7 Crouch, S. L., Computer simulation of mining in faulted ground. *Jnl. S. Afr. Inst. Min and Metall,* **79** (6), pp. 159–173, 1979

8 Diering, T. A. C., Simulation of mining in non-homogeneous ground using the displacement discontinuity method. *Jnl. S. Afr. Inst. Min and Metall,* **80,** No. 7, pp. 169–173, 1980

9 Wiles, T. D., Curran, J. H., A general 3-D displacement discontinuity method. *Proc. 4th Int. Conf. on Numerical Methods in Geomechanics,* ed. Eisenstein, A. A. Blakema, Rotterdam, **1,** pp. 103–109, 1982

10 Crawford, A. M., Curran, T. H., Higher order functional variation displacement discontinuity method. *Int. J. Rock Mech. Min. Sci. & Geomech.* Abstr., **19,** pp. 143–148, 1982

11 Deist, F. H. et al., A new digital method for three dimensional stress analysis in elastic media. Rock Mechanics **5,** 189–202, 1973

12 Deist, F. H., Georgiadis, E., A computer system for three-dimensional elastic analysis using a boundary element approach. Chamber of Mines of South Africa, *Research Report No. 43/76, Project No. GS 1510*

13 Rizzo, F. T., An integral equation approach to boundary value problems of classical elastostatics. Q. Appl. Math. **25,** 83–95, 1967

14 Cruse, T. A., Application of the boundary integral equation method to three-dimensional stress analysis. Computer & Structures, **3,** pp. 509–527, 1973

15 Lachat, J. C., A further development of the boundary integral technique for elastostatics. *Ph.D. Thesis,* University of Southampton, U.K.

16 Lachat, J. C., Watson, J. O., Effective numerical treatment of boundary integral equations: A formulation for three-dimensional elastostatics. Int. Jnl. for Num. Meth. in Eng. **10,** pp. 991–1005, 1976

17 Bannerjee, P. K., Butterfield, R., Boundary element method in Geomechanics. In: *Finite Elements in Geomechanics.* Ed. Gudekus, G., Chichester, Wiley 1977

18 Brady, B. H. G., Bray, J. W., The boundary element method for elastic analysis of tabular orebody extraction, assuming complete plane strain. Int. J. Rock Mech. Min. Sci. & Geomech. Abstr., **15,** pp. 29–37, 1978

19 Brady, B. H. G., A boundary element method for three-dimensional elastic analysis of tabular orebody extraction. Proc. 19th Rock Mech. Symp., pp. 431–438, 1978

20 Cruse, T. A., Formulation of boundary integral equations for three-dimensional elasto-plastic flow. Int. J. Solids & Structures, **7**

21 Hocking, G., Stress analysis of underground excavations incorporating slip and separation along discontinuities. In: *Recent Advances in Boundary Element Methods.* (Ed. Brebbia, C. A.) Pentech Press, Plymouth, pp. 195–214, 1978

22 Wardle, L. J., Crotty, J. M., Two Dimensional boundary integral equation analysis for non-homogeneous mining applications. In: *Recent Advances in Boundary Element Methods.* Ed. Brebbia, C. A., Pentech Press, Plymouth, pp. 233 – 251, 1978

23 Telles, J. C. F., Brebbia, C. A., Plasticity. In: *Progress in Boundary Element Methods.* Ed. Brebbia, C. A., Pentech Press, London 1981

24 Beer, G. et al., Efficient analysis in geomechanics. 4th Int. Conf. on Numerical Methods in Geomech. Ed. Eisenstein, A. A. Blakema, Rotterdam, **1,** pp. 5 – 13, 1982

25 Beer, G., Meek, J. L., Coupled Finite Element – Boundary Element analysis of infinite domain problems. Int. Conf. on Num. Meth. for Coupled Problems, ed. Bettess et al., Pineridge Press, Swansea, U.K., 1981

26 Sinha, K. P., Displacement discontinuity technique for analysis stresses and displacements due to mining in seam deposits. Ph.D. dissertation, University of Minnesota, 1979

27 Diering, T. A. C., An improved method for the determination by a MINSIM type of analysis of stresses and displacements around tabular excavations. Inl. S. Afr. Inst. Min Metall, **80,** No. 12, pp. 225 – 228, 1980

28 Wilson, R. B., Cruse, T. A., Efficient implementation of anisotropic three-dimensional boundary integral equation stress analysis. Int. J. Num. Meth. in Engng., **12,** 1383 – 1397

29 Banerjee, P. K., Integral equation methods for analysis of piecewise non-homogeneous three-dimensional elastic solids of arbitrary shape. Int. J. Mechanical Science, **18,** pp. 293 – 303, 1976

30 Brady, B. H. G., Bray, J. W., The boundary element method for determining stresses and displacements around long openings in a triaxial stress field. Int. J. Rock Mech. & Min. Sci., **15** (1), pp. 21 – 28, 1978

31 Hoek, E., Brown, E. T., Underground excavations in rock. Int. of Min. and Met., 44 Portland Place, London W1N 4BR, England

32 Massonet, C. E., Numerical use of integral procedures. Chapter 10 in *Stress Analysis.* Ed. Zienkiewicz, John Wiley, London 1965

33 Oliviera, E. R. A., Plane stress analysis by a general integral method. J. ASCE, **94,** No. EM1, PP. 79 – 101, 1968

34 Watson, J. O., Advanced implementation of the boundary element method for two and three dimensional elastostatics. Chapter 3 in *Developments in Boundary Element methods – 1.* Ed. Banerjee, Applied Science Publ., 1979

35 Diering, T. A. C., Further developments of the boundary element method with applications in mining. MSc. thesis, University of the Witwatersrand, Johannisburg, South Africa, 1981

36 Riccardella, P. C., An improved implementation of the boundary integral techniques for two-dimensional elasticity problems. Carnegie Inst. of Tech., Dept. Mech. Eng., Pittsburgh, Report No. SM-72-76, 1972

37 Riccardella, P. C., An implementation of the boundary integral technique for planar problems of elasticity and elasto-plasticity. Carnegie Inst. of Tech., Dept. Mech. Engng., Pittsburgh, Report No. SM-73-10, 1973

38 Crotty, J. M., Wardle, L. J., Program BITEM user's manual. Institute of earth resources, CSIRO Division of Applied Geomechanics, P.O. Box 54, Mount Waverley, Victoria, Australia, 1980

39 Wardle, L. J., Three dimensional boundary element program for mining applications: MIN3D2, Technical Report No. 116, CSIRO Div. of Applied Geomech., Australia, 1980

40 Watson, J. O., The solution of boundary integral equations of three-dimensional elastostatics for infinite regions. Proc. First Int. Seminar on Recent Advances in Boundary Element Methods, Univ. of Southampton, 1979

41 Lachat, J. C., Watson, J. O., Progress in the use of boundary integral equations, illustrated by examples. Computer Methods in Applied Mechanics and Engineering, **10** (3), pp. 273 – 289, 1977

42 Watson, J. O., Hermitian cubic boundary elements for plane problems of fracture mechanics, Res. Mechanica, 1981

43 Zienkiewicz, O. C., *The Finite Element Method* – Third Edition, McGraw-Hill 1977

44 Crotty, J. M., A block equation solver for large unsymmetric matrices arising in the Boundary Integral Equation Method. Int. J. Numer. Meth. in Engng., **18,** pp. 997 – 1017, 1982

45 Hocking, G., Stress analysis of the 1100 orebody at Mount Isa. Proc. 19th U.S. Symposium on Rock Mechanics, Stateline, Nevada, 1978

46 Crotty, J. M., Wardle, L. T., Integral equation analysis of piecewise homogeneous media with structural discontinuities. In preparation. C.S.I.R.O., Div. of Applied Geomechanics, Australia, 1982

47 Bolteus, L., Tullberg, O., BEMSTAT – A new type of boundary element program for two-dimensional elasticity problems. 3rd Int. Sem. on Recent Advances in Boundary Element Methods, Irvine, USA, 1981

48 Zienkiewicz, O. C., Pande, G. N., Time dependent multi-laminate model of rocks – A numerical study of deformation and failure of rock masses. Int. Jnl. Num. Anal. Meth. Geomech., **1**, pp. 219–247, 1977

49 Venturini, U. S., Brebbia, C. A., Boundary Element formulation for non-linear applications in geomechanics. 4th Int. Conf. on Recent Advances in Boundary Element Methods, ed. Brebbia, Southampton, England, 1982

50 Venturini, U., Brebbia, C. A., *The Boundary Element Method for the solution of no-tension materials in Boundary Element Methods.* Ed. Brebbia, C. A., Springer Verlag, Berlin, New York 1981

51 Zienkiewicz, O. C., Kelly, D. W., Bettess, P., The coupling of the finite element method and boundary solution methods. Int. J. Numer. Meth. Eng., **11** (2), pp. 355–375

52 Mustoe, G. G. W., Symmetric variational boundary integral and finite element procedures in continuum mechanics. Ph.D. dissertation, Dept. of Civil Eng., Univ. of Wales, Swanesa, U.K., 1980

53 Beer, G., Finite element, boundary element and coupled analysis of unbounded problems in elastostatics. Int. J. Numer. Meth. Engng., in press

54 Brady, B. H. G., Wassyng, A., A coupled finite element – boundary element method of stress analysis. Int. J. Rock Mech. Min. Sci. & Geomech. Abstr., **18**, pp. 475–485, 1981

55 Beer, G., BEFE – Users manual. Dept. of Civil Eng., University of Queensland, St. Lucia, Australia, 1982

56 Beer, G. et al., Prediction of the behaviour of shale hanging walls in deep underground excavations. 5th ISRM congress, Melbourne, Australia, 1983

57 Bettes, J. A., Economical solution technique for Boundary Integral matrices. Report C/R/399/81, Dept. of Civil Eng., University College of Swansea, U.K.

58 Brebbia, C. A., Walker, S., *Boundary Element techniques in engineering.* Newness-Butterworths, 1980

59 Mindlin, P. D., Force at a point in the interior of a semi-infinite solid. Physics, **7**, pp. 195–202, 1936

60 Telles, J. C. F., Brebbia, C. A., Boundary element solution for half-plane problems. Int. J. Solids Structures, **17**, No. 12, pp. 1149–1158, 1981

61 Wardle, L. J., Three-dimensional solutions for displacement discontinuities in cross-anisotropic media. CSIRO, Div. of Appl. Geomech., Technical paper No. 34, 1980

62 Wardle, L. J., Fundamental solutions and boundary integral formulation for tabular excavations in a non-homogeneous rock mass. CSIRO, Div. of Appl. Geomech., Techn. report No. 133, 1982

63 Wardle, L. J., Boundary element methods for solution of stress analysis problems in mining. CSIRO, Div. of Appl. Geomech., Techn. Report No. 102, 1980

64 Meek, J. L., *Matrix Structural Analysis.* McGraw-Hill Co, New York 1971

65 Hebblewhite, B. K., The use of modelling techniques for mine design. Australian Journal of Coal Mining Technology and Research, No. 1, 1982

66 Brebbia, C. A., Georgiou, P., Combination of boundary and finite elements in elastostatics. Appl. Math. Modelling, **3**, pp. 212–220, 1979

67 Banerjee, P. K., Butterfield, Boundary Element Methods in Engineering Science. McGraw-Hill, London 1981

68 Bywater, S., Cowling, R., Black, B. N., Stress Measurement and Analysis for Mine Planning. 5th ISRM Congress, Melbourne, Australia, 1983

69 Beer, G., BEFE – a combined oundary element finite element computer program. Adv. Eng. Software, Vol. 6, No. 2

# Chapter 9

# Finite Deflection of Plates

*by N. Kamiya and Y. Sawaki*

## 9.1 Introduction

The Boundary Element Method (BEM) is now an effective tool for the numerical analysis of nonlinear as well as linear problems. Some of those problems are physical or material nonlinear, such as elastoplasticity and creep in solid mechanics [1]. Another type of nonlinear problem are those concerned with geometrical nonlinearities, such as finite deformations. Among a variety of geometrically nonlinear behaviours in solid mechanics, the finite deflection of flat plates or shells is one of the most important problems from the standpoint of engineering practice.

Very little work has been carried out on the treatment of geometrically nonlinear equations using BEM. The conventional von Karman nonlinear differential equations for finite deflections of plates represent a complex situation involving coupling between inplane and out-of-plane deformations. So far the BEM appears to have been applied to find approximate solutions for the finite deflection of plates and shells [2−5]. In these studies, a pseudo-linear governing equation known as the Berger equation [6] has been used, which approximates nonlinear bending behaviour without introducing the above-mentioned coupling. The Berger decoupled field equation is not based on a sound physical assumption and is occasionally referred to as the Berger "hypothesis". While the equation may not theoretically be precise, several have shown its effectiveness for practical applications provided that the inplane displacements are constrained on the boundary.

The present chapter discusses the previous works and presents the general equations for finite deflection of thin elastic, flat plates based on the Kirchhoff-Love theory. The chapter also shows an integral equation formulation of the von Karman-type nonlinear governing equations and a short derivation of the approximate governing equation due to Berger and its corresponding integral form. Some results illustrate the application of the approximate integral formulation for the Berger equation. Finally, the chapter deals with the extension of the BEM to nonlinear bending problem with thin shallow shells and sandwich plates and shells. Some representative numerical results are included.

## 9.2 Geometrically Nonlinear Governing Equations [7, 8]

For convenience, the fundamental equations for the finite deflection of flat elastic plates of constant thickness will be summarized here. These equations are based on

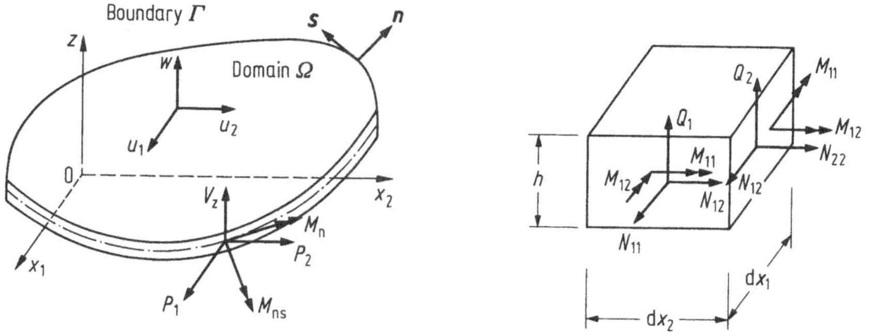


**Fig. 9.1.** Flat plate and its element. Coordinate and notations

the Kirchhoff-Love hypothesis and referred to the Cartesian coordinates $x_1$, $x_2$, $z$ shown in Fig. 9.1. The summation convention is used throughout ranging from 1 to 2. Linear and nonlinear terms are distinguished by the superfixes $l$ and $n$ respectively (for the complete notations, see Appendix III).

I) Fundamental definitions for plates

(1) Membrane strains:

$$e_{ij} = e_{ij}^l + e_{ij}^n \tag{9.1}$$

$$e_{ij}^l = \frac{1}{2}(u_{i,j} + u_{j,i}) \tag{9.2}$$

$$e_{ij}^n = \frac{1}{2} w_{,i}\, w_{,j} \tag{9.3}$$

(2) Stress resultants (membrane forces):

$$N_{ij} = \int_{-h/2}^{h/2} \sigma_{ij}\, dz = N_{ij}^l + N_{ij}^n \tag{9.4}$$

$$N_{ij}^l = D_{ijkl}\, e_{kl}^l \tag{9.5}$$

$$N_{ij}^n = D_{ijkl}\, e_{kl}^n \tag{9.6}$$

(3) Inplane boundary tractions:

$$p_i = N_{ij}\, n_j = p_i^l + p_i^n \tag{9.7}$$

$$p_i^l = N_{ij}^l\, n_j \tag{9.8}$$

$$p_i^n = N_{ij}^n\, n_j \tag{9.9}$$

(4) Bending and twisting moments:

$$M_{ij} = \int_{-h/2}^{h/2} \sigma_{ij}\, z\, dz = E_{ijkl}\, w_{,kl} \tag{9.10}$$

(5) Shear forces:

$$Q_i = Q_i^l + Q_i^n \tag{9.11}$$

$$Q_i^l = M_{ij,j} \tag{9.12}$$

$$Q_i^n = N_{ij}\, w_{,j} \tag{9.13}$$

(6) Normal and twisting moments on the edge:

$$M_n = M_{ij} n_i n_j \tag{9.14}$$

$$M_{ns} = -\varepsilon_{ij} M_{ik} n_k n_j \tag{9.15}$$

(7) Shear force on the edge:

$$V = Q_i n_i = V^l + V^n \tag{9.16}$$

$$V^l = M_{ij,j} n_i \tag{9.17}$$

$$V^n = N_{ij} w_{,j} n_i \tag{9.18}$$

(8) Kirchhoff effective shear force on the edge:

$$K_n = V + M_{ns,s} = K_n^l + K_n^n \tag{9.19}$$

$$K_n^l = Q_i^l n_i + M_{ns,s} \tag{9.20}$$

$$K_n^n = Q_i^n n_i \tag{9.21}$$

II) Equilibrium equations: Plate equilibrium is represented by the following two equations:

$$N_{ij,j} = 0 \tag{9.22}$$

and

$$M_{ij,ij} + (N_{ij} w_{,j})_{,i} + \bar{p} = 0 \tag{9.23}$$

or

$$-D \nabla^4 w + Q_{i,i}^n + \bar{p} = 0 \tag{9.24}$$

If we introduce a membrane stress function $F$, Eq. (A4), identically satisfying the first equilibrium Eq. (9.22), the governing equations can now be formulated in terms of the deflection $w$ and the stress function $F$. The resulting formulae are called *von Karman equations* for the finite deflection for plates (see Appendix II, Eqs. (A5) and (A6)). In what follows, we call Eqs. (9.22) to (9.24) *von Karman-type equations.*

III) Boundary conditions: Admissible boundary conditions associated with the above equilibrium equations are as follows:

(1) Mechanical boundary conditions on $\Gamma_1$:

$$\begin{aligned} p_i &= \bar{p}_i \quad (i = 1, 2) \\ M_n &= \bar{M}_n \\ K_n &= \bar{K}_n \end{aligned} \tag{9.25}$$

(2) Geometrical boundary conditions on $\Gamma_2$:

$$\begin{aligned} u_i &= \bar{u}_i \quad (i = 1, 2) \\ w &= \bar{w} \\ w_{,n} &= \bar{w}_{,n} \end{aligned} \tag{9.26}$$

where overscored values are prescribed on the boundary.

## 9.3 Integral Equation Formulation for von Karman-type Equations [9]

The finite deflection of thin elastic plates is governed by the equilibrium equations (9.22) and (9.23) or (9.24) and the boundary conditions, Eqs. (9.25) and (9.26). In order to formulate the integral equation for boundary element analyses, we will first consider the linear deformation case, i.e., infinitesimal or small bending theory. For this case, the nonlinear terms (superfix $n$) can be dropped out in the previous equations and the inplane and out-of-plane deformations become uncoupled. The resulting governing equations and boundary conditions are as follows:

(i) Inplane deformations

$$N^l_{ij,j}=0 \quad \text{in } \Omega \tag{9.27}$$

with

$$p_i=\bar{p}_i \quad \text{on } \Gamma_1 \quad \text{and} \quad u_i=\bar{u}_i \text{ on } \Gamma_2 \tag{9.28}$$

(ii) Out-of-plane deformations:

$$M_{ij,ij}+\bar{p}=0 \quad \text{in } \Omega \tag{9.29}$$

or

$$-D\nabla^4 w+\bar{p}=0 \quad \text{in } \Omega \tag{9.30}$$

with

$$M_n=\bar{M}_n, \quad K_n=\bar{K}_n \text{ on } \Gamma_1 \quad \text{and} \quad w=\bar{w}, \quad w_{,n}=\bar{w}_{,n} \text{ on } \Gamma_2 \tag{9.31}$$

In this case one can apply two well-known integral equation formulations, i.e., i) the two-dimensional plane stress integral formula, and ii) the linear plate bending integral expression. They are respectively,

$$\int_\Omega N^l_{ij}(u_k)_{,j}\, u^*_i \, d\Omega = \int_\Omega u_i\, N^{l*}_{ij}(u^*_k)_{,j}\, d\Omega + \int_\Gamma [p^l_i(u_k)\, u^*_i - u_i\, p^{l*}_i(u^*_k)]\, d\Gamma \tag{9.32}$$

and

$$-D\int_\Omega w^*(\nabla^2 w)\, d\Omega = -D\int_\Omega w(\nabla^4 w^*)\, d\Omega + \int_\Gamma [w^* K^l_n(w) - w^*_{,n} M_n(w) + w_{,n} M^*_n(w^*) - w\, K^{l*}_n(w^*)]\, d\Gamma - \int_\Gamma [w^* M_{ns}(w) - w\, M^*_{ns}(w^*)]_{,s}\, d\Gamma \tag{9.33}$$

where $u^*_i$ and $w^*$ are weight functions. They and their related kernels are denoted by an asterisk to indicate their specified character. Equation (9.33) was initially introduced by Bergman-Schiffer [10] for a smooth boundary and was extended by Bezine-Gamby [11] and Stern [12] to non-smooth boundaries with corners or points where the boundary conditions change. References [11, 12] discussed the problem of non-smooth boundary in great detail. For simplicity sake, we will assume that the boundary is smooth in what follows. Hence the last term in Eq. (9.33) can be neglected. Linear decoupled problem with both inplane and out-of-plane deformations are treated using Eqs. (9.32) and (9.33), and their effectiveness has been proven via several numerical applications.

In order to extend the above problem to the nonlinear coupled case, consider the following two integral expressions:

$$\int_\Omega N_{ij}(u_k, w)_{,j}\, u^*_i\, d\Omega \tag{9.34}$$

and

$$\int_{\Omega} [-D\nabla^4 w + Q_i^n(u_k, w)_{,i} + \bar{p}]\, w^*\, d\Omega \tag{9.35}$$

The field variables other than the displacements and moments $M_n$ can now be defined and composed of linear and nonlinear parts,

$$N_{ij}(u_k, w) = N_{ij}^l(u_k) + N_{ij}^n(w) \tag{9.36}$$

$$p_i(u_k, w) = p_i^l(u_k) + p_i^n(w) \tag{9.37}$$

$$K_n(u_k, w) = K_n^l(w) + K_n^n(u_k, w) \tag{9.38}$$

$$Q_i(u_k, w) = Q_i^l(w) + Q_i^n(u_k, w) \tag{9.39}$$

We can take the same weight functions $u_i^*$ for Eqs. (9.32) and (9.34), and $w^*$ for Eqs. (9.33) and (9.35), respectively. Consequently, the related kernels are identical to those used in the linear decoupled case,

$$p_i^*(u_k^*) = p_i^{l*}(u_k^*), \quad N_{ij}^*(u_k^*) = N_{ij}^{l*}(u_k^*), \quad K_n^*(w^*) = K_n^{l*}(w^*)\,. \tag{9.40}$$

We can now integrate Eqs. (9.34) and (9.35) by parts taking into consideration Eqs. (9.36) to (9.40). This gives

$$\int_{\Omega} u_i N_{ij}^*(u_k^*)_{,j}\, d\Omega + \int_{\Gamma} [p_i(u_k, w)\, u_i^* - u_i p_i^*(u_i^*)]\, d\Gamma = \int_{\Omega} N_{ij}^n(w)\, u_{i,j}^*\, d\Omega \tag{9.41}$$

and

$$-D\int_{\Omega} w\nabla^4 w^*\, d\Omega + \int_{\Omega} \bar{p}\, w^*\, d\Omega - \int_{\Gamma} [M_n(w)\, w_{,n}^* - w_{,n} M_n^*(w^*) + K_n(u_k, w)\, w^* - w\, K_n^*(w^*)]\, d\Gamma = \int_{\Omega} Q_i^n(u_k, w)\, w_{,i}^*\, d\Omega \tag{9.42}$$

Identical results could be obtained using weighted residuals for the governing equations and boundary conditions as shown in Reference [13]. The main difference between Eqs. (9.41)–(9.42) and Eqs. (9.32)–(9.33) consists of the additional nonlinear terms written on the former right hand sides. These terms are given by the integrals on the domain $\Omega$, with terms of $w$ in Eq. (9.41) and of $w$ and $u_k$ in Eq. (9.42) representing the coupling between inplane and out-of-plane deformations.

The weight functions introduced in Eqs. (9.32), (9.33) and (9.41), (9.42) are the respective fundamental or principal solutions of the following equations:

$$N_{ij}^*[u_k^*(P, Q)]_{,j} + \delta_m(P, Q) = 0 \tag{9.43}$$

and

$$\nabla^4 w^*(P, Q) + \delta(P, Q) = 0 \tag{9.44}$$

These solutions are known as the Kelvin solution for the two-dimensional infinite body in a state of plane stress and as the fundamental solution to the two-dimensional biharmonic equations for the out-of-plane deformation, respectively. Equation (9.43) represents the two-dimensional field at a point $P$ under unit inplane force concentrated at point $Q$ acting in the $m$-direction ($m = 1, 2$). Equation (9.44) corresponds to the deflection at $P$ due to a unit concentrated lateral load at $Q$. Their solutions are well-known and functions of the distance $r$

between the observation point $P$ and the force point $Q$. For

$$r = |\overline{PQ}| \tag{9.45}$$

we can write [11, 13],

$$u_k^* (P, Q)_m = \frac{1+\nu}{8\pi G}\left[\frac{3-\nu}{1+\nu}\log\frac{1}{r}\,\delta_{km} + r_{,k}\, r_{,m}\right] \tag{9.46}$$

$$p_k^* [u_k^* (P, Q)]_m = \frac{1+\nu}{4\pi r}\left[r_{,n}\left(\frac{1-\nu}{1+\nu}\,\delta_{km} + 2r_{,k}\, r_{,m}\right) - \frac{1-\nu}{1+\nu}\,(r_{,m}\, n_k - r_{,k}\, n_m)\right] \tag{9.47}$$

$$w^* (P, Q) = -\frac{1}{8\pi}\, r^2 \log r \tag{9.48}$$

where $u_{km}^*$ and $p_{km}^*$ denote the components of $u_k^*$ and $p_k^*$ due to the unit inplane force applied at $Q$ in the $m$-direction. The related kernels of Eq. (9.42), $w_{,n}^*$, $M_n^*$ and $K_n^*$, can be obtained by differentiating Eq. (9.48) (for further detail, see Ref. [11]).

Substituting the corresponding fundamental solutions into Eqs. (9.41) and (9.42) and computing the Cauchy principal value integrals, we finally have

$$- C(P)\, u_i(P) + \int_\Gamma \{p_k[u_m(Q)]\, u_k^* (P, Q)_{\,i} - u_k(Q)\, p_k^* [u_m^* (P, Q)]_{\,i}\}\, d\Gamma$$
$$= \int_\Omega N_{km}^n [w(\hat{Q})]\, u_k^* (P, \hat{Q})_{i,m}\, d\Omega \qquad (Q \in \Gamma,\ \hat{Q} \in \Omega) \tag{9.49}$$

and

$$D\, C(P)\, w(P) + \int_\Omega \bar{p}(\hat{Q})\, w^* (P, \hat{Q})\, d\Omega$$
$$- \int_\Gamma \{M_n[w(Q)]\, w^* (P, Q)_{,n} - w(Q)_{,n}\, M_n^* [w^* (P, Q)]$$
$$- K_n[u_k(Q), w(Q)]\, w^* (P, Q) + w(Q)\, K_n^* [w^* (P, Q)]\}\, d\Gamma$$
$$= \int_\Omega Q_k^n [u_k(\hat{Q}), w(\hat{Q})]\, w^* (P, \hat{Q})_{,k}\, d\Omega \qquad (Q \in \Gamma,\ \hat{Q} \in \Omega) \tag{9.50}$$

where the value of the coefficient $C(P)$ depends on whether the point $P$ belongs to the domain $\Omega$ or to the smooth boundary $\Gamma$, i.e.,

$$C(P) = \begin{cases} 1 & (P \in \Omega) \\ \frac{1}{2} & (P \in \Gamma) \end{cases} \tag{9.51}$$

An additional integral equation is needed for the analysis of bending problems. This equation can be derived by differentiating Eq. (9.50) with respect to the outward normal $n_0$ on the boundary point $P$, i.e.,

$$\tfrac{1}{2} D\, w_{,n_0}(P) + \int_\Omega \bar{p}(Q)\, w_{,n_0}^* (P, Q)\, d\Omega - \int_\Gamma \{M_n[w(Q)]\, w_{,nn_0}^* (P, Q) - w_{,n}(Q)$$
$$\times M_n^* [w^* (P, Q)]_{,n_0} - K_n[u_k(Q), w(Q)]\, w_{,n_0}^* (P, Q)$$
$$- w(Q)\, K_n^* [w^* (P, Q)]_{,n_0}\}$$
$$\times d\Gamma = \int_\Omega Q_k^n [u_k(\hat{Q}), w(\hat{Q})]\, w^* (P, \hat{Q})_{,kn_0}\, d\Omega$$
$$(P, Q \in \Gamma,\ \hat{Q} \in \Omega) \tag{9.52}$$

Equations (9.49), (9.50) and (9.52) constitute a set of simultaneous integral equations in terms of the following eight boundary quantities:

$$p_i, M_n, K_n, u_i, w \quad \text{and} \quad w_{,n} \qquad (i=1,2) \tag{9.53}$$

Of the above eight quantities, four need to be prescribed on the boundary points and the remaining four are to be determined.

The numerical procedure needed to solve Eqs. (9.49), (9.50) and (9.52) using BEM can now be applied. The boundary $\Gamma$ and the domain $\Omega$ are discretized into a finite number of one-dimensional elements and two-dimensional cells, i.e., boundary elements and domain cells. For each node taken on the boundary elements, the four equations (9.49), (9.50) and (9.52) are reduced to a system of algebraic equations in terms of nodal unknowns.

By using column vectors for the nodal unknowns, i.e.,

$$\mathbf{u}_1, \mathbf{u}_2, \mathbf{p}_1, \mathbf{p}_2, \mathbf{w}, \mathbf{w}_{,n}, \mathbf{M}_n \text{ and } \mathbf{K}_n \tag{9.54}$$

a system of simultaneous equations can be formulated as follows:

$$\mathbf{A}_1 \mathbf{u}_1 + \mathbf{B}_1 \mathbf{u}_2 + \mathbf{C}_1 \mathbf{p}_1 + \mathbf{D}_1 \mathbf{p}_2 = \mathbf{N}_1 \tag{9.55}$$

$$\mathbf{A}_2 \mathbf{u}_1 + \mathbf{B}_2 \mathbf{u}_2 + \mathbf{C}_2 \mathbf{p}_1 + \mathbf{D}_2 \mathbf{p}_2 = \mathbf{N}_2 \tag{9.56}$$

$$\mathbf{E}_1 \mathbf{w} + \mathbf{F}_1 \mathbf{w}_{,n} + \mathbf{G}_1 \mathbf{M}_n + \mathbf{H}_1 \mathbf{K}_n = \mathbf{N}_3 \tag{9.57}$$

$$\mathbf{E}_2 \mathbf{w} + \mathbf{F}_2 \mathbf{w}_{,n} + \mathbf{G}_2 \mathbf{M}_n + \mathbf{H}_2 \mathbf{K}_n = \mathbf{N}_4 \tag{9.58}$$

where $\mathbf{A}_i, \ldots, \mathbf{D}_i$, $\mathbf{E}_i, \ldots, \mathbf{H}_i$ $(i=1,2)$ are influence coefficient matrices and the right hand side vectors $\mathbf{N}_i$ $(i=1,\ldots,4)$ contain the nonlinear and inhomogeneous terms. To compute the nonlinear terms an iterative procedure is required. The $(t+1)$-st approximation is estimated using the $t$-th approximation for the right hand side terms. Calculation can then be repeated until one reaches the required accuracy. Equations (9.55) to (9.58) are required only on the boundary, i.e., $C(P)=\frac{1}{2}$ in Eqs. (9.49) and (9.50). However, to calculate the right hand side terms $\mathbf{N}_i$ one needs to compute the internal values of the displacements and their derivatives for which Eqs. (9.49) and (9.50) are used with $C(P)=1$. These integrations are usually performed numerically.

## 9.4 The Approximate Berger Equation [6]

In the previous section, the formulation of the boundary integral equations has been applied to the governing field equation due to von Karman. They are based on the Kirchhoff-Love assumption, and the resulting inplane and out-of-plane deformations are fully coupled, which makes their solution difficult. In this section, an approximate field equation to simplify the finite deflection analysis due to Berger [6] is presented.

The Berger equation was originally obtained from the variational formulation of the problem by neglecting without proper justification the second invariant of membrane strains in the potential energy expression for laterally loaded, homogeneous, isotropic, thin elastic plates. Since the resulting equation is uncoupled and quasi-linear it has been extensively applied in place of the complete equation derived in section 3. The equation has been used for dynamic as well as static problems, shallow shells, sandwich plates and several others [14–18]. It is important to notice that the Berger solution can give a fairly good approximation to the problem provided that the inplane displacements are restrained on the boundary.

If a thin elastic plate is subjected to a lateral load $\bar{p}(x_1, x_2)$ and a thermal field $T(x_1, x_2, z)$, the total potential energy can be expressed as

$$\Pi = \tfrac{1}{2} D \int_\Omega \left\{ (\nabla^2 w)^2 + \frac{12}{h^2} I_1^2 - 2(1-\nu)\left[\frac{12}{h^2} I_2 + w_{,11} w_{,22} - w_{,12}^2\right]\right\} d\Omega$$

$$- \int_\Omega \left[\bar{p} w + \frac{1}{1-\nu}(I_1 N_T - M_T \nabla^2 w)\right] d\Omega + \text{(Boundary terms)} \tag{9.59}$$

where

$$I_1 = u_{1,1} + u_{2,2} + \frac{1}{2}(w_{,1}^2 + w_{,2}^2) \tag{9.60}$$

$$I_2 = \left(u_{1,1} + \frac{1}{2} w_{,1}^2\right)\left(u_{2,2} + \frac{1}{2} w_{,2}^2\right) - \frac{1}{4}(u_{1,2} + u_{2,1} + w_{,1} w_{,2})^2 \tag{9.61}$$

and where

$$M_T = \int_{-h/2}^{h/2} E \alpha T z \, dz\,, \quad N_T = \int_{-h/2}^{h/2} E \alpha T \, dz \tag{9.62}$$

The complete von Karman-type equations can be derived from Eq. (9.59) using variational calculus. If we neglect $I_2$, i.e., the second invariant of the membrane strains, as suggested by Berger, and then apply variational calculus, the resulting equation is,

$$D \nabla^4 w - D \varkappa^2 \nabla^2 w + \frac{\nabla^2 M_T}{1-\nu} = \bar{p} \tag{9.63}$$

together with the identity,

$$I_1 - \frac{(1+\nu) N_T}{E h} = \frac{\varkappa^2 h^2}{12} = \text{const.} \tag{9.64}$$

Equation (9.63) is linear with respect to $w$, but the parameter $\varkappa^2$, sometimes called Berger constant, is dependent on the external force or temperature field. In other words, the second term in Eq. (9.63) represents nonlinear relationship between deformation and intensity of external actions.

The magnitude of the Berger constant $\varkappa^2$ can be estimated as follows. Let us integrate both sides of Eq. (9.64) over the whole two-dimensional plate domain $\Omega$, i.e.,

$$\int_\Omega \frac{\varkappa^2 h^2}{12} d\Omega = \int_\Omega \left[u_{1,1} + u_{2,2} + \frac{1}{2}(w_{,1}^2 + w_{,2}^2) - \frac{(1+\nu) N_T}{E h}\right] d\Omega = \frac{\varkappa^2 h^2}{12} A_\Omega \tag{9.65}$$

where the area of the domain $\Omega$ is denoted by $A_\Omega$. Consequently, we obtain,

$$\varkappa^2 = \frac{12}{h^2 A_\Omega} \int_\Omega \left[u_{1,1} + u_{2,2} + \frac{1}{2}(w_{,1}^2 + w_{,2}^2) - \frac{(1+\nu) N_T}{E h}\right] d\Omega\,. \tag{9.66}$$

## 9.5 Integral Formulation for the Berger Equation [2–5]

The integral formulae for the Berger equation can be easily obtained using a similar procedure as the one given for the linear problem. Let us write the second and third terms in Eq. (9.63) on the right hand side, which gives

$$D\nabla^4 w = \bar{p} - \frac{\nabla^2 M_T}{1-\nu} + D\,\varkappa^2 \nabla^2 w \tag{9.67}$$

Since now the left hand side contains only the biharmonic operator in terms of the deflection and a harmonic term in $M_T$ exists on the right hand side, the following generalized Green identity will hold for smooth boundary [19]:

$$D\int_\Omega w^* \left(\nabla^2 w + \frac{\nabla^2 M_T}{1-\nu}\right) d\Omega = D\int_\Omega \left[ w\,(\nabla^4 w^*) + \frac{M_T}{1-\nu}\nabla^2 w^* \right] d\Omega$$
$$- \int_\Gamma [w^* K_n^{lT}(w) - w^*_{,n} M_n^T(w) + w_{,n} M_n^*(w^*) - w K_n^{l*}(w^*)]\,d\Gamma \tag{9.68}$$

where $K_n^{l*}(w^*)$ and $M_n^*(w^*)$ denote related kernels, and $K_n^{lT}(w)$ and $M_n^T(w)$ correspond respectively to the Kirchhoff effective shear force in the linear deflection and the normal bending moment, under the action of mechanical and thermal fields,

$$K_n^{lT}(w) = K_n^l(w) - \left(\frac{M_T}{1-\nu}\right)_{,n} \tag{9.69}$$

$$M_n^T(w) = M_n(w) - \frac{M_T}{1-\nu} \tag{9.70}$$

By using the known fundamental solution (9.48) and applying Cauchy principal values, we arrive at the following equation:

$$D\,C(P)\,w(P) - \int_\Gamma \{w^*(P,Q)\,K_n^{lT}[w(Q)] - w^*_{,n}(P,Q)\,M_n^T[w(Q)]$$
$$+ w_{,n}(Q)\,M_n^*[w^*(P,Q)] - w(Q)\,K_n^{l*}[w^*(P,Q)]\}\,d\Gamma$$
$$= -\int_\Omega \left[ \{D\varkappa^2[\nabla^2 w(\hat{Q})] + \bar{p}(\hat{Q})\}\,w^*(P,\hat{Q}) - \frac{M_T}{1-\nu}(Q)\,\nabla^2 w^*(P,\hat{Q}) \right] d\Omega$$
$$(Q \in \Gamma,\ \hat{Q} \in \Omega) \tag{9.71}$$

where the coefficient $C(P)$ depends as previously on the position of the point $P$. Another equation for the normal derivative of the deflection on the boundary is obtained by differentiating Eq. (9.71) in the $n_0$ direction, i.e.,

$$\tfrac{1}{2} D\,w_{,n_0}(P) - \int_\Gamma \{w^*_{,n_0}(P,Q)\,K_n^{lT}[w(Q)] - w^*_{,nn_0}(P,Q)\,M_n^T[w(w(Q)]$$
$$- w_{,n}(Q)\,M_n^*[w^*(P,Q)]_{,n_0} - w(Q)\,K_n^{l*}[w^*(P,Q)]_{,n_0}\}\,d\Gamma$$
$$= \int_\Omega \left[ \{D\varkappa^2[\nabla^2 w(\hat{Q})] + \bar{p}(\hat{Q})\}\,w^*_{,n_0}(P,\hat{Q}) - \frac{M_T}{1-\nu}(\hat{Q})\,\nabla^2 w^*_{,n_0}(P,\hat{Q}) \right] d\Omega$$
$$(P, Q \in \Gamma,\ \hat{Q} \in \Omega) \tag{9.72}$$

The discretization of Eqs. (9.71) and (9.72) is performed as in section 3. Once this is done in $\Omega$ and on $\Gamma$, the following set of simultaneous equations in terms of the boundary unknowns $\mathbf{w}$, $\mathbf{w}_{,n}$, $\mathbf{M}_n^T$ and $\mathbf{K}_n^{lT}$ can be obtained;

$$\mathbf{E}_1' \, \mathbf{w} + \mathbf{F}_1' \, \mathbf{w}_{,n} + \mathbf{G}_1' \, \mathbf{M}_n^T + \mathbf{H}_1' \, \mathbf{K}_n^{lT} = \mathbf{N}_3' \tag{9.73}$$

$$\mathbf{E}_2' \, \mathbf{w} + \mathbf{F}_2' \, \mathbf{w}_{,n} + \mathbf{G}_2' \, \mathbf{M}_n^T + \mathbf{H}_2' \, \mathbf{K}_n^{lT} = \mathbf{N}_4' \tag{9.74}$$

These are similar to Eqs. (9.57) and (9.58) although their influence coefficient matrices are different. The right hand side column vectors $\mathbf{N}_3'$ and $\mathbf{N}_4'$ contain the nonlinear and inhomogeneous parts of the original governing equation (9.63). Once the right hand side vectors are computed one can find the values of the two unknowns among the four boundary quantities $w$, $w_{,n}$, $M_n^T$ and $K_n^{lT}$. However, it must be pointed out that since Eqs. (9.73) and (9.74) are formulated in terms of $K_n^{lT}$ instead of $K_n^T [= K_n^l + K_n^n - M_{T,n}/(l-\nu)]$, the Kirchhoff effective shear force $K_n^T$ is not expressed properly and hence the free boundary cannot be modelled in this way. To estimate the nonlinear terms, one starts with an initial solution, usually the linear one.

As indicated above, the Berger equation gives a fairly accurate approximation for restrained boundary conditions. For such a case, the Berger constant can be easily estimated. Over the boundary, if

$$u_1 = u_2 = 0 \quad \text{on } \Gamma$$

we obtain from Eq. (9.66)

$$\varkappa^2 = \frac{12}{h^2 A_\Omega} \int_\Omega \left[ \frac{1}{2} (w_{,1}^2 + w_{,2}^2) - \frac{(1+\nu)\, N_T}{E\, h} \right] d\Omega \tag{9.75}$$

If in addition to the condition, either

$$w = 0 \quad \text{or} \quad w_{,n} = 0 \tag{9.76}$$

is forced on the boundary, the constant is further reduced to

$$\varkappa^2 = -\frac{12}{h^2 A_\Omega} \int_\Omega \left[ \frac{1}{2} w \, \nabla^2 w + \frac{(1+\nu)\, N_T}{E\, h} \right] d\Omega \tag{9.77}$$

This equation can be integrated numerically using the distributions of the deflection $w$ and the thermal resultant $N_T$ over the domain $\Omega$.

It should be pointed out that with the exception of the determination of Berger constant and the resulting iterative procedure, the numerical scheme and the system of simultaneous equations are similar to those used in the linear bending problem.

## 9.6 Numerical Examples

In this section some numerical results obtained using the integral formulation of Berger equation are presented.

The plate boundary and domain are discretized into constant elements and cells. In this way a set of algebraic simultaneous equations is obtained from Eqs. (9.73) and (9.74) using numerical integration. Calculation of the integral related to the Berger constant can also be carried out numerically for a given distribution of the deflection and thermal field. An iterative procedure is applied until the required convergence is achieved.

To check the validity of the proposed approximate procedure, some simple circular and square plate problems have been solved. For these problems, the nonlinear solution can be obtained using conventional methods such as Runge-Kutta-Gill (RKG) method or classical energy method due to their simple geometry. The plates are considered to be subjected to uniformly distributed lateral load

$$\bar{p} = \text{const}$$

or a distributed thermal field given by (say, for circular plates),

$$T(R, z) = \left[T_0 + T_1\left(1 - \frac{R^2}{a^2}\right)\right]\left(1 + \frac{2}{3}\frac{z}{h}\right), \quad R^2 = x_1^2 + x_2^2$$

where $T_0$ and $T_1$ are constants. The temperature is assumed to vary linearly within the thickness of the plate.

Figures 9.2 to 9.5 show results for the maximum deflection of circular and square plates; obtained by using 24 boundary elements and 72 cells in the case of circular plate, and 48 boundary elements and 64 cells for square plates [2, 3]. The boundary conditions considered are indicated in each of the figure captions. Poisson ratio has been taken $\nu = 0.3$.

The present boundary element solutions (marked by small circles in figures) are compared against the more rigorous results obtained using the von Karman equation. A set of results obtained using coarse discretization (12 boundary elements) is also included in Fig. 9.3 and plotted as black small circles.

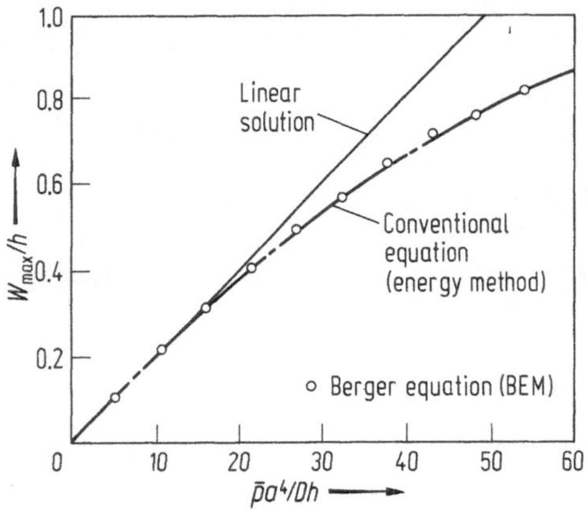


**Fig. 9.2.** Maximum deflections of immovably clamped square plate under uniform lateral load

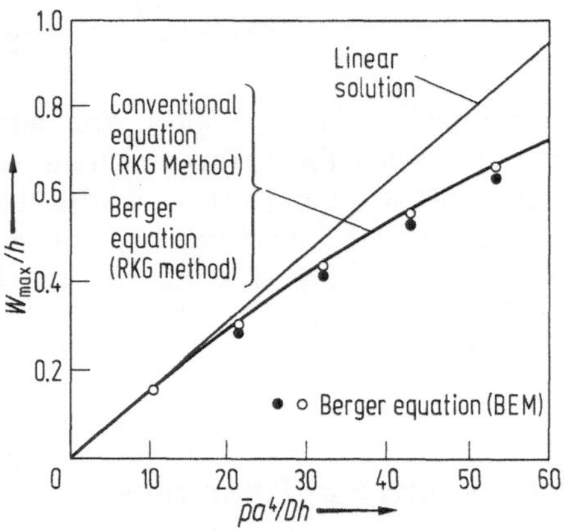


**Fig. 9.3.** Maximum deflections of immovably clamped circular plate under uniform lateral load

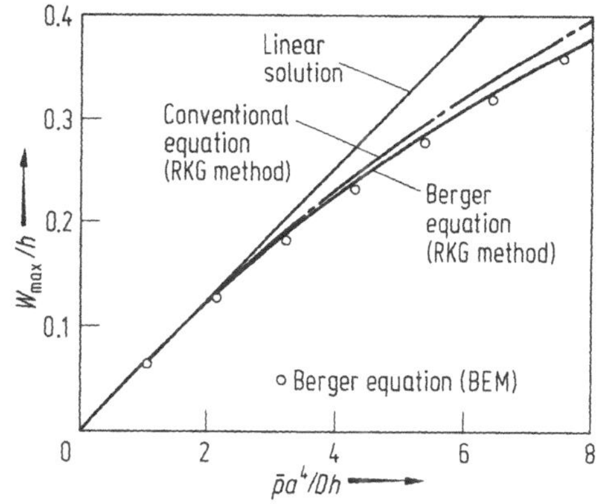


**Fig. 9.4.** Maximum deflections of immovably supported, rotation free circular plate under uniform lateral load

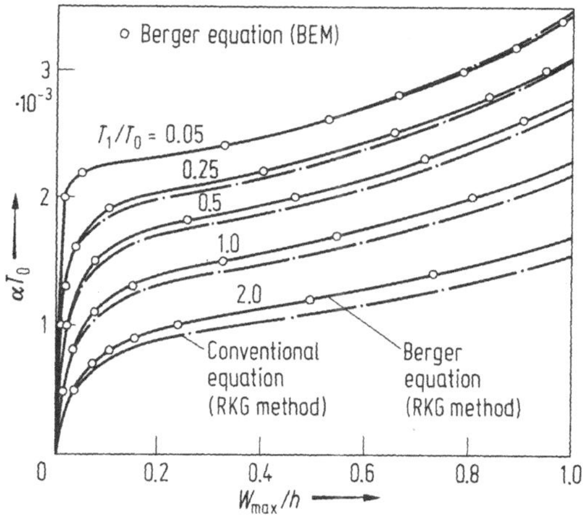


**Fig. 9.5.** Maximum deflections of immovably clamped heated circular plate

The Berger solution for 24 boundary elements agrees well with the more accurate results. A significant improvement in the accuracy of the solution is noticeable when comparing this solution with the one obtained using the coarse mesh (Fig. 9.3). It should be pointed out that it is always difficult to carry out rigorous nonlinear analysis for plates with arbitrary plans, hence the importance of boundary element formulations to solve this problem.

Figure 9.5 shows the deflections of a plate heated by inhomogeneous temperature distribution. The boundary element solution coincides with the RKG solution of the Berger equation for the whole range of the temperature parameter $T_1/T_0$. The difference between the results obtained using the Berger or the complete von Karman formulations, i.e., the uncoupled and coupled solutions, varies with the magnitude of the parameter $T_1/T_0$ and the stage of deformation. Nevertheless, the difference does not seem significant from the practical point of view and the Berger solution may trace well the coupled results even for highly nonlinear behaviour.

## 9.7 Nonlinear Shallow Shell and Sandwich Plate/Shell Problems [4, 5]

There have been many applications of Berger original idea to approximate the finite deflection of shallow shells and sandwich structures. In this section, we will outline briefly how this can be extended to find an integral formulation for problems with shallow shells and sandwich plates and shells.

*Shallow Elastic Shells* [4]

The governing field equations for finite deflections of elastic shells become more complicated even if we follow the conventional Kirchhoff-Love assumption and use small curvature theory. The assumptions produce a fully coupled system of equations.

To generalize the method the Berger hypothesis has been applied to shallow elastic shells [14, 15]. Consider a shallow elastic shell of constant thickness and of double curvature $k_1$ and $k_2$ referred to the orthogonally curvilinear coordinate system $x_1$, $x_2$, $z$ (Fig. 9.6). The coordinate axes $x_1$ and $x_2$ are along the principal axes of curvature in the middle plane of the shell. Through a similar procedure to flat plates, i.e., by disregarding the second invariant of the membrane strains and then minimizing the potential energy of the shell, we can arrive to the following equation

$$D\,\nabla^4 w - D\,\varkappa^2(\nabla^2 w + k_1 + k_2) + \frac{\nabla^2 M_T}{1-\nu} = \bar{p} \tag{9.78}$$

with the identity

$$I_1 - \frac{(1+\nu)\,N_T}{E\,h} = \frac{\varkappa^2 h^2}{12} = \text{const.} \tag{9.79}$$

where the first invariant of the membrane strain is

$$I_1 = u_{1,1} + u_{2,2} + \tfrac{1}{2}(w_{,1}^2 + w_{,2}^2) - (k_1 + k_2)\,w \tag{9.80}$$

Notice that we are assuming that lateral load $\bar{p}$ and a thermal field are acting on the shell. Equations (9.78) and (9.80) differ from the corresponding Eqs. (9.63) and (9.60) for flat plates only by the additional terms representing the total curvature $k_1 + k_2$. Therefore, we can formulate integral equations similarly as in section 5.

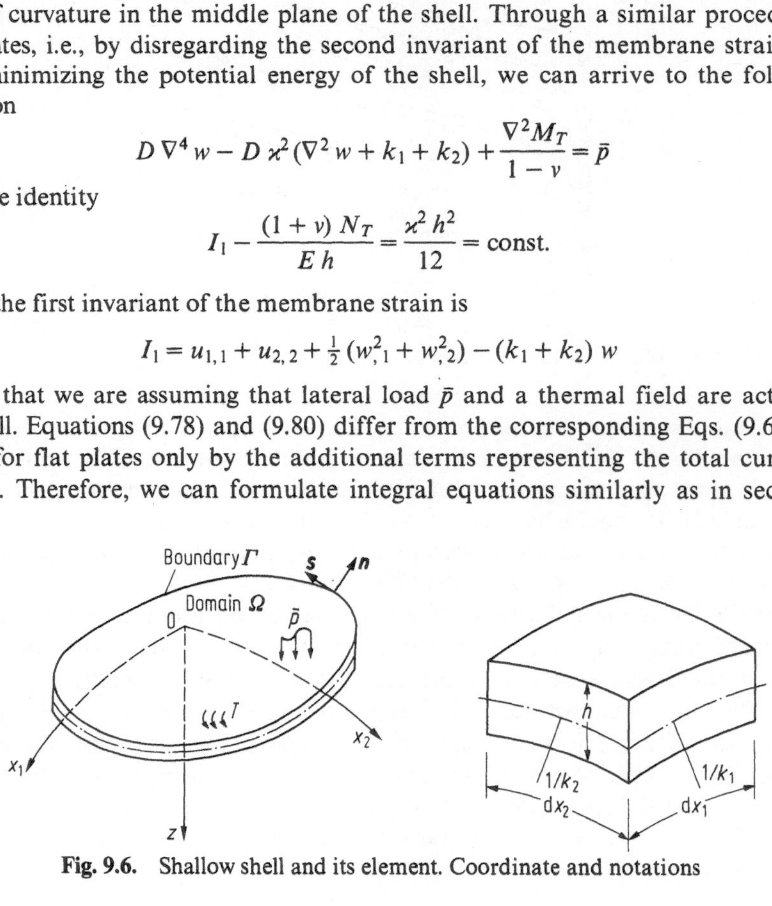


**Fig. 9.6.** Shallow shell and its element. Coordinate and notations

The final integral formulae are of the same form as those in Eqs. (9.71) and (9.72) except that the term $\nabla^2 w$ has to be replaced by $\nabla^2 w + k_1 + k_2$. It must be pointed out that the Berger constant for shallow shells is also affected by the curvature.

The results obtained for axisymmetric finite deflections of circular spherical caps under constant lateral load and distributed temperature are shown in Figs. 9.7 and 9.8. Berger results coincide with the corresponding RKG solutions for a whole range of total curvature.

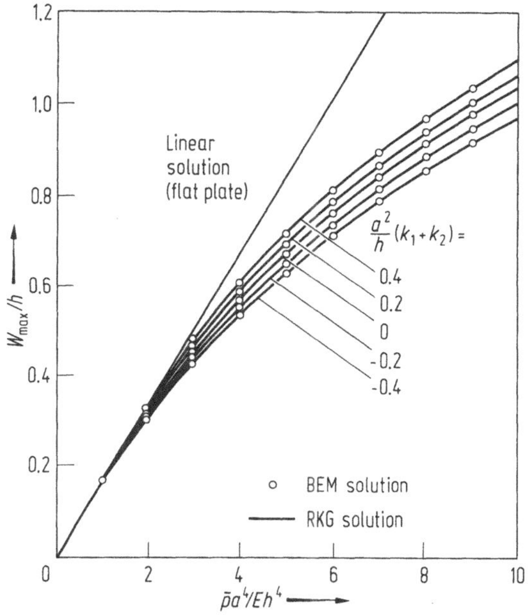


Fig. 9.7. Maximum deflections of immovably clamped spherical cap under uniform lateral load

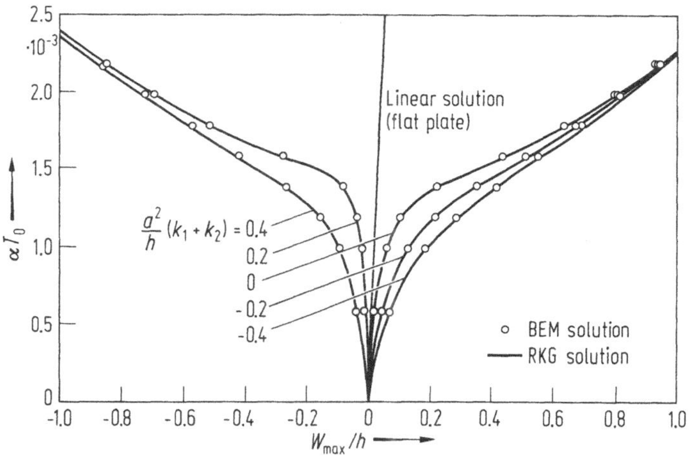


Fig. 9.8. Maximum deflections of immovably clamped heated spherical cap

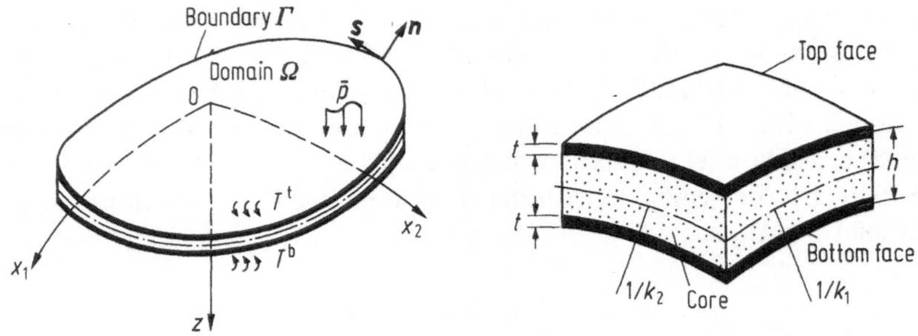


**Fig. 9.9.** Sandwich shell and its element. Coordinate and notations

*Sandwich Plates and Shells* [5]

Consider now three-layered isotropic sandwich plate and shell as shown in Fig. 9.9, subjected to both lateral load and temperature difference between top and bottom faces. We assume that the core withstands shear deformation, whereas the thin faces withstand bending deformation only. The governing field equations for finite deflections of sandwich plates and shells are complicated because of an additional effect due to the shear deformation occurring in the core. The Berger method was extended via a slight modification to derive their approximate governing field equations [17, 18]. For three-layered sandwich plates and shallow sandwich shells with isotropic faces and core, the field equations obtained by the Berger procedure can be reduced to a single differential equation in terms of deflection alone. Supplementary constraint condition; that is, invariance of a particular parameter, is also necessary.

The final equation is

$$\nabla^4 w - \frac{\varkappa^2}{A\varkappa^2+1}(\nabla^2 w + k_1 + k_2) - \frac{1}{D_s(A\varkappa^2+1)}[\bar{p} - A\nabla^2\bar{p} + B\nabla^2 f - AD_s\varkappa^2\nabla^2(k_1+k_2)] = 0 \tag{9.81}$$

with the identity

$$I_1^m - (1+\nu^f)\,\alpha^f T^m = \frac{\varkappa^2 h^2}{4} = \text{const} \tag{9.82}$$

where

$$I_1^m = u_{1,1}^m + u_{2,2}^m + \tfrac{1}{2}(w_{,1}^2 + w_{,2}^2) - (k_1+k_2)\,w \tag{9.83}$$

$$u_1^m = \tfrac{1}{2}(u_1^t + u_1^b), \quad u_2^m = \tfrac{1}{2}(u_2^t + u_2^b) \tag{9.84}$$

$$T^m = \tfrac{1}{2}(T^t + T^b), \quad f = T^t - T^b \tag{9.85}$$

and where

$$D_s = \frac{E^f t h^2}{2[1-(\nu^f)^2]}, \quad A = \frac{D_s}{G^c h}, \quad B = \frac{D_s(1+\nu^f)\,\alpha^f}{h}$$

The derived approximate field equation (9.81) is slightly complicated compared with those for single-layer flat plates or shallow shells, but nonlinearity of deformation rests on only a parameter $\varkappa^2$ defined as Eq. (9.82) as for previous formulae. Such formulation very much facilitates the subsequent integral formulation and calculation.

Since the principal term of Eq. (9.81) is the biharmonic representation of the deflection, we may follow a procedure analogous to that of section 7.1, to obtain the following integral equations:

$$
\begin{aligned}
C(P)\, w(P) &+ \int_{\Gamma} \{K_n^s(Q)\, w^*(P, Q) - M_n^s(Q)\, w^*_{,n}(P, Q) + M_n^*[w^*(P, Q)]\, w_{,n}(Q) \\
&- K_n^{l*}[w^*(P, Q)]\, w(A)\}\, d\Gamma = -\int_{\Omega} \left[ \frac{\varkappa^2}{A\varkappa^2 + 1} \{[\nabla^2 w^*(P, \hat{Q})]\, w(w(\hat{Q}) \right. \\
&+ w^*(P, \hat{Q})\, [k_1(\hat{Q}) + k_2(\hat{Q})]\} + \frac{1}{D_s(A\varkappa^2 + 1)} \{\bar{p}(\hat{Q}) - A\nabla^2 \bar{p}(\hat{Q}) \\
&\left. + B\nabla^2 f(\hat{Q}) - AD_s \varkappa^2 \nabla^2 [k_1(\hat{Q}) + k_2(\hat{Q})]\}\, w^*(P, \hat{Q}) \right] d\Omega
\end{aligned}
$$

and

$$(Q \in \Gamma, \quad \hat{Q} \in \Omega) \qquad (9.86)$$

$$
\begin{aligned}
\tfrac{1}{2} w_{,n_0}(P) &+ \int_{\Gamma} \{K_n^s(Q)\, w^*_{,n_0}(P, Q) - M_n^s(Q)\, w^*_{,nn_0}(P, Q) \\
&+ M^*_{n,n_0}[w^*(P, Q)]\, w_{,n}(Q) - K^{l*}_{n,n_0}[w^*(P, Q)]\, w(Q)\}\, d\Gamma \\
= -\int_{\Omega} &\left[ \frac{\varkappa^2}{A\varkappa^2 + 1} \{[\nabla^2 w^*_{,n_0}(P, \hat{Q})]\, w(\hat{Q}) + w^*_{,n_0}(P, \hat{Q})\, [k_1(\hat{Q}) + k_2(\hat{Q})]\} \right. \\
&+ \frac{1}{D_s(A\varkappa^2 + 1)} \{\bar{p}(\hat{Q}) - A\nabla^2 \bar{p}(\hat{Q}) + B\nabla^2 f(\hat{Q}) - AD_s \varkappa^2 \nabla^2 [k_1(\hat{Q}) + k_2(\hat{Q})] \\
&\left. \times w^*_{,n_0}(P, \hat{Q}) \right] d\Omega \qquad (P, Q \in \Gamma, \hat{Q} \in \Omega) \qquad (9.87)
\end{aligned}
$$

with

$$C(P) = \begin{cases} 1 & (P \in \Omega) \\ \frac{1}{2} & (P \in \Gamma) \end{cases}$$

where

$$K_n^s = K_n(w) + \frac{\varkappa^2}{A\varkappa^2 + 1} w_{,n} \qquad (9.88)$$

$$M_n^s = M_n(w) + \frac{\varkappa^2}{A\varkappa^2 + 1} w \qquad (9.89)$$

From these formulae the discretized algebraic equations can be obtained as usual and a numerical iterative scheme can be carried out in order to obtain the solution.

Some numerical results are shown in Figs. 9.10 to 9.13 for clamped circular flat sandwich plates and square shallow sandwich shells under uniform lateral load and specified temperature difference between both faces. The following geometrical and material parameters are used in the calculation:

$$t = 0.635 \text{ mm}, \quad h = 17.13 \text{ mm}, \quad \nu^f = 0.3$$

$$E^f = 0.72 \times 10^5 \text{ MPa}, \quad G_0^c = 0.414 \times 10^2 \text{ MPa}$$

Temperature distribution for circular sandwich plates:

$$T^t = \frac{2}{3} T_0 \left\{1 + \left(1 - \frac{R^2}{a^2}\right)\right\}, \quad T^b = 2T^t, \quad a = 254 \text{ mm} \quad \text{(radius)}$$

and for square sandwich shells:

$$T^t = \frac{2}{3} T_0 \left\{1 + \left(1 - \frac{x_1^2}{a^2}\right)\left(1 - \frac{x_2^2}{a^2}\right)\right\}, \quad T^b = 2T^t, \quad 2a = 508 \text{ mm} \quad \text{(edge length)}$$

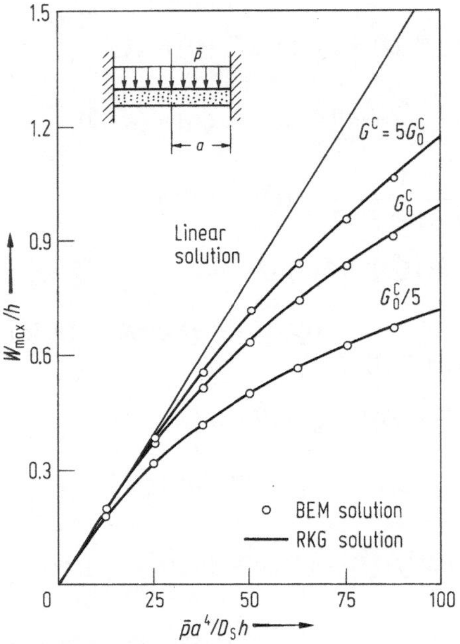


**Fig. 9.10.** Maximum deflections of flat circular sandwich plate under uniform lateral load

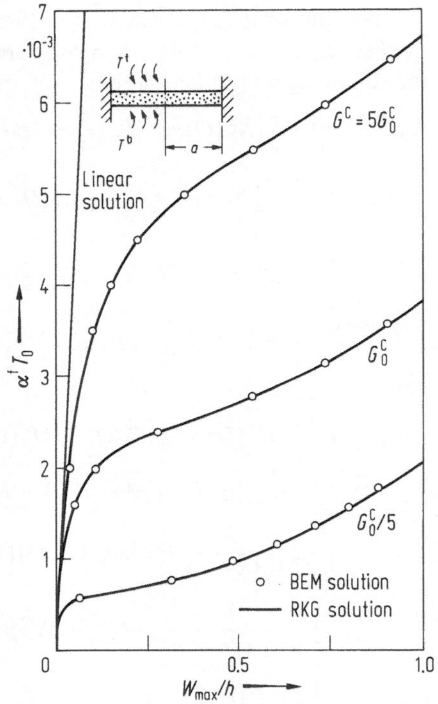


**Fig. 9.11.** Maximum deflections of heated flat circular sandwich plate

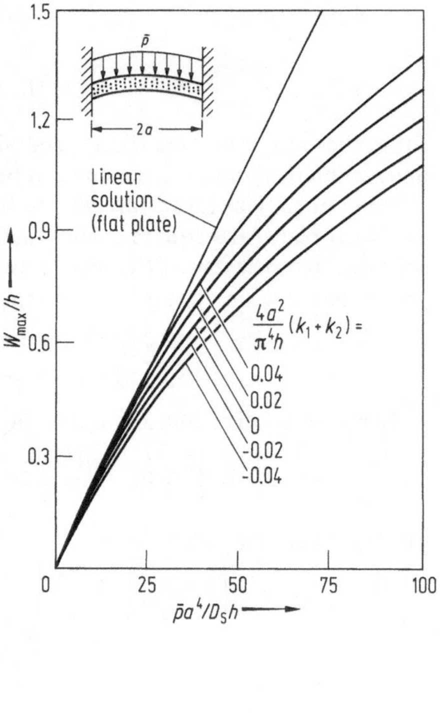


**Fig. 9.12.** Maximum deflections of shallow square sandwich shell under uniform lateral load ($G^c = G_0^c$)

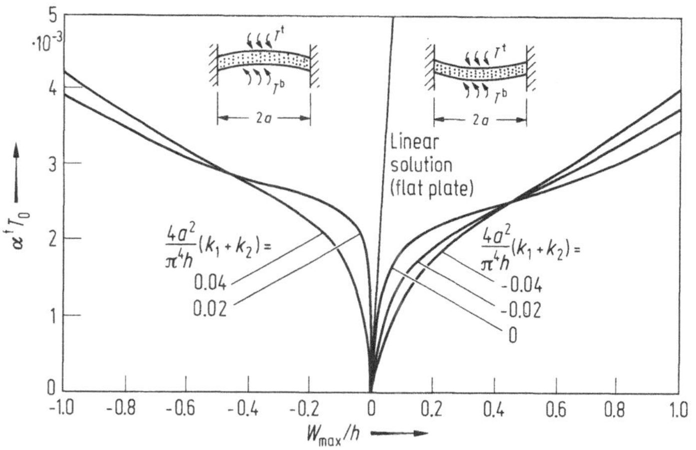


**Fig. 9.13.** Maximum deflections of heated shallow square sandwich shell ($G^c = G_0^c$)

# 9.8 Concluding Remarks

In this chapter, the application of BEM to solve finite deflection of elastic thin plates, shallow shells and sandwich plates and shells has been presented. This is an important problem in structural and solid mechanics for which integral equation formulation can be conveniently applied.

Integral equation formulations based on the von Karman theory have been described as well as the so-called Berger approximation. The former gives a coupled system of field equations which is reduced to an uncoupled system for the latter.

Some numerical solutions are presented for the equation under the Berger hypothesis. This approximation gives fairly good results provided that the boundary conditions are suitable and represents a simple way of simulating the complex nonlinear behaviours of plates and shells.

Although the results presented in this chapter were restricted to simple geometrical shapes of plates and shells, external force distribution, special boundary condition – thus for the case of Berger equation – piecewise constant elements and smooth boundaries; the potentialities of the method are demonstrated. Further studies should be carried out to implement the technique in a more general form. Moreover, other integral formulations could be studied in addition to those given by the Berger and von Karman field equations.


## Acknowledgement

The authors wish to express their cordial thanks to Dr. C. A. Brebbia of Southampton University and Institute of Computational Mechanics for his valuable discussions and suggestions.

## References


1 Brebbia, C. A., *Progress in Boundary Element Methods.* Pentech Press, London 1981
2 Kamiya, N., Sawaki, Y., An integral equation approach to finite deflection of elastic plates. Int. J. Non-Linear Mech., **17**, pp. 187–194, 1982
3 Kamiya, N., Sawaki, Y., Nakamura, Y., Fukui, A., An approximate finite deflection analysis of a heated elastic plate by the boundary element method. Appl. Math. Modell., **6**, pp. 23–27, 1982
4 Kamiya, N., Sawaki, Y., A simplified nonlinear bending analysis of flat plates and shallow shells by boundary element approach based on Berger equation. Proc. Int. Conf. Num. Meth., pp. 289–297, 1982
5 Kamiya, N., Sawaki, Y., Nakamura, Y., Boundary element nonlinear bending analysis of clamped sandwich plates and shells. Proc. Fourth Int. Conf. on Boundary Element Methods in Engineering, Brebbia, C. A., ed., pp. 515–525, 1982
6 Berger, H. M., A new approach to the analysis of large deflections of plates. J. Appl. Mech., **22**, pp. 465–472, 1955
7 Timoshenko, S., Woinowsky-Krieger, S., *Theory of Plates and Shells.* McGraw-Hill, New York 1959
8 Washizu, K., *Variational Methods in Elasticity and Plasticity.* Pergamon Press, 1968
9 Kamiya, N., Sawaki, Y., Integral formulation for nonlinear bending of plates – Formulation by weighted residual method. Zeit. ang. Math. Mech., **62**, pp. 651–655, 1982
10 Bergman, S., Schiffer, M., *Kernel Functions and Elliptic Differential Equations in Mathematical Physics.* Academic Press, New York 1953
11 Bezine, G. P., Gamby, D. A., A new integral equation formulation for plate bending problems. *Recent Advances in Boundary Element Methods.* Brebbia, C. A., ed., Pentech Press, pp. 327–342, London 1978
12 Stern, M., A general boundary integral formulation for the numerical solution of plate bending problems. Int. J. Solids Struct., **15**, pp. 769–782, 1979
13 Brebbia, C. A., *The Boundary Element Method for Engineers.* Pentech Press, London 1978
14 Nash, W. A., Modeer, J. R., Certain approximate analyses of the nonlinear behaviours of plates and shallow shells. Proc. Symp. Theory of Thin Elastic Shells (IUTAM), pp. 331–354, 1960
15 Nowinski, J. L., Ismail, I. A., Certain approximate analyses of large deflections of cylindrical shells. Zeit. ang. Math. Phys., **15**, pp. 449–456, 1964
16 Pal, M. C., Large deflections of heated circular plates. Acta Mech., **8**, pp. 82–103, 1969
17 Kamiya, N., Governing equations for large deflections of sandwich plates. AIAA J., **14**, pp. 250–253, 1976
18 Kamiya, N., Analysis of the large thermal bending of sandwich plates by a modified Berger method. J. Strain Anal., **13**, pp. 17–22, 1978.
19 Kamiya, N., Sawaki, Y., Nakamura, Y., Thermal bending analysis by boundary integral equation method. Mech. Res. Comm., **8**, pp. 369–373, 1981


## Appendices

**Appendix I.** Non-zero components of bending and stiffness coefficients $D_{ijkl}$ and $E_{ijkl}$:

$$D_{1111} = D_{2222} = \frac{E\,h}{1-\nu^2}, \quad D_{1122} = D_{2211} = \frac{E\,h}{1-\nu}, \quad D_{1212} = D_{1221} = D_{2112} = D_{2121} = G\,h \tag{A 1}$$

$$E_{1111} = E_{2222} = -D, \quad E_{1122} = E_{2211} = -\nu D, \quad E_{1212} = E_{1221} = E_{2112} = E_{2121} = -\frac{D(1-\nu)}{2} \tag{A 2}$$

**Appendix II.** Alternative expressions for basic quantities:

$$V^l = -D\nabla^2 w_{,n}, \quad M_{ns} = -D(1-\nu)\, w_{,ns}, \quad K_n^l = -D\,[\nabla^2 w + (1-\nu)\, w_{,s}]_{,n} \tag{A 3}$$

Membrane stress function $F$:

$$N_{11} = F_{,22}, \quad N_{22} = F_{,11}, \quad N_{12} = -F_{,12} \tag{A 4}$$

von Karman equations:

$$D\nabla^4 w = \bar{p} + F_{,22}\, w_{,11} + F_{,11}\, w_{,22} - 2F_{,12}\, w_{,12} \tag{A 5}$$

$$\nabla^4 F = E\,h\,(w_{,12}^2 - w_{,11}\, w_{,22}) \tag{A 6}$$

**Appendix III. Notations**

| | |
|---|---|
| $a$ | linear dimension of plate or shell |
| $D$ | bending rigidity [$= E\,h^3/12\,(1-\nu^2)$] |
| $D_{ijkl}$ | elastic stiffness coefficient |
| $E$ | Young's modulus |
| $e_{ij}$ | membrane strain |
| $E_{ijkl}$ | bending coefficient |
| $F$ | membrane stress function |
| $G$ | shear modulus |
| $h$ | thickness or core thickness |
| $I_1, I_2$ | first and second invariants of membrane strains |
| $K_n$ | Kirchhoff effective shear force on the boundary |
| $k_i$ | principal curvature |
| $M_{ij}$ | bending and twisting moments |
| $M_n$ | normal bending moment on the boundary |
| $M_{ns}$ | twisting moment on the boundary |
| $M_T$ | thermal bending resultant |
| $\mathbf{n}$ | outward normal |
| $n_i$ | direction cosine of outward normal [$= \cos(\mathbf{x}_i, \mathbf{n})$] |
| $N_T$ | thermal stress resultant |
| $N_{ij}$ | stress resultant (membrane force) |
| $\bar{p}$ | lateral load |
| $p_i$ | inplane boundary traction |
| $Q_i, V$ | shear forces |
| $r$ | distance between observation and force points |
| $\mathbf{s}$ | boundary tangent |
| $t$ | thickness of face |
| $T$ | temperature |
| $u_i$ | inplane displacement |
| $w$ | deflection |
| $(x_1, x_2, z)$ | rectangular Cartesian or orthogonally curvilinear coordinate system |
| $\alpha$ | linear thermal expansion coefficient |
| $\Gamma$ | boundary curve |
| $\Gamma_1, \Gamma_2$ | boundaries where mechanical and geometrical boundary conditions are specified, respectively |

| | |
|---|---|
| $\varepsilon_{ij}$ | permutation symbol |
| $\delta$, $\delta_i$ | Dirac delta function |
| $\delta_{ij}$ | Kronecker delta |
| $\varkappa^2$ | Berger constant |
| $\Omega$ | domain |
| $\nu$ | Poisson ratio |
| $\Pi$ | potential energy |
| $\sigma_{ij}$ | stress |
| $\nabla^2$ | Laplace operator in two-dimensional space |
| $\nabla^4$ | biharmonic operator in two-dimensional space |

*Subscripts*

| | |
|---|---|
| $i, j, k \ldots$ (except $n$, $s$) | 1, 2 (summation convention for repeated indices) |
| $(\,)_{,i}$ | $\partial(\,)/\partial x_i$ |
| $(\,)_{,n}$ | $\partial(\,)/\partial n$ normal derivative on the boundary |
| $(\,)_{,s}$ | $\partial(\,)/\partial s$ tangential derivative on the boundary |

*Superscripts*

| | |
|---|---|
| $b$ | refer to bottom face |
| $c$ | refer to core |
| $f$ | refer to face |
| $l$ | refer to linear term |
| $m$ | refer to average value on top and bottom faces |
| $n$ | refer to nonlinear term |
| $t$ | refer to top face |
| * | weight function or its related functions |

# Chapter 10

# Trefftz Method

*by Ismael Herrera*

## 10.1 Introduction

In recent years, by a boundary method, it is usually understood a numerical procedure in which a subregion or the entire region, is left out of the numerical treatment, by use of available analytical solutions (or, more generally, previously computed solutions). Boundary methods reduce the dimensions involved in the problem leading to considerable economy in the numerical work and constitute a very convenient manner of treating adequately unbounded regions by numerical means. Generally, the dimensionality of the problem is reduced by one, but even when part of the region is treated by finite elements, the size of the discretized domain is reduced [1–2].

There are two main approaches for the formulation of boundary methods; one is based on boundary integral equations and the other one, on the use of complete systems of solutions. In numerical applications, the first one of these methods has received most of the attention [3–4]. This is in spite of the fact that the use of complete systems of solutions presents important numerical advantages; e.g., it avoids the introduction of singular integral equations and it does not require the construction of a fundamental solution. The latter is especially relevant in connection with complicated problems, for which, it may be extremely laborious to build up a fundamental solution. This is illustrated by the fact that there are methods for synthesizing fundamental solutions starting from plane waves, which can be shown to be a complete system [5].

The use of complete systems of solutions is frequently associated with the name of Trefftz [6]. The idea of his method [7] consists in looking for approximate solutions, from among the appropriate class of functions that satisfy exactly the differential equation, but do not necessarily satisfy the prescribed boundary conditions. Although Trefftz's original formulation was linked to a variational principle, this is not required. Indeed, complete systems of solutions can be used to treat differential equations which are linear but otherwise arbitrary [8].

The method has been used in many fields. For example, applications to Laplace's equation are given by Mikhlin [9], to the biharmonic equation by Rektorys [10] and to elasticity by Kupradze [11]. Also many scattered contributions to the method can be found in the literature. Special mention is made here of work by Amerio, Fichera, Kupradze, Picone and Vekua [12–16]. Colton [17, 18] has constructed families of solutions which are complete for parabolic equations.

However, systematic analysis applicable to arbitrary linear differential equations were lacking. Motivated by this situation the author started a systematic research of the subject. The aim of the study has been two-fold; firstly, to clarify the theoretical foundations required for using complete systems of solutions in a reliable manner, and secondly, to expand the versatility of such methods, making them applicable to any problem which is governed by partial differential equations which are linear.

Considerable progress has been made [8, 19]. The systematic development includes: a) approximating procedures and conditions for their convergence [5, 8, 20–22]; b) formulation of variational principles [23–28]; and c) development of complete systems of solutions [29–33]. In addition, the algebraic frame-work [34] in which the theory has been constructed has been used for developing biorthogonal systems of solutions [35].

Recently, the theory has been extended to free-boundary problems [36, 37]. These are non-linear even when the governing differential equations are linear. Boundary integral equations have been applied to some of these problems [38]. For seepage problem, numerical experiments using Trefftz method were reported [22], but a theoretical analysis was not included. Many others have been treated using finite element approximations; for example, contact problems in elasticity [39, 40].

Systems of solutions which are complete, frequently preserve this property when the region of definition of the problem is changed. This is the case, for example, of the systems given in Tables 10.1 and 10.2 (see Section 10.6). It is important to develop procedures for constructing such systems and to formulate criteria to elucidate their completeness. This subject is being studied at present and a preliminary survey appeared recently [33].

## 10.2 Scope

To fix ideas we first consider a simple example. Take Laplace or Poisson's equation in a bounded region $\Omega$, illustrated in Fig. 10.1, and subjected to

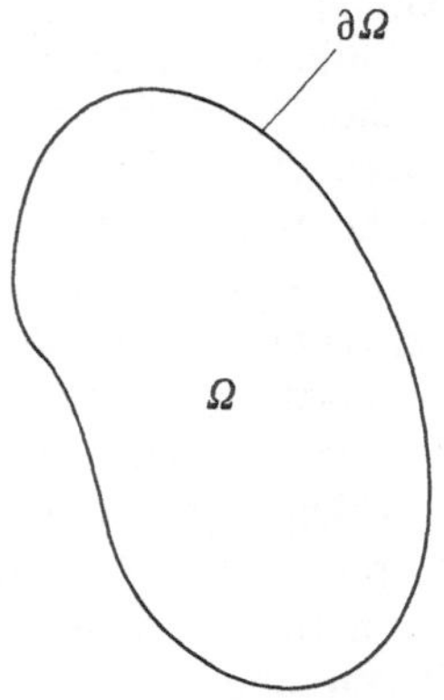


**Fig. 10.1**

boundary conditions of Dirichlet type:

$$\Delta u = f_\Omega; \quad \text{in } \Omega \tag{10.1a}$$

and

$$u = f_{\partial\Omega}; \quad \text{on } \partial\Omega \tag{10.1b}$$

where $f_\Omega$ and $f_{\partial\Omega}$ are given functions.

In general, the application of boundary methods requires transforming equation (10.1a) into a homogeneous equation. This can be achieved by introducing a particular solution $U$ of equation (10.1a); i.e.

$$\Delta U = f_\Omega; \quad \text{in } \Omega \tag{10.2a}$$

In applications the construction of function $U$ is not difficult, because it is not required to satisfy any prescribed boundary conditions. For example, when a fundamental solution is available, it can be obtained by quadrature.

In addition, let $V$ be a function such that

$$V = f_{\partial\Omega}; \quad \text{on } \partial\Omega \tag{10.2b}$$

Then, Dirichlet problem (10.1) is equivalent to

$$\Delta(u - U) = 0; \quad \text{in } \Omega \tag{10.3a}$$

and

$$u = V; \quad \text{on } \partial\Omega \tag{10.3b}$$

In order to formulate this problem precisely, it is necessary to define a space $D$ of admissible functions. Consider Sobolev space $H^s(\Omega)$, where $s$ is any real number ($-\infty < s < \infty$). As it is well known the trace operator (i.e. the boundary values) is not defined for some elements of $H^s(\Omega)$ when $s \leqq 1/2$ [41, 42]. However, there is a wide class of functions of $H^s(\Omega)$ for which this trace is defined and belongs to $H^{s-1/2}(\partial\Omega)$. Thus, define

$$D^e = \bigcup_s h^s(\Omega) \tag{10.4}$$

and

$$D = \left\{ u \in D^e \,|\, \gamma_0 u \in \bigcup_s H^s(\partial\Omega) \right\} \tag{10.5}$$

where $\gamma_0$ stands for the trace of $u$ on $\partial\Omega$. In general, for simplicity the symbol $\gamma_0$ will be omitted when it is clear from the context that we refer to the boundary values. It can be noticed that the linear space $D$ defined by (10.5) is not closed. Indeed, a metric is not defined in the whole space.

Let

$$N_P = \{u \in D \,|\, \Delta u = 0 \quad \text{in } \Omega\} \tag{10.6}$$

and

$$I = \{u \in D \,|\, u = 0 \quad \text{on } \partial\Omega\} \tag{10.7}$$

Then, Dirichlet problem can be formulated as a problem of linear restrictions. Given any $U \in D$ and $V \in D$ (these functions can be taken as data of the problem), find and element $u \in D$ such that

$$u - U \in N_P \quad \text{and} \quad u - V \in I \tag{10.8}$$

The first of equations (10.8) is equivalent to (10.3a), while the second one to (10.3b).

A first advantage of formulating the problem in this manner is connected with its existence properties. Clearly equation (10.3b) is equivalent to $u - U = V - U$ on $\partial\Omega$. By well known results on the existence of solution [41], this problem possesses a unique solution. Indeed, given $U \in D$ and $V \in D$ there are real numbers $r$ and $s$ such that $U \in H^r(\Omega)$ and the trace $\gamma_0(V - U) \in H^s(\partial\Omega)$. Then, $u - U \in H^{s+1/2}(\Omega)$. Therefore, $u = U + (u - U)$ belongs to $H^t(\Omega)$ where $t = \min\{r, s + 1/2\}$. This shows $u \in D$.

The above discussion also shows that there is no lack of generality by restricting attention to the homogeneous case; i.e.

$$\Delta u = 0; \quad \text{on } \Omega \tag{10.9a}$$

and

$$u = f_{\partial\Omega}; \quad \text{on } \partial\Omega \tag{10.9b}$$

The boundary method to be applied depends on the continuity of the solutions on their boundary values. In principle it can be applied when the space of admissible functions $D$ is given by (10.5). However, this would lead to consider inner products in the space of boundary values $H^s(\partial\Omega)$ with arbitrary $s$, which may be inconvenient in numerical applications. It is preferable to keep the computations in $\mathcal{L}^2(\partial\Omega) = H^0(\partial\Omega)$, which, as will be seen, leads to least-squares fitting. This, can be achieved if attention is restricted to functions with boundary values belonging to $H^0(\partial\Omega) = \mathcal{L}^2(\partial\Omega)$. When this condition is incorporated in the definition of the space of admissible functions, one gets

$$D = \{u \in D^e \mid \gamma_0 u \in H^0(\partial\Omega)\} \tag{10.10}$$

This is again a linear space which is not closed.

In addition, in many applications it is necessary to compute the normal derivative $\partial u/\partial n$ on the boundary $\partial\Omega$. Similar considerations lead to require that $\partial u/\partial n$ belong to $H^0(\partial\Omega) = \mathcal{L}^2(\partial\Omega)$. When these two requirements are incorporated in the definition of the space of admissible functions, equation (10.5) becomes

$$D = \{u \in D^e \mid \gamma_0 u \in H^0(\partial\Omega), \gamma_1 u \in H^0(\partial\Omega)\} \tag{10.11a}$$

Here, as it is costumary, $\gamma_1 u$ stands for the trace of the normal derivative on $\partial\Omega$. This is again a linear space.

General results on the existence and continuity properties of solutions of elliptic equations [41], imply that any harmonic function $u$ whose trace $\gamma_0 u$ belongs to $H^0(\partial\Omega)$, necessarily is a member of $H^{1/2}(\Omega)$. Therefore $N_P \subset H^{1/2}(\Omega)$ in this case. Even more, due to the continuity properties just mentioned, $N_P$ is a closed subspace of $H^{1/2}(\Omega)$. This will be represented by $N^{1/2}(\Omega)$. Thus

$$N_P = N^{1/2}(\Omega) \tag{10.11b}$$

when $D$ is defined by equation (10.10). Similarly, when equation (10.11a) holds corresponding properties imply that

$$N_P = N^{3/2}(\Omega) \tag{10.11c}$$

where $N^{3/2}(\Omega)$ is the subspace of harmonic functions belonging to $H^{3/2}(\Omega)$ which can also be shown to be closed.

If $\mathscr{B}=\{w_1, w_2, \ldots\} \subset N^{3/2}$ is a system of harmonic functions in $\Omega$ which spans $N^{3/2}$ then it also spans $N^{1/2}$, because $N^{3/2} \subset N^{1/2}$ is dense in $N^{1/2}$. Thus, let $\mathscr{B}=\{w_1, w_2, \ldots\} \subset N^{3/2}$ be such a system. Then, given $u \in N^{1/2}$ there is a sequence of approximations

$$u^N = \sum_{n=1}^{N} a_n^N w_n; \quad N = 1, 2, \ldots \tag{10.12}$$

such that

$$u^N \to u \quad \text{in} \quad H^{1/2}(\Omega) \tag{10.13}$$

Notice that $a_n^N$ depends on the number of terms $N$, of the approximation. This is essential in order for (10.13) to hold. When $a_n^N$ is independent of $N$, (10.12) becomes a series and the approximation by a series can only be granted when the system of functions $\mathscr{B}$ is orthogonal. This fact explains some of the difficulties that were encountered in applications to electromagnetic field studies [43].

In order for representation (10.12) to be useful, it will be required to have a procedure for deriving the coefficients $a_n^N$ from boundary data only. This is, indeed, possible. General results on the existence and continuity properties of solutions of elliptic equations [41] show that the range of the traces $\gamma_0 u$ when $u \in N^{1/2}(\partial\Omega)$ is $H^0(\partial\Omega)$. Given $u \in N^{1/2}(\Omega)$ if the coefficients $a_n^N$ are chosen so that

$$\gamma_0 u^N \to \gamma_0 u \quad \text{on} \quad H^0(\partial\Omega) = \mathscr{L}^2(\partial\Omega) \tag{10.14}$$

then (10.13) holds necessarily.

If $\mathscr{B}=\{w_1, w_2, \ldots\} \subset N^{3/2}(\Omega)$ spans $N^{3/2}(\Omega)$, then the continuity properties of elliptic equations imply that the associated system of traces $\{\gamma_0 w_1, \gamma_0 w_2, \ldots\}$ span $H^1(\partial\Omega)$, which in turn implies that

$$\{\gamma_0 w_1, \gamma_0 w_2, \ldots\} \quad \text{spans} \quad H^0(\partial\Omega) \tag{10.15}$$

These remarks show that the coefficients $a_n^N$ can be chosen so that (10.14) holds. Clearly, to this end it will be sufficient to take $\gamma_0 u^N$ as the projection of the boundary values $\gamma_0 u \in H^0(\partial\Omega)$ on the subspace spanned by $\{w_1, \ldots, w_N\}$. Therefore, the coefficients can be computed by standard procedures for projecting on a subspace. This yields a least-squares method because only $\mathscr{L}^2(\partial\Omega) = H^0(\partial\Omega)$ inner products are being used.

This procedure leads to the system of equations:

$$\sum_{n=1}^{N} M_{nm} a_n^N = c_m; \quad m = 1, \ldots, N \tag{10.16}$$

for the coefficients $a_n^N$ occurring in equation (10.12). Here

$$M_{nm} = \int_{\partial\Omega} w_n w_m \, dx \tag{10.17a}$$

and

$$c_m = \int_{\partial\Omega} f_{\partial\Omega} w_m \, dx \tag{10.17b}$$

In order for this system to be invertible it is required that the system of traces $\{w_1, \ldots, w_N\} \subset H^0(\partial\Omega)$ be linearly independent. Then, the matrix $M_{nm}$ can be seen to be Hermitian.

Computation of the boundary values may need a special device. If the normal derivatives $\partial u/\partial n$ are required on the boundary, in the case of Dirichlet problem, they can not be obtained from the approximating sequence $u^N$, directly. Indeed, as mentioned, from $u^N \to f_{\partial\Omega}$ on $H^0(\partial\Omega)$, one can only grant that $u^N \to u$ in $H^{1/2}(\Omega)$. Hence $\frac{\partial u^N}{\partial n} \to \frac{\partial u}{\partial n}$ in the metric of $H^{-1}(\partial\Omega)$ [41]; thus, $\frac{\partial u^N}{\partial n}$ diverges in $H^0(\partial\Omega)$ in general. As a matter of fact, $\frac{\partial u}{\partial n} \in H^{-1}(\partial\Omega)$ is only defined as a continuous functional on $H^1(\partial\Omega)$. The procedure to be presented can be extended to supply a method for computing this functional, but for simplicity such extension is not included here.

This can be avoided if the space of admissible functions $D$ is restricted. Let it be defined by equations (10.11). In this case the boundary data is more smooth, necessarily; indeed, $f_{\partial\Omega} \in H^1(\partial\Omega)$ and $\frac{\partial u}{\partial n} \in H^0(\partial\Omega)$.

Define the bilinear functional

$$\langle A\, u, v\rangle = \int_{\partial\Omega} \left\{ v \frac{\partial u}{\partial n} - u \frac{\partial v}{\partial n} \right\} dx \tag{10.18}$$

This is well-defined for every $u \in D$ and $v \in D$, because their traces belong to $H^0(\partial\Omega)$. Observe that

$$\langle A\, u, v\rangle = (v, \partial u/\partial n)_\partial - (u, \partial v/\partial n)_\partial \tag{10.19}$$

where we have written $(,)_\partial$ for the inner product in $H^0(\partial\Omega)$. Also $\langle A\, u, v\rangle = 0$ whenever $u \in N_P$ and $v \in N_P$. This yields the reciprocity relation

$$(v, \partial u/\partial n)_\partial = (u, \partial v/\partial n)_\partial \tag{10.20}$$

Using it, the following approximating sequence will be constructed

$$\sum_{n=1}^{N} b_n^N w_n \to \frac{\partial u}{\partial n} \quad \text{in} \quad H^0(\partial\Omega) \tag{10.21}$$

When $u \in N^{3/2}(\Omega)$ the trace $\gamma_1 u$ (i.e. the normal derivative on the boundary) spans

$$\{1\}^\perp \subset H^0(\partial\Omega) = \mathscr{L}^2(\partial\Omega) \tag{10.22}$$

Here, $\{1\}^\perp$ stands for the orthogonal complement in $H^0(\partial\Omega)$ of the constant function. Thus, $\frac{\partial u}{\partial n} \in \{1\}^\perp$, if and only if, $\frac{\partial u}{\partial n} \in H^0(\partial\Omega)$ and

$$\int_{\partial\Omega} \frac{\partial u}{\partial n} dx = 0\,. \tag{10.23}$$

Taking $\mathscr{B} = \{w_1, w_2, \ldots\} \subset N^{3/2}(\Omega)$ as before, and in view of (10.15), given any $u \in N^{3/2}(\Omega)$, it is possible to construct an approximation satisfying (10.21). Again, for such representation to be useful it will be required to have a procedure for computing $b_n^N$ using boundary data only. This will be based on the use of the reciprocity relation (10.20). It has interest to observe a general feature of representation (10.21); namely, that the normal derivative $\partial u/\partial n$ of the sought solution is not approximated by the normal derivatives of the basic system $\mathscr{B}$, but by its boundary values, instead. The coefficients $b_n^N$ will be chosen so that

$$\left\| \frac{\partial u}{\partial n} - \sum_{n=1}^{N} b_n^N w_n \right\|^2 \tag{10.24}$$

is minimized.

This leads to take the projection of $\partial u/\partial n$ on the space spanned by $\{w_1, \ldots, w_N\} \subset H^0(\partial\Omega)$. This requires the orthogonality condition

$$\left( \frac{\partial u}{\partial n} - \sum_{n=1}^{N} b_n^N w_n, w_m \right) = 0\,, \qquad m = 1, \ldots, N \tag{10.25}$$

to be satisfied. Expanding (10.25), one gets

$$\sum_{n=1}^{N} K_{nm}\, b_n^N = d_m \tag{10.26}$$

where

$$K_{nm} = \int_{\partial\Omega} w_n\, w_m\, dx\,; \qquad n, m = 1, \ldots, N \tag{10.27 a}$$

and

$$d_m = \int_{\partial\Omega} \frac{\partial u}{\partial n} w_m\, dx = \int_{\partial\Omega} f_{\partial\Omega} \frac{\partial w_m}{\partial n} dx\,; \qquad m = 1, \ldots, N \tag{10.27 b}$$

Observe that the use of the reciprocity relation (10.20) has permitted to express $d_m$ in terms of boundary data only.

An additional point must be mentioned. In order for the approximating sequence $\sum_{n=1}^{N} b_n^N w_n$ to be convergent, it is necessary that the solution $\frac{\partial u}{\partial n} \in H^0(\partial\Omega)$. This is granted if $f_{\partial\Omega} \in H^1(\partial\Omega)$. Alternatively, this condition can be expressed in matrix form. Let $\mathbf{K}^N$ be the $N \times N$ square matrix whose elements are given by (10.27 a). Similarly $\mathbf{d}^N$ is the $1 \times N$ vector defined by (10.27 b). Assume, the system of traces $\{w_1, \ldots, w_N\} \subset H^1(\partial\Omega)$ is linearly independent, which is required in order for the system (10.26) to be invertible, and denote by $(\mathbf{K}^N)^{-1}$ the inverse of $\mathbf{K}^N$. Then, the sequence of real numbers

$$\left\| \sum_{n=1}^{N} b_n^N w_n \right\|^2 = \mathbf{d}^N \cdot (\mathbf{K}^N)^{-1} \cdot \mathbf{d}^N \geqq 0\,, \qquad N = 1, 2, \ldots \tag{10.28}$$

is non-negative and increasing. Convergence, of the approximating sequence is granted when the sequence (10.28) is bounded. The meaning of this condition is more easily understood by observing that when the system of traces $\{w_1, w_2, \ldots\}$ is orthonormal (i.e. $K_{nm} = \delta_{nm}$), in which case the coefficients $d_n$ are independent of $N$, it becomes

$$\sum_{n=1}^{\infty} d_n^2 < \infty \tag{10.29}$$

The treatment of Neuman problem is similar. Let the space of admissible functions be given again by equations (10.11 a). Then equation (10.9 b) is replaced by

$$\frac{\partial u}{\partial n} = g_{\partial\Omega}\,; \qquad \text{on } \partial\Omega \tag{10.30}$$

where the boundary values $g_{\partial\Omega} \in \{1\}^{\perp} \subset H^0(\partial\Omega)$. The previous argument still holds if (10.17) is replaced by

$$M_{nm} = \int_{\partial\Omega} \frac{\partial w_n}{\partial n} \frac{\partial w_m}{\partial n} dx \tag{10.31 a}$$

and

$$c_m = \int_{\partial\Omega} g_{\partial\Omega} \frac{\partial w_m}{\partial n} dx \tag{10.31 b}$$

In this case $u^N \to u$ in $H^{3/2}(\Omega)$; therefore, also $u^N \to u$ in $H^{1/2}(\Omega)$. It must be observed that this assertion is not strictly true because the solution of Neuman's problem contains an undetermined constant. To remove it one can take $\mathscr{B} = \{1, w_1, w_2, \dots\} \subset N^{3/2}(\Omega)$ and require

$$\int_{\partial\Omega} w_j \, dx = 0\,; \quad j = 1, 2, \dots \tag{10.32}$$

Then

$$\int_{\partial\Omega} u \, dx = 0 \tag{10.33}$$

In general, if the normal derivative $\dfrac{\partial u^N}{\partial n} \to g_{\partial\Omega}$ in $H^0(\partial\Omega)$, then $u^N \to u$ in $H^{3/2}(\Omega)$; hence, on the boundary $u^N \to u$ in $H^1(\partial\Omega)$, which implies $u^N \to u$ in $H^0(\partial\Omega)$. Thus, the boundary values (i.e. $\gamma_0 u$ on $\partial\Omega$), which in case of Neuman problem are not known beforehand, can be derived from the approximating sequence directly. However, the use of the reciprocity relation (10.20) offers an alternative for computing them. Indeed, one simply has to replace equations (10.21) and (10.27), by

$$u^N = \sum_{n=1}^{N} b_n^N \frac{\partial w_n}{\partial n} \to u\,, \quad \text{in } H^0(\partial\Omega) \tag{10.34}$$

$$K_{nm} = \int_{\partial\Omega} \frac{\partial w_n}{\partial n} \frac{\partial w_m}{\partial n} dx \tag{10.35 a}$$

and

$$d_m = \int_{\partial\Omega} u \frac{\partial w_m}{\partial n} dx = \int_{\partial\Omega} g_{\partial\Omega} w_m \, dx \tag{10.35 b}$$

Again, equations (10.26) have to be satisfied. When this is the case the solution $u$ in (10.34) fulfills (10.33). This method can be used to accelerate the convergence of the approximating sequence on the boundary. As a matter of fact, when the system of equations (10.12), (10.16) and (10.31) is applied, the norm $\left\| \dfrac{\partial u^N}{\partial n} - g_{\partial R} \right\|$ in the $\mathscr{L}^2(\partial\Omega)$ sense, is minimal; however, $\| u^N - u \|$ in $\mathscr{L}^2(\partial\Omega)$ in general is not minimal. When equations (10.26), (10.34) and (10.35) are applied, on the contrary, $\| u^N - u \|$ in the $\mathscr{L}^2(\partial\Omega)$ sense, is minimal; i.e. in the first case, the approximation of the boundary data is optimal, while by the second method, the approximation of the unknown boundary values is optimal. In applications, generally, the latter would be preferable.

Generally, when dealing with partial differential equations only some boundary values of the functions and their derivatives are relevant in the discussion of the problems. For example, for Laplace equation these are the function $u$ and its normal derivative $\partial u/\partial n$. For Elasticity the displacements $\mathbf{u}$ and tractions $\mathbf{T}(\mathbf{u})$. When a boundary value problem is formulated, only one part of this boundary information is prescribed and the other part must be derived after the solution has been obtained. For Dirichlet problem, for example, $u$ is prescribed and $\partial u/\partial n$ is derived. The converse has to be done in the case of Neuman problem. Approximating sequences for the complementary boundary values which depend on reciprocity relations, such as (10.20) can be derived for very general classes of differential equations. The reciprocity relations can be obtained from corresponding Green's formulas. For example, from

$$\int_{\Omega} \{v\,\Delta u - u\,\Delta v\}\,dx = \int_{\partial\Omega} \left\{v\,\frac{\partial u}{\partial n} - u\,\frac{\partial v}{\partial n}\right\} dx \tag{10.36}$$

one obtains

$$\int_{\partial\Omega} v\,\frac{\partial u}{\partial n}\,dx = \int_{\partial\Omega} u\,\frac{\partial v}{\partial n}\,dx \tag{10.37}$$

when $u$ and $v$ are harmonic in $\Omega$. Equation (10.37) can be recognized as (10.20). The procedure used to derive approximations (10.21) and (10.34) can be traced back to a group of italian mathematicians [12–14] and was discussed extensively by Kupradze [15]. The author has introduced an abstract formulation which permits extending this procedure to problems with prescribed jumps [34] (applications to elasticity are given in [5]). This is linked to a systematic classification of boundary values and will be explained in Section 3.

The possibility of applying the boundary method here explained depends on the availability of a system of solutions $\mathscr{B} = \{w_1, w_2, \ldots\} \subset N^{3/2}(\Omega)$ of Laplace equation which spans $N^{3/2}(\Omega)$. In this connection, there are two general categories of theoretical questions which must be analyzed in order to increase the flexibility and versatility of the procedure. These are: criteria for deciding when a system $\hat{\mathscr{B}}$ is complete and methods for constructing complete systems which can be applied to many problems.

Regarding the first one, we have seen that what is required is that the system $\mathscr{B} = \{w_1, w_2, \ldots\}$ spans $N^{3/2}(\Omega)$. However, in applications it is frequently difficult to verify this in a direct manner and it is necessary to use alternative criteria; these can be established by analyzing the spaces spanned by the boundary values. For example, for Laplace equation, given a system of functions $\mathscr{B} = \{w_1, w_2, \ldots\}$ defined in $\Omega$, let us denote by $\hat{w}_\alpha = [w_{\alpha 1}, w_{\alpha 2}]$ the system of traces $w_{\alpha 1} = \gamma_0\, w_\alpha$ and $w_{\alpha 2} = \gamma_1\, w_\alpha$. In addition $\hat{\mathscr{B}} = \{\hat{w}_1, \hat{w}_2, \ldots\}$, $\mathscr{B}_1 = \{w_{11}, w_{21}, \ldots\}$ and $\mathscr{B}_2 = \{w_{12}, w_{22}, \ldots\}$. For example, when the region $\Omega$ is a circle (the unit circle for definiteness), by separation of variables one obtains (in polar coordinates)

$$\mathscr{B} = \{1;\, r^n \cos n\,\theta,\, r^n \sin n\,\theta;\; n = 1, 2, \ldots\} \tag{10.38}$$

This system is made of harmonic polynomials

$$\mathscr{B} = \{1,\, x^2 - y^2,\, x\,y, \ldots\} \tag{10.39}$$

which can be recognized as Re $z^n$ and Im $z^n$ $(n = 0, 1, \ldots)$.

Setting $r=1$ in (10.38), one obtains the system of traces

$$\mathscr{B}_1 = \{\cos n\,\theta, \sin n\,\theta;\ n=0,1,\ldots\} \tag{10.40 a}$$

and

$$\mathscr{B}_2 = \{-n\sin n\,\theta, n\cos n\,\theta;\ n=0,1,\ldots\} \tag{10.40 b}$$

Denote by $N_1$ and $N_2$ the spaces spanned in the $\mathscr{L}^2(\partial\Omega)$ metric by the traces $\gamma_0 u$ and $\gamma_1 u$, respectively, when $u$ ranges over $N^{3/2}(\Omega)$. Clearly, $N_1 = \mathscr{L}^2(\partial\Omega) = H^0(\partial\Omega)$ while $N_2 = \{1\}^\perp \subset \mathscr{L}^2(\partial\Omega)$, Here, the orthogonal complement $\{1\}^\perp$ is taken in the $\mathscr{L}^2(\partial\Omega)$ inner product.

Let $\mathscr{B} \subset N^{3/2}(\Omega)$ be a system such that

$$\operatorname{span}\mathscr{B}_1 = N_1 = \mathscr{L}^2(\partial\Omega) \quad \text{and} \quad \operatorname{span}\mathscr{B}_2 = N_2 = \{1\}^\perp \tag{10.41}$$

where the spans are taken in the $\mathscr{L}^2(\partial\Omega)$ sense.

For simplicity, assume that the constant function $w_0 = 1$ is a member of $\mathscr{B}$, so that

$$\mathscr{B} = \{1\} \cup \mathscr{B}' \tag{10.42}$$

where $\mathscr{B}' = \{w_1, w_2, \ldots\}$. It will also be assumed that

$$\int_{\partial\Omega} w_\alpha\, dx = 0\,; \qquad \alpha = 1, 2, \ldots \tag{10.43}$$

Any harmonic function $u \in N^{3/2}(\Omega)$ can be written uniquely as

$$u = a_0 + u' \tag{10.44}$$

where $a_0$ is the constant

$$a_0 = \int_{\partial\Omega} u\, dx\,, \quad \text{while} \int_{\partial\Omega} u'\, dx = 0 \tag{10.45}$$

In view of (10.41) and $\gamma_1 w_0 = 0$, it is clear that

$$\operatorname{span}\mathscr{B}_2' = \{1\}^\perp \tag{10.46}$$

Also $\gamma_1 u' \in \{1\}^\perp$, since $u'$ is harmonic in $\Omega$, so that $\gamma_1 u'$ is in the $\mathscr{L}^2(\Omega)$ – span of $\mathscr{B}_2'$. This shows that there is a sequence $v^N$ of linear combination of $\mathscr{B}'$ such that

$$\gamma_0 v^N \underset{N\to\infty}{\longrightarrow} \gamma_0 u'\,, \qquad \text{in } \mathscr{L}^2(\partial\Omega) \tag{10.47}$$

In view of (10.43), the second of conditions (10.45) and continuity properties [41] of solutions of elliptic equations, it is clear that $v^N \to u'$ in the metric of $H^{3/2}(\Omega)$. Therefore, the linear combination $u^N = a_0 + v^N$ of elements of $\mathscr{B} \subset N^{3/2}(\Omega)$, is such that $u^N \to u$ in $H^{3/2}(\Omega)$. This shows that

$$\operatorname{span}\mathscr{B} = N^{3/2}(\Omega) \tag{10.48}$$

where the span is taken in the $H^{3/2}(\Omega)$ metric. Thus, in this case we have derived the completeness of the system $\mathscr{B} \subset N^{3/2}(\Omega)$ from the fact that the system of traces $\mathscr{B}_1$, spans the same space as the traces of harmonic functions (i.e. solutions of the homogeneous equation) in $N^{3/2}(\Omega)$. Similar results hold in a more general context.

Let $\hat{\mathscr{H}} = H^0(\partial\Omega) \oplus H^0(\partial\Omega)$ be the space of pairs $\hat{u} = [u_1, u_2]$ with $u_1 \in H^0(\partial\Omega)$ and $u_2 \in H^0(\partial\Omega)$, provided with the usual inner product

$$((\hat{u}, \hat{v})) = (u_1, v_1)_\partial + (u_2, v_2)_\partial \tag{10.49}$$

Denote by $\hat{\mathcal{N}} \subset \hat{\mathcal{H}}$ the image of $N^{3/2}(\Omega)$ under the mapping $u \rightarrow \hat{u} = [\gamma_0 u, \gamma_1 u] \in \hat{\mathcal{H}}$. It can be shown that $\hat{\mathcal{N}} \subset \hat{\mathcal{H}}$ is closed in the metric of $\hat{\mathcal{H}}$. Notice that the reciprocity relation (10.20) becomes (we assume Hilbert-spaces are being taken with real coefficients):

$$(v_1, u_2) = (v_2, u_1) \ \forall \ \hat{u} \in \hat{\mathcal{N}} \ \& \ \hat{v} \in \hat{\mathcal{N}} \tag{10.50}$$

A system of function $\mathscr{B} \subset N^{3/2}(\Omega)$ will be said to be $T$-complete* if, for every $\hat{u} \in \hat{\mathcal{H}}$, one has

$$(w_1, u_2) = (w_2, u_1) \ \forall \ w \in \mathscr{B} \Rightarrow \hat{u} \in \hat{\mathcal{N}} \tag{10.51}$$

Using this notation the following characterization of complete systems holds [19, 34].

**Theorem 10.1.** *Let $\mathscr{B} \subset N^{3/2}(\Omega)$. Then the following assertions are equivalent:*

(i) *$\mathscr{B} \subset N^{3/2}(\Omega)$ spans $N^{3/2}(\Omega)$ in the metric of $H^{3/2}(\Omega)$;*
(ii) *$\hat{\mathscr{B}} \subset \hat{\mathcal{H}}$ spans $\hat{\mathcal{N}}$ in the metric of $\hat{\mathcal{H}}$;*
(iii) *$\hat{\mathscr{B}} \subset \hat{\mathcal{N}}$ is a T-complete system;*
(iv) *Equations* (10.41) *are satisfied when the spans are taken in the $\mathscr{L}^2(\partial\Omega)$ sense.*

An advantage of having a system which satisfies any of the criteria (i) to (iv), is that the same system can be used for both a Dirichlet and a Neuman problem. Indeed, the same $T$-complete system can be used for any linear boundary condition which is prescribed point-wise. Such condition can be written as

$$a_1 u + a_2 \partial u/\partial n = f_{\partial R}, \qquad \text{on } \partial\Omega \tag{10.52}$$

The arguments presented previously, can be extended to this case by introduction of more general Green's formulas. This will be discussed in Section 3.

It has interest to observe that it is possible to develop systems which are complete in regions which are, to a large extent, arbitrary. For example, the system of harmonic polynomials given by (10.38) and (10.39), is $T$-complete in any bounded and simply connected region [29]. Also, the system

$$\{\text{Log}\, r, \ \text{Re}\, z^{-n}, \ \text{Im}\, z^{-n}; \ n = 1, 2, \ldots\} \tag{10.53}$$

is $T$-complete in the exterior of any simply connected and bounded region which contains the origin.

To develop general criteria establishing conditions under which a system which is complete in a region is also complete in another one, is quite valuable. Especially if such criteria are applicable to a wide class of partial differential equations. For this purpose the notion of $T$-completeness is useful.

## 10.3 Green's Formulas

The development of Green's formulas for general classes of partial differential equations is a classical topic of the theory of partial differential equations [41]. A

* Trefftz-complete. Previously, such systems had been called $c$-complete by the author.

theory which permits obtaining such formulas systematically and which in some respects enlarges the kind of problems that can be treated in this manner, has been developed by the author [19, 28, 34]. The fundamental notions are closely related with simplectic geometry [44].

Basically, what is done is to characterize the space of boundary values which are relevant for each differential equation or system of such equations. Then such space is decomposed into two subspaces. With every Green's formula there is associated such a decomposition and conversely with every decomposition there is a unique Green's formula. A procedure for reconstructing the Green's formula when the decomposition is known, is established [34].

We consider a bilinear functional $P$ defined on an arbitrary linear space $D$; it will be denoted by $P: D \to D^*$ because it can be thought as an operator defined on the linear space $D$ and taking values on its algebraic dual $D^*$ (this is the space of linear functionals defined on $D$) [45]. The value of such bilinear functional at elements $u \in D$ and $v \in D$, will be denoted by $\langle P u, v \rangle$. The transposed bilinear functional of $P: D \to D^*$, will be $P^*: D \to D^*$; thus

$$\langle P^* u, v \rangle = \langle P v, u \rangle \tag{10.54}$$

The theory is applicable to general non-symmetric linear operators, although its application to formally symmetric ones is simpler, because it does not require the introduction of a formal adjoint. Here, attention is restricted to such operators. Given an operator $P: D \to D^*$, we define the antisymmetric bilinear form

$$A = P - P^* \tag{10.55}$$

The operator $A$, given by (10.55) plays a central role in the theory. Firstly, we are going to use it, to define the relevant boundary values. For this purpose, we consider the null subspace $N_A$ of $A$; i.e.

$$N_A = \{u \in D \mid A u = 0\} \tag{10.56}$$

With reference to the reduced wave equation

$$\Delta u + k^2 u = 0\,, \qquad \text{on } R \tag{10.57}$$

as an example (recall that Laplace equation corresponds to the case $k = 0$), consider the bilinear functional $P: D \to D^*$, given by

$$\langle P u, v \rangle = \int_R v\, (\Delta u + k^2 u)\, dx \tag{10.58}$$

Then $A = P - P^*$ is

$$\langle A u, v \rangle = \int_{\partial R} \left\{ v \frac{\partial u}{\partial n} - u \frac{\partial v}{\partial n} \right\} dx \tag{10.59}$$

The null subspace $N_P$, is the linear subspace of functions which satisfy (10.57).

There are many ways of taking the linear space $D$. A convenient one is by means of equation (10.11a). This defines a linear subspace, but we do not introduce a topology in it. We notice that the null subspace $N_P$ is well defined, if (10.57) is interpreted in the sense of distributions [41]. Also the bilinear form $A: D \to D^*$, given by (10.59); however, the operator $P: D \to D^*$, given by (10.58) is not. Many technical difficulties are avoided by leaving the operator $P$ out of the discussion.

It is easy to see that

$$N_A = \left\{u \in D \mid u = \frac{\partial u}{\partial n} = 0\,, \quad \text{on } \partial\Omega\right\} \tag{10.60}$$

Due to (10.60), the relevant boundary values for Laplace and reduced wave equations (10.60), will be $u$ and $\frac{\partial u}{\partial n}$, on $\partial\Omega$. We notice that given $u \in D$ and $v \in D$, one has that

$$u = v; \quad \frac{\partial u}{\partial n} = \frac{\partial v}{\partial n}, \quad \text{on } \partial\Omega \tag{10.61}$$

if and only if $u - v \in N_A$; i.e. two functions $u \in D$ and $v \in D$ have the same relevant boundary values, if and only if, $u - v \in N_A$.

Similar notions can be applied to any linear differential equation. Let us consider the biharmonic equation:

$$\Delta^2 u = 0\,; \qquad \text{on } \Omega \tag{10.62}$$

which occurs, for example, in connection with incompressible flows at low Reynolds numbers. Define

$$\langle P\,u, v\rangle = \int_\Omega v\,\Delta^2 u\,dx \tag{10.63}$$

Then

$$\langle A\,u, v\rangle = \int_{\partial\Omega} \left\{v\,\frac{\partial \Delta u}{\partial n} - \Delta u\,\frac{\partial v}{\partial n} + \Delta v\,\frac{\partial u}{\partial n} - u\,\frac{\partial \Delta v}{\partial n}\right\} dx \tag{10.64}$$

Again, a convenient definition of the space $D$ is (see equation (10.4)):

$$D = \left\{u \in D^e \mid u, \frac{\partial u}{\partial n}, \Delta u \text{ and } \frac{\partial \Delta u}{\partial n} \text{ belong to } H^0(\partial\Omega)\right\} \tag{10.65}$$

Then, $A$ as given by (10.64) is well defined, and $N_P$ can be taken as the linear subspace of $D$ which satisfies (10.62) in the sense of distributions. The operator $P: D \to D^*$, given (10.63), is not defined for this space $D$, and we leave it out from our discussion.

The null subspace $N_A$, is

$$N_A = \left\{u \in D \mid u = \frac{\partial u}{\partial n} = \Delta u = \frac{\partial \Delta u}{\partial n} = 0, \text{ on } \partial\Omega\right\} \tag{10.66}$$

The classification of boundary values induced by (10.66), is characterized by quadruplets of functions $u$, $\frac{\partial u}{\partial n}$, $\Delta u$, $\frac{\partial \Delta u}{\partial n}$; recall that these functions yield enough information to have $u$ and its derivatives up to order 3 determined.

The homogeneous stationary Stokes equations are

$$\nu \Delta \mathbf{u} - \nabla p = 0 \tag{10.67 a}$$

$$\nabla \cdot \mathbf{u} = 0 \tag{10.67 b}$$

where $\nu$ is the viscosity. In this case, it is convenient to define the bilinear form $P: D \to D^*$, by

$$\langle P\,\hat{u}, \hat{v}\rangle = \int_\Omega \{\mathbf{v} \cdot (\nu \Delta \mathbf{u} - \nabla p) + q \nabla \cdot \mathbf{u}\}\,dx \tag{10.68}$$

Here, $\hat{u}$ stands for a pair of functions; $\mathbf{u}$ which is vector valued and defined in $\Omega$, and $p$ scalar valued and also defined in $\Omega$. With $\hat{v}$, we have associated the pair $\mathbf{v}, q$. Then

$$\langle A\,\hat{u}, \hat{v}\rangle = \int_{\partial\Omega} \left\{ \mathbf{v} \cdot \left( v \frac{\partial \mathbf{u}}{\partial n} - p\,\mathbf{n} \right) - \mathbf{u} \cdot \left( v \frac{\partial \mathbf{v}}{\partial n} - q\,\mathbf{n} \right) \right\} dx\,. \tag{10.69}$$

Elements of the linear space $D$ will be pairs $\hat{u} = [\mathbf{u}, p]$ such that the traces $\mathbf{u}$ and $v \dfrac{\partial \mathbf{u}}{\partial n} - p\,\mathbf{n}$ are well defined and span $\mathbf{H}^0(\partial\Omega)$. One must also require that the set of functions $N_P \subset D$ which satisfy Stokes equations (10.67) in the sense of distributions be well defined. In general, $P: D \to D^*$ may not be defined in this space. The null subspace

$$N_A = \left\{ \hat{u} \in D \mid \mathbf{u} = v \frac{\partial \mathbf{u}}{\partial n} - p\,\mathbf{n} = \mathbf{0}, \text{ on } \partial\Omega \right\}. \tag{10.70}$$

The classification of boundary values induced by (10.70) is characterized by the values of $\mathbf{u}$ and $v \dfrac{\partial \mathbf{u}}{\partial n} - p\,\mathbf{n}$ on the boundary $\partial\Omega$.

As it has been seen in the specific examples given thus far, in general, it is not necessary to define on operator $P: D \to D^*$ in order for the theory to be applicable. Thus, in what follows, it will simply be assumed that there is available an antisymmetric bilinear form $A: D \to D^*$.

A subspace $I \subset D$ is said to be regular for $A$, when

(i) For every $u \in I$ and $v \in I$,

$$\langle A\,u, v\rangle = 0\,; \tag{10.71}$$

i.e. $I$ is a commutative subspace for $A$.

(ii)

$$I \supset N_A \tag{10.72}$$

We have seen that the null subspace $N_A$, induces a classification of $D$ which defines what could be properly called, the boundary values which are relevant for the differential equation considered. In the light of this fact, condition (ii) implies that a regular subspace is characterized by boundary values, only.

To illustrate this fact, assume, $I \subset D$ is a regular subspace. In connection with the examples given previously, let $u \in D$ and $v \in D$, be such that

$$u = v\,; \quad \frac{\partial u}{\partial n} = \frac{\partial v}{\partial n}\,, \quad \text{on } \partial\Omega \tag{10.73}$$

when the reduced wave equation is considered; or

$$u = v\,; \quad \frac{\partial u}{\partial n} = \frac{\partial v}{\partial n}\,; \quad \Delta u = \Delta v\,; \quad \frac{\partial \Delta u}{\partial n} = \frac{\partial \Delta v}{\partial n}\,; \quad \text{on } \partial\Omega \tag{10.74}$$

for the biharmonic equation. Then, there are only two mutually exclusive possibilities

a) $u$ and $v$ belong to $I$, or
b) neither $u$, nor $v$ belongs to $I$.

A corresponding proposition holds for $\hat{u} \in D$ and $\hat{v} \in D$, in connection with Stokes equations, when it is assumed that

$$\mathbf{u} = \mathbf{v}; \quad \nu \frac{\partial \mathbf{u}}{\partial n} - p\,\mathbf{n} = \nu \frac{\partial \mathbf{v}}{\partial n} - q\,\mathbf{n}; \quad \text{on } \partial\Omega \tag{10.75}$$

The following statement summarizes this discussion. A regular subspace, is a commutative subspace which is defined through boundary values only.

Examples of regular subspaces for the reduced wave equation are

$$I_1 = \{u \in D \mid u = 0, \text{ on } \partial\Omega\} \tag{10.76 a}$$

$$I_2 = \left\{u \in D \;\middle|\; \frac{\partial u}{\partial n} = 0, \text{ on } \partial\Omega\right\} \tag{10.76 b}$$

and

$$I_3 = \left\{u \in D \;\middle|\; \alpha \frac{\partial u}{\partial n} + \beta u = 0, \text{ on } \partial\Omega\right\} \tag{10.76 c}$$

where $\alpha^2 + \beta^2 \neq 0$.

Many examples of regular subspaces can be given for the biharmonic equation; an interesting set of such subspaces is

$$I_1 = \left\{u \in D \;\middle|\; u = \frac{\partial u}{\partial n} = 0, \text{ on } \partial\Omega\right\}, \tag{10.77 a}$$

$$I_2 = \{u \in D \mid u = \Delta u = 0, \text{ on } \partial\Omega\} \tag{10.77 b}$$

and

$$I_3 = \left\{u \in D \;\middle|\; \frac{\partial u}{\partial n} = \frac{\partial \Delta u}{\partial n} = 0, \text{ on } \partial\Omega\right\} \tag{10.77 c}$$

More general examples were given previously [8].

For Stokes problem we have the following regular suspaces

$$I_1 = \{\hat{u} \in D \mid \mathbf{u} = \mathbf{0}, \text{ on } \partial\Omega\} \tag{10.78 a}$$

$$I_2 = \left\{\hat{u} \in D \;\middle|\; \nu \frac{\partial \mathbf{u}}{\partial n} - p\,\mathbf{n} \in H^0(\partial\Omega)\right\} \tag{10.78 b}$$

Of course, many more can be given.

Of special interest is the case when a regular subspace $I \subset D$, has the following additional property

(iii) For every $u \in D$

$$\langle A u, v \rangle = 0 \;\forall\; v \in I \Rightarrow u \in I \tag{10.79}$$

A regular subspace, which enjoys (iii) is called completely regular.

It is not difficult to verify that in all the examples given in equations (10.76) through (10.78) the subspaces are, actually, completely regular.

Given an antisymmetric bilinear from $A: D \to D^*$, a pair of subspaces $\{I_1, I_2\}$ is said to be a canonical decomposition of $D$ for $A$, when

(i) $I_1$ and $I_2$ are regular subspaces; and

(ii) $D = I_1 + I_2$. (10.80)

It has been shown [28, 34] that when $\{I_1, I_2\}$ is a canonical decomposition of $D$, then $I_1$ and $I_2$ are necessarily, completely regular and

$$N_A = I_1 \cap I_2 \tag{10.81}$$

Now, condition (10.80) is equivalent to the requirement that given any $u \in D$, one can find elements $u_1 \in I_1$ and $u_2 \in I_2$ such that

$$u = u_1 + u_2 \tag{10.82}$$

In the presence of equation (10.81), this representation of $u$, is unique except for elements of subspace $N_A$; more precisely, if $u_1' \in I_1$ and $u_2' \in I_2$ are such that

$$u = u_1' + u_2' \tag{10.83}$$

then $u_1 - u_1' \in N_A$ and $u_2 - u_2' \in N_A$. Taking into account that $N_A$ is the set of functions with vanishing boundary values, it is seen that the boundary values of $u_1$ and $u_2$ are uniquely defined. Thus, when a canonical decomposition $\{I_1, I_2\}$ is available, representation (10.82) supplies a convenient manner of dividing the information on the boundary values of the function $u$ into two parts, $u_1 \in I_1$ and $u_2 \in I_2$, which is useful in the formulation of many boundary value problems.

For the reduced wave equation, the pair $\{I_1, I_2\}$, defined by (10.76 a) and (10.76 b), constitutes a canonical decomposition of the space $D$, with respect to $A$, as defined by (10.59). In this case, the representation (10.82), breaks the boundary information in the following manner

$$u = u_2; \quad \frac{\partial u}{\partial n} = \frac{\partial u_1}{\partial n}; \quad \text{on } \partial\Omega \tag{10.84}$$

The pair $\{I_1, I_3\}$, given by (10.76 a) and (10.76 c), is also a canonical decomposition, whenever $\alpha \neq 0$. In this case, if $u = u_1 + u_3$, with $u_1 \in I_1$ and $u_3 \in I_3$, then the boundary values are given by

$$u = u_1 + u_3; \quad \frac{\partial u}{\partial n} = \frac{\partial u_1}{\partial n} + \frac{\partial u_3}{\partial n}; \quad \text{on } \partial\Omega \tag{10.85}$$

If we define

$$I_4 = \left\{ u \in D \;\middle|\; \gamma \frac{\partial u}{\partial n} + \delta u = 0, \text{ on } \partial\Omega \right\} \tag{10.86}$$

it is easy to see that $\{I_3, I_4\}$ is a canonical decomposition, whenever $\alpha\delta - \beta\gamma \neq 0$. Clearly, the previous ones are particular cases of this more general canonical decomposition.

For the biharmonic equation, the following pair is a canonical decomposition

$$I_1 = \left\{ u \in D \;\middle|\; u = \frac{\partial u}{\partial n} = 0, \text{ on } \partial\Omega \right\} \tag{10.87 a}$$

$$I_2 = \left\{ u \in D \;\middle|\; \Delta u = \frac{\partial \Delta u}{\partial n} = 0, \text{ on } \partial\Omega \right\} \tag{10.87 b}$$

Also

$$I_1 = \{ u \in D \mid u = \Delta u = 0, \text{ on } \partial\Omega \} \tag{10.88 a}$$

$$I_2 = \left\{ u \in D \;\middle|\; \frac{\partial u}{\partial n} = \frac{\partial \Delta u}{\partial n} = 0, \text{ on } \partial\Omega \right\} \tag{10.88 b}$$

Finally, for Stokes problems one has

$$I_1 = \{\hat{u} \in D \mid \mathbf{u} = \mathbf{0}, \text{ on } \partial\Omega\} \tag{10.89 a}$$

$$I_2 = \left\{\hat{u} \in D \;\middle|\; v\frac{\partial \mathbf{u}}{\partial n} - p\,\mathbf{n} = \mathbf{0}, \text{ on } \partial\Omega\right\} \tag{10.89 b}$$

Of course many more can be constructed.

In many boundary value problems the prescribed boundary data is given by means of one of the elements in (10.82), for example $u_1$, and the complementary boundary information $u_2$, can only be obtained after the boundary value problem has been solved. In Dirichlet problem for example, $u$ is prescribed on $\partial\Omega$ and the derived boundary information $\frac{\partial u}{\partial n}$ on $\partial\Omega$, is obtained, only after the problem has been solved.

The notion of Green's formula is closely related with that of canonical decomposition. Some auxiliary notions are required in order to introduce abstract Green's formulas.

Given the bilinear form $B: D \to D^*$, let

$$N_B = \{u \in D \mid B\,u = 0\} \tag{10.90}$$

be the null subspace of $B$. Then, if

$$D = N_B + N_{B^*} \tag{10.91}$$

where $B^*: D \to D^*$ is the transposed bilinear form of $B$, one says that $B$ and $B^*$ can be varied independently. When $B$ and $B^*$ can be varied independently and

$$A = B - B^* \tag{10.92}$$

equation (10.92) is called a Green's formula. It can be shown [34] that in this case $B: D \to D^*$ is necessarily a boundary operator.

There is a general result of the theory according to which there is a one-to-one correspondence between canonical decompositions $\{I_1, I_2\}$ and Green's formulas. This is established as follows:

(i) Given a Green's formula, define

$$I_1 = N_{B^*}; \qquad I_2 = N_B \tag{10.93}$$

then $\{I_1, I_2\}$ is a canonical decomposition.

(ii) Given a canonical decomposition $\{I_1, I_2\}$, let $B: D \to D^*$ be defined by

$$\langle B\,u, v\rangle = \langle A\,u_1, v_2\rangle \tag{10.94}$$

Here, the representation (10.82) of every element $u \in D$ of the space, in terms of its components $u_1 \in I_1$ and $u_2 \in I_2$, has been used.

To illustrate these notions, in the case of Laplace and reduced wave equation, we notice that if we define

$$\langle B\,u, v\rangle = \int_{\partial R} v\,\frac{\partial u}{\partial n}\,dx \tag{10.95}$$

then (10.91) and (10.92) are fulfilled. Also, the canonical decomposition $\{I_1, I_2\}$, given by (10.76 a, b) satisfies (10.93).

In the case of the biharmonic equation, the canonical decomposition (10.87), is associated with

$$\langle B\, u, v\rangle = \int_{\partial R} \left\{ v \frac{\partial \Delta u}{\partial n} - \Delta u \frac{\partial v}{\partial n} \right\} dx \tag{10.96}$$

The canonical decomposition (10.88), on the other hand, yields

$$\langle B\, u, v\rangle = \int_{\partial R} \left\{ v \frac{\partial \Delta u}{\partial n} + \Delta v \frac{\partial u}{\partial n} \right\} dx \tag{10.97}$$

Finally, for Stokes equations, the canonical decomposition (10.89), is associated with

$$\langle B\, \hat{u}, \hat{v}\rangle = \int_{\partial R} \mathbf{v} \cdot \left( \nu \frac{\partial \mathbf{u}}{\partial n} - p\, \mathbf{n} \right) dx \tag{10.98}$$

## 10.4 Illustration of Green's Formulas

In this section general examples of Green's formulas are presented. Many of the operators listed are formally symmetric in the classical sense; others can be included due to the extension of this concept introduced in the algebraic theory of boundary value problems [19, 34] which supplies the basic frame-work for this chapter.

*Elliptic Equations*

This subject is classical. The reader is referred to the book by Lions and Magenes [41]. The extension of such formulas to problems with prescribed jumps can be done along the lines presented in Section 5. A general discussion of Green's formulas from the point of view of the algebraic theory will appear soon [19].

*Time Dependent Problems*

For a discussion of the spaces which are suitable for the formulation of this class of problems, the reader is referred to the second volume of the treatise by Lions and Magenes [41]. In this Section we simply assume that the linear space of functions $D$ is such that the operators to be considered are well defined.

Two examples will be given: the heat and the wave equations. These can be associated with formally symmetric operators, in the sense of the algebraic theory [19, 34], using Gurtin's convolutions [46, 47]. The basic ideas can be applied to more general problems. For each one of these operators we give only one Green's formula; of course, many more can be constructed.

*(i) The Heat Equation.* Consider the cylinder $\Omega \times [0, T]$ (Fig. 10.2). Let the linear space $D$, be made of functions defined on $\Omega \times [0, T]$. The operator $P : D \to D^*$, is defined by

$$\langle P\, u, v\rangle = \int_{\Omega} v * \left( \frac{\partial u}{\partial t} - \Delta u \right) dx \tag{10.99}$$

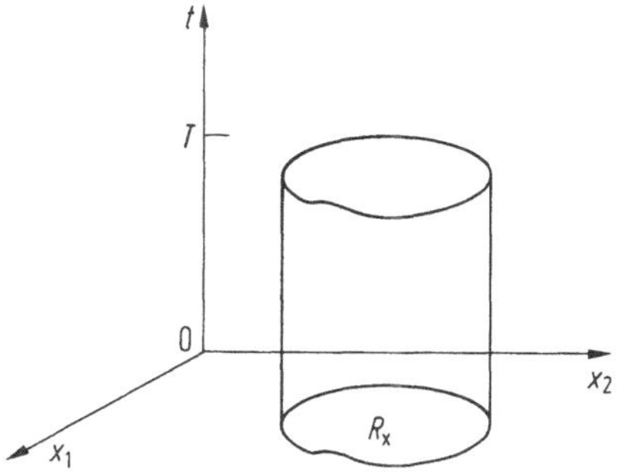


**Fig. 10.2**

where the notation

$$u * v = \int_0^T u(T-t)\, v(t)\, dt \tag{10.100}$$

is used. Let $A = P - P^*$, then

$$A = B - B^* \tag{10.101}$$

where

$$\langle B u, v \rangle = \int_{\partial\Omega} u * \frac{\partial v}{\partial n} dx - \int_{\Omega} v(T)\, u(0)\, dx \tag{10.102}$$

*(ii) The Wave Equation.* The incorporation of this equation in the frame-work here presented is similar.

Taking the region $\Omega \times [0, T]$ and the linear space of functions $D$, as explained before, define $P: D \to D^*$ by

$$\langle P u, v \rangle = \int_{\Omega} v * \left( \frac{\partial^2 u}{\partial t^2} - \Delta u \right) dx \tag{10.103}$$

with the convention (10.100). A Green's formula for this operator is obtained taking $B: D \to D^*$ as

$$\langle B u, v \rangle = \int_{\partial\Omega} u * \frac{\partial v}{\partial n} dx - \int_{\Omega} \left\{ v(T) \frac{\partial u}{\partial t}(0) + \frac{\partial v}{\partial t}(T)\, u(0) \right\} dx \tag{10.104}$$

Formulas (10.102) and (10.104) are suitable for application to initial value problems when the function $u$ is prescribed on the lateral boundary of the space-time cylinder (Fig. 10.2). More general boundary conditions can be treated by using the Green's formula of the Laplace operator, associated with the canonical decomposition defined by (10.76 c) and (10.86).

*Elasticity*

Let the elastic tensor $C_{ijpq}$ be $C^\infty(\Omega)$, satisfy the usual symmetry conditions [48]

$$C_{ijpq} = C_{pqij} = C_{jipq} \tag{10.105}$$

and be strongly elliptic; i.e.

$$C_{ijpq}\, \xi_i\, n_j\, \xi_p\, \eta_q > 0 \quad \text{whenever} \quad \| \xi \| \neq 0, \ \| \eta \| \neq 0 \tag{10.106}$$

*(i) Static and Periodic Motions.* Let $D = H^s(\Omega) = H^s(\Omega) \oplus H^s(\Omega) \oplus H^s(\Omega)$, $s \geqq 2$. Define

$$\tau_{ij}(u) = C_{ijpq} \frac{\partial u_p}{\partial x_q}, \quad \text{on } \Omega \tag{10.107}$$

$$\mathscr{S}_i(u) = \frac{\partial \tau_{ij}}{\partial x_j} + \varrho\, \omega^2 u_i, \quad \text{on } \Omega \tag{10.108}$$

where summation convention is understood. Here the density $\varrho$, is a function of position belonging to $C^\infty(\Omega)$ while $\omega$ is a constant. The case $\omega = 0$, is associated with elastostatics.

Let $P: D \to D^*$ be

$$\langle P\, u, v \rangle = \int_\Omega v_i\, \mathscr{S}_i(u)\, dx \tag{10.109}$$

Then, $A = P - P^*$ is given by

$$\langle A\, u, v \rangle = \int_{\partial\Omega} \{v_i\, T_i(u) - u_i\, T_i(u)\}\, dx \tag{10.110}$$

where

$$T_i(u) = \tau_{ij}(u)\, n_j\,. \tag{10.111}$$

An operator $B: D \to D^*$ that decomposes $A$ is

$$\langle B\, u, v \rangle = -\int_{\partial\Omega} u_i\, T_i(v)\, dx\,. \tag{10.112}$$

There are many more.

*(ii) Dynamics.* Let $D$ be a suitable linear space of functions defined on $\Omega \times [0, T]$. Define

$$\langle P\, u, v \rangle = \int_\Omega v_i * \left(\varrho \frac{\partial^2 u_i}{\partial t^2} - \mathscr{S}_i\, u\right) dx \tag{10.113}$$

where the conventions (10.100) and (10.108) (with $\omega = 0$) are used. $A = P - P^*$ is given by

$$\langle A\, u, v \rangle = \int_{\partial\Omega} \{u_i * T_i(v) - v_i * T_i(u)\}\, dx \tag{10.114}$$

$$+ \int_\Omega \varrho \left\{v_i(0) \frac{\partial u_i}{\partial t}(T) + \frac{\partial v_i}{\partial t}(0)\, u_i(T) - u_i(0) \frac{\partial v_i}{\partial t}(T) - \frac{\partial u_i}{\partial t}(0)\, v_i(T)\right\} dx$$

Many operators that decompose $A$ can be constructed. One such operator is

$$\langle B\, u, v \rangle = \int_{\partial\Omega} u_i * T_i(v)\, dx - \int_\Omega \left\{u_i(0) \frac{\partial v_i}{\partial t}(T) - \frac{\partial u_i}{\partial t}(0)\, v_i(T)\right\} dx \tag{10.115}$$

## 10.5 Green's Formulas in Discontinuous Fields

An advantage of introducing abstract boundary operators is the large class of problems that can be formulated using them; a very general example is the problem of connecting or matching [28]. This is an abstract version of problems formulated in discontinuous fields with prescribed jump conditions.

Such problems occur in many applications. In potential theory, for example, the jumps of the function and its normal derivative are usually prescribed, while in Elasticity the prescribed functions are the jumps of the displacements and the tractions. Variational principles for some of these problems were developed by Prager [49] and Nemat-Nasser [50, 51] presented more recent surveys. Here, general Green's formulas for such problems are developed which are applicable irrespectively of the specific operators.

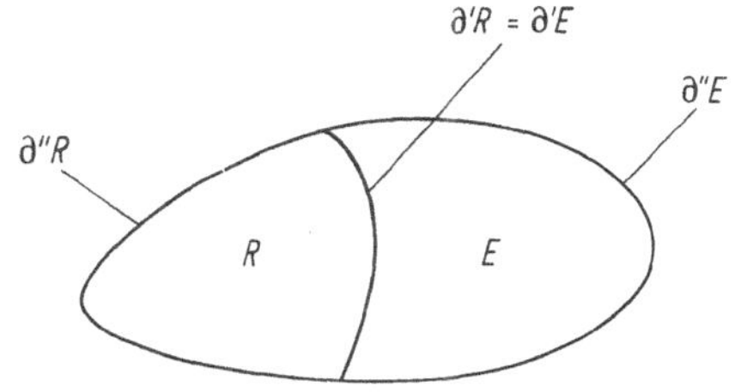


**Fig. 10.3**

Consider two neighboring regions $R$ and $E$ (Fig. 10.3), let $\partial' R = \partial' E$ be the common boundary separating them; in addition, $\partial'' R$ and $\partial'' E$ will be the remaining parts of the boundaries of $R$ and $E$, respectively. Let $D_R$ and $D_E$ be two linear spaces; in the applications to be made their elements will be functions defined on $R$ and $E$, respectively. Consider the product space $\hat{D} = D_R \oplus D_E$; elements $\hat{u} \in \hat{D}$ are pairs $\hat{u} = \{u_R, u_E\}$ where $u_R \in D_R$ while $u_E \in D_E$. Given operators $P_R\colon D_R \to D_R^*$ and $P_E\colon D_E \to D_E^*$, define $\hat{P}\colon \hat{D} \to \hat{D}^*$ by

$$\langle \hat{P}\,\hat{u}, \hat{v}\rangle = \langle P_R\, u_R, v_R\rangle + \langle P_E\, u_E, v_E\rangle \tag{10.166}$$

This additive property is usually satisfied when $\hat{P}\colon \hat{D} \to \hat{D}^*$ is defined by means of an integral on the region $R \cup E$. From (10.116) it follows that

$$\langle \hat{A}\,\hat{u}, \hat{v}\rangle = \langle A_R\, u_R, v_R\rangle + \langle A_E\, u_E, v_E\rangle \tag{10.117}$$

The symbol $\hat{N}_A$ will be used for the null subspace of $\hat{A}\colon \hat{D} \to \hat{D} \to \hat{D}^*$. A linear subspace $\hat{S} \subset \hat{D}$ will be considered. Elements $\hat{u} = \{u_R, u_E\} \in \hat{S}$ will be said to be smooth. When $\hat{u} = \{u_R, u_E\}$ is smooth, $u_R \in D_R$ and $u_E \in D_E$ will be said to be smooth extension of each other.

Let $\hat{S} \subset \hat{D} = D_R \oplus D_E$ be a linear subspace. Then $\hat{S}$ is said to be a smoothness relation if every $u_R \in D_R$ possesses at least one smooth extension $u_E \in D_E$ and conversely. A smoothness relation $\hat{S}$ is said to be regular or completely regular for $\hat{P}$, when as a subspace, it is regular or completely regular for $\hat{P}$, respectively. Therefore, a smoothness relation $\hat{S}$ is regular when

$$\text{a)} \qquad \hat{S} \supset \hat{N}_A \tag{10.118a}$$

and

$$\text{b)} \quad \langle \hat{A}\,\hat{u}, \hat{v}\rangle = 0 \;\forall\; \hat{u} \in \hat{S} \;\text{ and }\; \hat{v} \in \hat{S} \tag{10.118b}$$

Similarly, it is completely regular when

$$\langle \hat{A}\,\hat{u}, \hat{v}\rangle = \langle A_R\, u_R, v_R\rangle + \langle A_E\, u_E, v_E\rangle = 0 \quad \forall\; \hat{v} \in \hat{S} \Leftrightarrow \hat{u} \in \hat{S} \tag{10.119}$$

The mapping $\tau: \hat{D} \to \hat{D}$ defined by $\tau\,\hat{u} = \{u_R, -u_E\}$, for every $u = \{u_R, u_E\} \in \hat{D}$ will be used in the following discussion. Given $\hat{S}$, let $\tau\,\hat{S}$ be the image of $\hat{S}$ under this mapping; i.e.

$$\tau\,\hat{S} = \{\hat{u} = \tau\,\hat{v} \in \hat{D} \,|\, \hat{v} \in \hat{S}\} \tag{10.120}$$

Given $\hat{u} = \{u_R, u_E\}$, let be $\hat{u}' = \{u'_R, u'_E\}$, where $u'_R$ and $u'_E$ are smooth extensions of $u_E$ and $u_R$, respectively. Define

$$\bar{u} = \tfrac{1}{2}(\hat{u} + \hat{u}') = \tfrac{1}{2}\{u_R + u'_R, u_E + u'_E\} \tag{10.121 a}$$

and

$$[\hat{u}] = \hat{u}' - \hat{u} = \{u'_R - u_R, u'_E - u_E\} \tag{10.121 b}$$

Then

$$\hat{u} = \bar{u} - \tfrac{1}{2}[\hat{u}] \tag{10.122}$$

and it can be seen that $\hat{u} \in \hat{S}$ while $[\hat{u}] \in \tau\,\hat{S}$. Therefore

$$\hat{D} = \hat{S} + \tau\,\hat{S} \tag{10.123}$$

From (10.123), it follows that when $\hat{S} \subset \hat{D}$ is a regular smoothness condition, then the pair $\tau\,\hat{S}$, $\hat{S}$ is a canonical decomposition of $\hat{D}$ with respect to $\hat{P}: \hat{D} \to \hat{D}^*$.

Application of formula (10.96), shows that the relation

$$\hat{P} - \hat{P}^* = \hat{A} = \hat{J} - \hat{J}^* \tag{10.124}$$

is a Green's formula when $\hat{J}: \hat{D} \to \hat{D}^*$ is defined by

$$\langle \hat{J}\,\hat{u}, \hat{v} \rangle = -\tfrac{1}{2}\langle \hat{A}\,[\hat{u}], \bar{v} \rangle \tag{10.125}$$

This is called jump operator [19, 34] because it characterizes the jumps since

$$\hat{J}\,\hat{u} = \hat{J}\,\hat{v} \Leftrightarrow \hat{u} - \hat{v} \in \hat{S} \tag{10.126}$$

by virtue of the second of equations (10.93).

To apply these results to potential theory and reduced wave equation, given $\varrho$ and non-zero functions $k_R$ and $k_E$, define (Fig. 10.3)

$$\mathcal{L} u = \nabla \cdot k\,\nabla u + \varrho\,u \tag{10.127}$$

$$\langle P_R\,u_R, v_R \rangle = \int_R v\,\mathcal{L}\,u\,dx + \int_{\partial_1 R} u\,k\,\frac{\partial v}{\partial n}\,dx - \int_{\partial_2 R} v\,k\,\frac{\partial u}{\partial n}\,dx \tag{10.128}$$

and $P_E: D_E \to D_E^*$ replacing $R$ by $E$. Integrating by parts it is seen that

$$\langle \hat{A}\,\hat{u}, \hat{v} \rangle = \int_{\partial' R} k_R\left(v_R\frac{\partial u_R}{\partial n} - u_R\frac{\partial v_R}{\partial n}\right)dx + \int_{\partial' E} k_E\left(v_E\frac{\partial u_E}{\partial n} - u_E\frac{\partial v_E}{\partial n}\right)dx \tag{10.129}$$

Observe that the unit normal vector $\mathbf{n}$ is taken pointing outwards from the region of integration. Equation (10.129) implies that

$$\hat{N}_A = \{\{u_R, u_E\} \in \hat{D} \,|\, u_R = u_E = \partial u_E/\partial n = \partial u_R/\partial n = 0, \quad \text{on} \quad \partial_3 R\} \tag{10.130}$$

Smoothness conditions can be defined in many alternative manners. One which is suitable in many applications (in flow through porous media, for example) is

$$\hat{S} = \{\hat{u} \in \hat{D} \,|\, u_R = u_E,\, k_R\,\partial u_R/\partial n = k_E\,\partial u_E/\partial n, \quad \text{on} \quad \partial' R\} \tag{10.131}$$

Given $\hat{u} = \{u_R, u_E\} \in \hat{D}$, let be $\hat{u}' = \{u'_R, u'_E\}$, where $u'_R$ and $u'_E$ are smooth extensions of $u_E$ and $u_R$, respectively. Write

$$[\hat{u}] = \{[\hat{u}]_R, [\hat{u}]_E\}; \quad \bar{u} = \{\bar{u}_R, \bar{u}_E\} \tag{10.132}$$

Applying definitions (10.121) and (10.131), it is seen that

$$[\hat{u}]_R = u'_R - u_R = u_E - u_R, \quad \text{on } \partial' R \tag{10.133 a}$$

$$\frac{\partial [\hat{u}]_R}{\partial n} = \frac{\partial u'_R}{\partial n} - \frac{\partial u_R}{\partial n} = \frac{k_E}{k_R}\frac{\partial u_E}{\partial n} - \frac{\partial u_R}{\partial n}, \quad \text{on } \partial' R \tag{10.133 b}$$

where the normal derivative is taken pointing outwards from $R$. When $\hat{v} \in \hat{S}$, equation (10.129) reduces to

$$\langle \hat{A}\,\hat{u}, \hat{v} = \int_{\partial' R} \left\{ k\,[\hat{u}] \frac{\partial v}{\partial n} - v \left[ \hat{k} \frac{\partial \hat{u}}{\partial n} \right] \right\} dx \tag{10.134}$$

Here, as in what follows, the components ($R$ or $E$) to be used when carrying out the integration are indicated by the subindex under the integral sign. Also the notation

$$\left[ \hat{k} \frac{\partial \hat{u}}{\partial n} \right]_R = k_R \frac{\partial}{\partial n} [\hat{u}]_R = k_E \frac{\partial u_E}{\partial n} - k_R \frac{\partial u_R}{\partial n} \tag{10.135}$$

was introduced. Application of (10.134) in (10.125) yields

$$2 \langle \hat{J}\,\hat{u}, \hat{v} \rangle = \int_{\partial' R} \left\{ \bar{v} \left[ \hat{k} \frac{\partial \hat{u}}{\partial n} \right] - k \overline{\frac{\partial v}{\partial n}} [\hat{u}] \right\} dx \tag{10.136}$$

Observe that equations (10.121) imply

$$[\hat{u}]_R = u_E - u_R, \quad 2\bar{v} = v_R + v_E \quad \text{and} \quad 2k_R \overline{\frac{\partial v}{\partial n}} = k_E \frac{\partial v_E}{\partial n} + k_R \frac{\partial v_R}{\partial n} \tag{10.137}$$

In view of equations (10.128) and (10.137), the Green's formula

$$\int_R \{v \mathscr{L} u - u \mathscr{L} v\}\, dx = \langle (\hat{B} + \hat{J})\,\hat{u}, \hat{v} \rangle - \langle (\hat{B} + \hat{J})\,\hat{v}, \hat{u} \rangle \tag{10.138}$$

is clear. Here

$$\langle \hat{B}\,\hat{u}, \hat{v} \rangle = \int_{\partial_1(R \cup E)} u\, k \frac{\partial v}{\partial n}\, dx - \int_{\partial_2(R \cup E)} v\, k \frac{\partial u}{\partial n}\, dx \tag{10.139}$$

In a similar fashion for static and quasi-static elasticity, when the smoothness criterium consists of continuity of displacements and tractions, one obtains for the jump operator [28]

$$2 \langle \hat{J}\,\hat{u}, \hat{v} \rangle = \int_{\partial' R} \{\bar{v}_i [T_i(\mathbf{u})] - [\hat{u}_i]\, \overline{T_i(\mathbf{v})}\}\, dx \tag{10.140}$$

This yields corresponding Green's formulas.

The formulation of Greens's formulas in discontinuous fields here presented is applicable to arbitrary formally symmetric operators which are linear. Thus, for example, the biharmonic equation or Stoke's problem are included. Green's formulas for two phases systems have also been derived in this manner [28].

## 10.6 *T*-Complete Systems

With every operator $P: D \to D^*$, we can associate a linear subspace $I_P \subset D$, defined by

$$I_P = N_P + N_A \tag{10.141}$$

this equation implies that every element $u \in I_P$, can be written as

$$u = u_P + u_A \tag{10.142}$$

with $u_P \in N_P$ while $u_A \in N_A$; since $u_A$ vanishes on the boundary, we see that a function $u$ belongs to $I_P$, if and only if, there is a solution $u_P$ of the homogeneous partial differential equation such that the boundary values of $u$ and $u_P$ coincide. As illustration, in the example given previously of the reduced wave equation, a function $v \in I_P$, if and only if, there is a solution $u \in D$ of the homogeneous equation such that $v = u$ and $\dfrac{\partial v}{\partial n} = \dfrac{\partial u}{\partial n}$, on the boundary $\partial\Omega$.

It can be shown [28, 34] that $I_P$, as defined by (6.1), is always regular. Due to this fact the concept of *T*-complete system will be useful. Let $I_P \subset D$, be regular, and $\mathscr{B}$ be a subset of $I_P$, then we say that $\mathscr{B} \subset I_P$ is *T*-complete for $I_P$, when for every $u \in D$

$$\langle A\,u, w\rangle = 0 \quad \forall\, w \in \mathscr{B} \Rightarrow u \in I_P \tag{10.143}$$

Under very general conditions $N_P \subset I_P$ is *T*-complete for $I_P$ [19, 28, 34]. For the representation of solutions it is, however, of greater interest to have denumerable subsets $\mathscr{B} \subset N_P$ which are *T*-complete. Examples of such systems are given in Tables 10.1 and 10.2. It has interest to mention that for the reduced wave equations the author has shown that a system of plane waves, which have a very simple structure, is *T*-complete in any bounded and simply connected region [5].

In these tables $J_n(r)$ and $H_n^{(1)}(r)$ are Bessel and Hankel functions of the first class [52, 53]. $P_n^q$ is the associated Legendre function, while $j_n$ and $h_n^1$ are the spherical Bessel and Hankel functions [52]. We recall, in addition, that the *T*-complete systems given in Tables 1 and 2 for Laplace equation in a bounded region are harmonic polynomials expressed in polar and spherical coordinates. Observe that the detailed shape of $\Omega$ is arbitrary.

**Table 10.1.** *T*-complete systems in two dimensions

| Bounded $\Omega$ | $\Omega$ = Exterior of a bounded region |
|---|---|
| Laplace Equation | |
| $\{1, r^n \cos n\,\theta, r^n \sin n\,\theta\}$ | $\{\mathrm{Ln}\, r^{-n} \cos n\,\theta, r^{-n} \sin n\,\theta\}$ |
| Reduced Wave Equation $\Delta u + u = 0$ | |
| $\{J_0(r), J_n(r) \cos n\,\theta, J_n(r) \sin n\,\theta\}$ | $\{H_0^{(1)}(r), H_n^{(1)}(r) \cos n\,\theta, H^{(1)}(r) \sin n\,\theta\}$ |
| $n = 1, 2, \ldots$ | |

**Table 10.2.** *T*-complete systems in three dimensions

| Bounded $\Omega$ | $\Omega$ = Exterior of a bounded region |
|---|---|
| Laplace Equation | |
| $\{r^n P_n^q(\cos\theta)\, e^{iq\varphi}\}$ | $\{r^{-n-1} P_n^q(\cos\theta)\, e^{iq\varphi}\}$ |
| Reduced Wave Equation | |
| $\{j_n(r)\, P_n^q(\cos\theta)\, e^{iq\varphi}\}$ | $\{h_n^{(1)}(r)\, P_n^q(\cos\theta)\, e^{iq\varphi}\}$ |
| $n = 0, 1, 2, \dots; \quad -n \leqq q \leqq n$ | |

## 10.7 Hilbert-Space Formulation

Associated with every Green's formula or equivalently, with every canonical decomposition, there is a Hilbert-space formulation.

For this purpose, we focus our attention in boundary values; i.e. we identify functions possessing the same boundary values. More precisely, two functions $u$ and $v$ of $D$, are identified whenever $u - v \in N_A$. The resulting space $\mathscr{D}$ is called the quotient space; i.e.

$$\mathscr{D} = D/N_A \tag{10.144}$$

Thus, for example

(i) For Laplace and reduced wave equation, $\mathscr{D}$ is made of pairs of functions $u, \dfrac{\partial u}{\partial n}$, defined on the boundary $\partial R$ and square integrable there. Indeed

$$\mathscr{D} = \left\{\left[u, \frac{\partial u}{\partial n}\right] \middle| \; u \in H^0(\partial R), \frac{\partial u}{\partial n} \in H^0(\partial R)\right\} \tag{10.145}$$

(ii) Biharmonic equation

$$\mathscr{D} = \left\{\left[u, \frac{\partial u}{\partial n}, \Delta u, \frac{\partial \Delta u}{\partial n}\right] \middle| \; \text{Each one of } u, \frac{\partial u}{\partial n}, \Delta u, \frac{\partial \Delta u}{\partial n} \in H^0(\partial R)\right\}$$

(iii) Stokes equation

$$\mathscr{D} = \left\{\left[\mathbf{u}, v\frac{\partial \mathbf{u}}{\partial n} - p\,\mathbf{n}\right] \middle| \; \mathbf{u} \in \mathbf{H}^0(\partial R), v\frac{\partial \mathbf{u}}{\partial n} - p\,\mathbf{n} \in H^0(\partial R)\right\} \tag{10.146}$$

In each of these examples, one can give to $\mathscr{D}$, the structure of a Hilbert space. Possible choices for the corresponding inner products are

$$\text{(i)} \quad \int_{\partial\Omega} \left\{u\,v + \frac{\partial u}{\partial n}\frac{\partial v}{\partial n}\right\} dx \tag{10.147a}$$

$$\text{(ii)} \quad \int_{\partial\Omega} \left\{u\,v + \frac{\partial u}{\partial n}\frac{\partial v}{\partial n} + \Delta u\,\Delta v + \frac{\partial \Delta u}{\partial n}\frac{\partial \Delta v}{\partial n}\right\} dx \tag{10.147b}$$

$$\text{(iii)} \quad \int_{\partial\Omega} \left\{\mathbf{u}\cdot\mathbf{v} + \left(v\frac{\partial \mathbf{u}}{\partial n} - p\,\mathbf{n}\right)\cdot\left(v\frac{\partial \mathbf{v}}{\partial n} - q\,\mathbf{n}\right)\right\} dx \tag{10.147c}$$

With these inner products, the linear space $\mathscr{D}$ is isomorphic to the following Hilbert spaces:

$$\text{(i)} \quad H^0(\partial\Omega) \oplus H^0(\partial\Omega) \tag{10.148a}$$

$$\text{(ii)} \quad H^0(\partial\Omega) \oplus H^0(\partial\Omega) \oplus H^0(\partial\Omega) \oplus H^0(\partial\Omega) \tag{10.148b}$$

$$\text{(iii)} \quad H^0(\partial\Omega) \oplus H^0(\partial\Omega) \tag{10.148c}$$

Now, given any canonical decomposition $\{I_1, I_2\}$ it is possible to chose the Hilbert-space structure so that the associated operator $B: D \to D^*$ (equation 10.94) is given by

$$\langle B\,u, v\rangle = (u_1, v_2) \tag{10.149}$$

Thus, for example, when the inner product (10.147a) is used, equation (10.149) yields the operator $B$ associated with the canonical decomposition given by (10.76a and b). The same happens if this decomposition is replaced by (10.76a and c). When one uses the inner product (10.147b), equation (10.149) supplies the operator $B: D \to D^*$ associated with any canonical decomposition corresponding to the biharmonic equation; for example, those given by equations (10.87) or (10.88). For Stokes problem the inner product can be (10.147c) and a possible canonical decomposition is defined by (10.89).

## 10.8 Representation of Solutions

For the formulation of the general boundary value problem to be considered here, we assume there is a canonical decomposition $\{I_1, I_2\}$, and an operator $B: D \to D^*$ such that (10.92) is a Green's formula. Using the representation (10.82), we formulate the problem as follows; find $u \in N_P$, such that

$$u_1 = U_1 \tag{10.150}$$

where $U_1$ is a given element of $I_1$.

Let $\mathscr{N}_P = N_P/N_A \subset \mathscr{D} = \hat{\mathscr{H}}$, be the linear space generated by the boundary values of solutions of the homogeneous equation. Then every $u \in \mathscr{N}_P$ can be written as

$$u = u_1 + u_2; \tag{10.151}$$

where $u_1 \in \mathscr{I}_1 = I_1/N_A$ while $u_2 \in \mathscr{I}_2 = I_2/N_A$. Let $\mathscr{N}_1 \subset \mathscr{I}_1$ be the range of values taken by $u_1$, in (10.151), when $u$ ranges over $\mathscr{N}_P$. Similarly, let $\mathscr{N}_2 \subset \mathscr{I}_2$, be the range of values taken by $u_2$, in (10.151), when $u$ ranges over $\mathscr{N}_P$.

Given a system of functions $\mathscr{B} = \{w_1, w_2, \ldots\} \subset N_P$, write

$$w_\alpha = w_{\alpha 1} + w_{\alpha 2} \tag{10.152}$$

We denote

$$\mathscr{B}_1 = \{w_{11}, w_{21}, w_{31}, \ldots\} \subset \mathscr{I}_1; \quad \mathscr{B}_2 = \{w_{12}, w_{22}, w_{32}, \ldots\} \subset \mathscr{I}_2 \tag{10.153}$$

Clearly, we will be able to approximate the boundary values of every solution of (10.150), if and only if

$$\operatorname{span} \mathscr{B}_1 = \bar{\mathscr{N}}_1 \tag{10.154}$$

Here, the bar refers to the closure of $\mathscr{N}_1$.

A result similar to Theorem 10.1, holds in this more general context. Assume $I_P = N_P + N_A$ is completely regular and $\mathscr{B} \subset N_P$. Then the following statements are equivalent:

$$\begin{aligned}&\text{(i)} \quad \hat{\mathscr{B}} \subset \hat{\mathscr{N}}_P \subset \hat{\mathscr{H}} \text{ spans } \hat{\mathscr{N}}_P \text{ in the metric of } \hat{\mathscr{H}};\\ &\text{(ii)} \quad \hat{\mathscr{B}} \subset \hat{\mathscr{N}}_P \text{ is } T\text{-complete; and}\\ &\text{(iii)} \quad \text{span } \mathscr{B}_1 = \bar{\mathscr{N}}_1 \text{ while span } \mathscr{B}_2 = \bar{\mathscr{N}}_2\end{aligned} \tag{10.155}$$

Therefore, when $\mathscr{B}$ is $T$-complete, it is possible to construct approximating sequences

$$u^N = \sum_{n=1}^{N} a_n^N w_n; \quad N = 1, 2, \ldots \tag{10.156}$$

such that $u_1^N \to U_1$, whenever $U_1 \in \bar{\mathscr{N}}_1$; therefore, if the problem (10.150) has a solution $u$, then

$$u^N \to u \tag{10.157}$$

The convergence in (10.157), is in any metric in which the solution of the problem, depends continuously on the boundary data $U_1$.

When a Green's formula is available, the results of Section 7, yield an efficient procedure to compute the complementary boundary data. In this case, for every $w_\alpha \in \mathscr{B}$, we have

$$(w_{\alpha 2}, U_1) = (w_{\alpha 2}, u_1) = (w_{\alpha 1}, u_2) \tag{10.158}$$

which gives $(w_{\alpha 1}, u_2)$ in terms of the boundary data $U_1$. This gives the approximating sequence

$$u_2^N = \sum_{n=1}^{N} b_n^N w_{n1} \tag{10.159}$$

where the coefficients $b_n^N$ satisfy, for every fixed $N$, the system of equations

$$(w_{m2}, U_1) = \sum_{n=1}^{N} b_n^N (w_{m1}, w_{n1}) \tag{10.160}$$

This generalizes the results of Section 2.

Observe that the values of $u_2$ are approximated by linear combinations of $w_{n1}$. This implies, for example, that in applications to problems formulated in discontinuous fields, with precribed jump conditions, the averages of the functions across discontinuities are approximated by the jumps of the basic systems (a specific application of this kind is given in [5]).

## References


1 Zienkiewicz, O. C., *The Finite Element Method in Engineering Science.* McGraw-Hill, New York 1977

2 Zienkiewicz, O. C., Kelly, D. W., Bettess, P., The coupling of the finte element method and boundary solution procedures. Int. J. Num. Meth. Eng., **11,** pp. 355–375, 1977

3 Brebbia, C. A., *The Boundary Element Method for Engineers.* Pentech Press, London 1978

4 Brebbia, C. A., Boundary element methods. Proc. Third Int. Seminar, Irvine, Ca., 1981; Proc. Fourth Int. Seminar, Southampton 1982, Springer-Verlag, Berlin-Heidelberg, New York

5 Sánchez-Sesma, F. J., Herrera, I., Avilés, J., Boundary methods for elastic wave diffraction-application to scattering of SH waves by surface irregularities. Bull. Seism. Soc. Am., **72** (2), pp. 473–490, 1982

6 Rektorys, K., *Survey of Applicable Mathematics.* ILIFFE Books Ltd, London 1969
7 Trefftz, E., Ein Gegenstück zum Ritz'schen Verfahren, Proc. Second Int. Congress Appl. Mech., Zürich 1926
8 Herrera, I., Boundary Methods for Fluids. *Finite Elements for Fluids.* **IV**, Gallagher, R. H. (Ed.), John Wiley & Sons Limited, Chapter 19, pp. 403–432, 1982
9 Mikhlin, S. G., Variational methods in mathematical physics. Pergamon Press, 1964
10 Rektorys, K., Variational methods in mathematics, science and engineering. D. Reidel Pub., Co., 1977
11 Kupradze, V. D. et al., Three dimensional problems of the mathematical theory of elasticity and thermoelasticity. North Holland, 1979
12 Amerio, L., Sul calculo delle autosoluzioni dei problemi al contorno per le equazioni lineari del secondo ordine di tipo ellitico. Rend. Acc. Lincei, **1**, pp. 352–359 and 505–509, 1946
13 Fichera, G., Teoremi di completezza sulla frontiera di un dominio per taluni sistema di funzioni. Ann. Mat. Pura e Appl, **27**, pp. 1–28, 1948
14 Picone, M., Nouvi metodi risolutivi per i problemi d'integrazione delle equazioni lineari a derivati parziali e nuova applicazione delle transformate multipla di Laplace nel caso delle equazioni a coefficienti constanti. Atti Acc. Sc. Torino, **76**, pp. 413–426, 1940
15 Kupradze, V. D., On the approximate solution of problems in mathematical physics. Russian Math. Surveys, **22**, (2), pp. 58–108, 1967, (Uspehi Mat. Nauk, **22** (2), pp. 59–107, 1967)
16 Vekua, I. N., New methods for solving elliptic equations. North Holland Pub. Co., 1967
17 Colton, D., Watzlawek, W., Complete families of solutions to the heat equation and generalized heat equation in $R^n$. Jour. Differential Equations, **25** (1), pp. 96–107, 1977
18 Colton, D., The approximation of solutions to initial boundary value problems for parabolic equations in one space variable. Quart. Appl. Math., **34** (4), pp. 377–386, 1976
19 Herrera, I., *Boundary Methods.* An Algebraic Theory, Pitman Publishing Co., 1984
20 Sabina, F. J., Herrera, I., England, R., Theory of connectivity: Applications to scattering of seismic waves. SH-wave motion, Proc. Second Int. Conference on Microzonation, San Francisco, Ca., 1979
21 England, R., Sabina, F. J., Herrera, I., Scattering of SH-waves by surface cavities of arbitrary shape using boundary methods. Physics of the Earth and Planetary Interiors, **21**, pp. 148–157, 1980
22 Herrera, I., Boundary methods in flow problems. Proc. Third International Conference on Finite Elements in Flow Problems, Banff, Canada, **1**, pp. 30–42, 1980 (Invited General Lecture)
23 Herrera, I., General variational principles applicable to the hybrid element method. Proc. Nat. Acad. Sci. USA, **74**, pp. 2595–2597, 1977
24 Herrera, I., Theory of connectivity for formally symmetric operators. Proc. Nat. Acad. Sci. USA, **74**, pp. 4722–4725, 1977
25 Herrera, I., On the variational principles of mechanics. Trends in Applications of Pure Mathematics to Mechanics, II, Zorsky, H. (Ed.), Pitman Publishing Limited, pp. 115–128, 1979
26 Herrera, I., Theory of connectivity: A systematic formulation of boundary element methods. Applied Mathematical Modelling, **3**, pp. 151–156, 1979
27 Herrera, I., Theory of connectivity: A unified approach to boundary methods. *Variational Methods in the Mechanics of Solids.* Nemat-Nasser, S. (Ed.), Pergamon Press, Oxford and New York, pp. 77–82, 1980
28 Herrera, I., Variational principles for problems with linear constraints. Prescribed jumps and continuation type restrictions. Jour. Inst. Maths. Applics., **25**, pp. 67–96, 1980
29 Herrera, I., Sabina, F. J., Connectivity as an alternative to boundary integral equations. Construction of bases. Proc. Nat. Acad. Sci. USA, **75** (5), pp. 2059–2063, 1978
30 Herrera, I., Boundary methods. A criterion for completeness. Proc. Nat. Acad. Sci. USA, **77** (8), pp. 4395–4398, (1980
31 Gourgeon, H., Herrera, I., Boundary methods. C-complete systems for the biharmonic equation. *Boundary Element Methods.* Brebbia, C. A. (Ed.), Springer-Verlag, Berlin, pp. 431–441, 1981

32 Herrera, I., Gourgeon, H., Boundary methods. C-complete systems for Stokes problems. Computer Methods in Applied Mechanics and Engineering, **30**, pp. 225–241, 1982
33 Herrera, I., Boundary methods: Development of complete systems of solutions. *Finite Element Flow Analysis*. Kawai, T. (Ed.), University of Tokyo Press, pp. 897–906, 1982
34 Herrera, K., An algebraic theory of boundary value problems. Kinam, **3** (2), pp. 161–230, 1981
35 Herrera, I., Spence, D. A., Framework for biorthogonal Fourier series. Proc. Nat. Acad. Sci. USA, **78** (12), pp. 7240–7244, 1981
36 Alduncin, G., Herrera, I., Solution of free boundary problems using C-complete systems. *Boundary Element Methods in Engineering*. Brebbia, C. A. (Ed.), Springer-Verlag, Berlin, Heidelberg, New York, pp. 34–42, 1982
37 Alduncin, G., Herrera, I., Contribution to free boundary problems using boundary elements. Trefftz Approach, Communicaciones Técnicas, IIMAS-UNAM, 1983 (Computers Methods in Applied Mechanics and Engineering, submitted)
38 Liggett, J. A., Liu, P. L.-F., The boundary integral equation method applied to flow in porous media. Allen and Unwin, 1982
39 Kikuchi, N., Oden, J. T., Contact problems in elasticity. TICOM Report 79-8, 1979
40 Oden, J. T., Kim, S. J., Interior penalty methods for finite element approximations of the Signorini problem in elastostatics. Comp. & Math. with Applics., **8**, pp. 35–56, 1982
41 Lions, J. L., Magenes, E., *Non-Homogeneous Boundary Value Problems and Applications*. 3 Volumes, Springer-Verlag, New York-Heidelberg-Berlin, 1972
42 Oden, J. T., Reddy, J. N., An introduction to the mathematical theory of finite elements. Pure & Applied Mathematics Series, J. Wiley & Sons, New York, London, Sydney, Toronto, 1976
43 Bates, R. H. T., Analytic constraints on electro-magnetic field computations. IEEE Trans. on Microwave Theory and Techniques, **23**, pp. 605–623, 1975
44 Abraham, R., Marsden, J. E., Foundations of mechanics. The Benjamin Cummins Publishing Co. Inc., pp. 159–187, 1978
45 Herrera, I., A general formulation of variational principles. Instituto de Ingeniería, UNAM, **E-10**, 1974
46 Gurtin, M. E., Variational principles for linear initial value problems. Quart. Appl. Math. **22**, pp. 252–256, 1964
47 Herrera, I., Bielak, J., A simplified version of Gurtin's variational principles. Arch. Rat. Mechanics and Analysis, **53** (2), pp. 131–149, 1974
48 Gurtin, M., The linear theory of elasticity. In: *Encyclopedia of Physics*. VI a/2, Springer-Verlag, Berlin, pp. 1–295, 1972
49 Prager, W., Variational principles of linear elastodynamics for discontinuous displacements, strains and stresses, in recent progress in applied mechanics. In: *The Folke-Adqvist Volume*. Broberg, B. Hult, J. & Niordson, F. (Ed.), Almgvist and Wiksell, Stockholm, pp. 463–474, 1967
50 Nemat-Nasser, S., General variational principles in non-linear and linear elasticity with applications, in mechanics today. Nemat-Nasser, S. (Ed.), Pergamon, **1**, pp. 214–261, 1972
51 Nemat-Nasser, S., On variational methods in finite and incremental elastic deformation problems with discontinuous fields. Quart. Appl. Math., **30** (2), pp. 143–156, 1972
52 Jackson, J. D., *Classical Electrodynamics*. Wiley, New York, pp. 65, 69, 86, 541, 1962
53 Morse, P., Feshbach, H., *Methods of Theoretical Physics*. McGraw-Hill, New York, 1953

# SUBJECT INDEX

# Lecture Notes in Engineering

Editors: **C. A. Brebbia, S. A. Orszag**

Volume 1
**J. C. F. Telles**

## The Boundary Element Method Applied to Inelastic Problems

1983. IX, 243 pages. ISBN 3-540-12387-3

Volume 2
**B. Amadei**

## Rock Anisotropy and the Theory of Stress Measurements

1983. XVII, 478 pages. ISBN 3-540-12388-1

Volume 3

## Computational Aspects of Penetration Mechanics

Proceedings of the Army Research Office Workshop on Computational Aspects of Penetration Mechanics held at the Ballistic Research Laboratory at Aberdeen Proving Ground, Maryland, 27–29 April, 1982
Editors: **J. Chandra, J. E. Flaherty**
1983. VII, 221 pages. ISBN 3-540-12634-1

Volume 4
**W. S. Venturini**

## Boundary Element Method in Geomechanics

1983. 114 figures. VIII, 246 pages. ISBN 3-540-12653-8

Volume 5
**M. Manzoor**

## Heat Flow Through Extended Surface Heat Exchangers

1984. VII, 286 pages. ISBN 3-540-13047-0

Volume 6
**M. B. Allen III**

## Collocation Techniques for Modeling Compositional Flows in Oil Reservoirs

1984. VI, 210 pages. ISBN 3-540-13096-9

Springer-Verlag
Berlin
Heidelberg
New York
Tokyo